19/2011

Dear John:
Happy B-Day! Hope
this book can give good info
how to make you next 77!
Hugs
J.J.

Happy Birthday, John!
So blessed to know you and
have the chance to work with you!
Helen.

Hi John.
You look great
at your age.
Happy happy
Birthday
Cheers
Helen Clements

Dear John,
Happy Birthday!
Wish you the
best health,
and happiness!
Jane
Orlov

W9-AAZ-607

TRANSCEND

TRANSCEND

NINE STEPS TO LIVING WELL FOREVER

RAY KURZWEIL AND
TERRY GROSSMAN, MD

RODALE

© 2009 by Ray Kurzweil and Terry Grossman, MD

All rights reserved. No part of this publication may be reproduced or transmitted in any form or by any means, electronic or mechanical, including photocopying, recording, or any other information storage and retrieval system, without the written permission of the publisher.

Rodale books may be purchased for business or promotional use or for special sales. For information, please write to:

Special Markets Department, Rodale Inc., 733 Third Avenue, New York, NY 10017

Printed in the United States of America

Rodale Inc. makes every effort to use acid-free ∞, recycled paper ♻.

Illustrations on pages 72, 207, and 313 by www.CartoonStock.com

All other illustrations by Laksman Frank

Page 19 images © 2006–2008 Visualbiotech, Lausanne, Switzerland, All Rights Reserved. www.visualbiotech.ch.

All other photographs by Mitch Mandel

Interior design by Christopher Rhoads

Library of Congress Cataloging-in-Publication Data

Kurzweil, Ray.
    Transcend : nine steps to living well forever / Ray Kurzweil and Terry Grossman.
       p.   cm.
    Includes bibliographical references.
    ISBN-13: 978–1–60529–956–3 hardcover
    ISBN-10: 1–60529–956–1 hardcover
    1. Longevity.   2. Health.   3. Well-being.   I. Grossman, Terry.   II. Title.
RA776.75.K876 2009
612.6'8—dc22   2009003923

Distributed to the trade by Macmillan

2 4 6 8 10 9 7 5 3 1 hardcover

LIVE YOUR WHOLE LIFE™

We inspire and enable people to improve their lives and the world around them

For more of our products visit **rodalestore.com** or call 800-848-4735

To Sonya, Ethan, and Amy, who
inspire me to live forever.
—Ray

To Karen . . . forever.
—Terry

CONTENTS

ACKNOWLEDGMENTS

Ray—I would like to express appreciation to my mother, Hannah, who taught me the power of ideas; to my wife of 33 years, Sonya, who lovingly shares with me all of the big and little challenges of life and gives me confidence in the future; and to my kids, Ethan and Amy, and my daughter-in-law, Rebecca, who inspire me every day.

Terry—My contribution to this book was only possible thanks to the help and support of my family: my parents, Louis and Irene Grossman, for teaching me to think for myself (and then not giving up on me after realizing what they had done); my children, Abraya and Jay Johnson and Sam Grossman, apples (and grape seed skins) of my eye; my grandchildren, Harrison and Lucette, whose growth and successes I look forward to celebrating throughout this remarkable century ahead (and beyond); and Karen, wife, best friend, muse, sounding board, in-house psychologist . . . and more.

We would both like to thank our support team of many multitalented individuals who provided fellowship, friendship, love, and support and were instrumental in helping bring this project to fruition:

Shannon Welch, our highly skilled editor at Rodale, who helped guide this project and went well beyond the "call of duty" by agreeing to serve as the female model in the book; Karen Rinaldi, Rodale's publisher, who provided many resources, support, and guidance; and the rest of our outstanding Rodale team: Chris Rhoads, associate art director; Andy Carpenter, art director; Mitch Mandel, photographer; Troy Schnyder, photo assistant; Beth Tarson, senior publicist; Beth Davey, publicity director; Hope Clarke, senior project editor; and Meredith Quinn, editorial assistant.

Loretta Barrett, our literary agent, who has devotedly and expertly guided this along with our other book projects.

Tom Garfield, who performed outstanding service in helping us with extensive research and wordsmithing.

Sarah Brangan, who also assisted with research and provided invaluable help in developing the recipes, including testing all of them.

Lolita Hanks, FNP, who provided valuable assistance with research.

Laksman Frank, our talented artist and graphic designer, who developed the illustrations.

Our research and communications team, who helped us find, assemble, and communicate the myriad facts in this rapidly changing field: Amara Angelica, Sarah Black, Sarah Reed, Celia Black-Brooks, Emily Brangan, Aaron Kleiner, Kathryn Myronuk, Nanda Barker-Hook, Peter St. Onge, Paula Florez, Diane Henry, and Robyn Stephens.

Our technical team, who helped us communicate with each other: Ken Linde, Sandi Dube, and Matt Bridges.

Our administrative and business team, who kept this project organized and on track: Aaron Kleiner, Denise Scutellaro, Joan Walsh, Maria Ellis, Bob Beal, Casey Beal, Mary Lou Sousa, Ardis Hoffman, Kelli Krill, and Cara Vandel.

Martine Rothblatt, whose dedication and commitment to all of the ideas in this book helped to inspire our work.

Donna Rae Smith, RN, for helping to develop and refine our fitness program.

A special thank you to our many readers who have helped us greatly over the years with suggestions, support, and corrections.

Our peer expert readers, who provided invaluable service in assuring the accuracy of the material in this book through fastidious reading of the manuscript and meticulous feedback: Michael Catalano, MD; Paul Dragul, MD; Karen Kurtak, L. Ac., Dipl. Ac.; Lee Light, MD; Joel Miller, MD; Tadashi Mitsuo, MD; Glenn Rothfeld, MD; and Kazuo Tsubota, MD. Special thanks finally to Aubrey de Grey, PhD, both for his seminal contributions to the field of longevity medicine by way of SENS (Strategies for Engineered Senescence) as well as for making time in his already overbooked schedule to read through and comment on the entirety of the manuscript. The above individuals provided many ideas and corrections that we were able to make thanks to their expert guidance. For any mistakes that remain, the authors take sole responsibility.

INTRODUCTION

Until quite recently, progress in health and medicine was hit-or-miss. We'd find something without having a good understanding of how it worked. *Oh, here is a substance that lowers blood pressure. We have no idea why it works.* We'd "discover" drugs to perform desirable functions, often with many severe side effects, but we lacked the means to *design* medical interventions for a carefully targeted purpose.

For example, in the mid-1990s, Pfizer developed a novel type of blood pressure medicine by the name of sildenafil. It blocked an enzyme that helps regulate bloodflow through arteries. By blocking this enzyme, they hoped to relax the arteries and lower blood pressure. Sildenafil made it all the way to human clinical trials, but it just didn't lower blood pressure enough for a marketable blood pressure medication. The study investigators told all the patients that the clinical trial was over and asked them to send their samples back.

Almost all the women did so, but a significant portion of the men in the study did not. Follow-up phone interviews revealed that many of the men had experienced a completely unexpected side effect. You undoubtedly now know sildenafil by its brand name: Viagra. Its developers weren't working on a drug to treat erectile dysfunction. They set out to develop a marketable blood pressure medication but got lucky and ended up with a blockbuster that made billions of dollars.

Such a random and haphazard approach to medical discovery is typical. But now this situation is changing—and very rapidly. With the completion of the Human Genome Project just a few years ago and with newly discovered means of modifying how our adult genes function, we have moved from the old paradigm, in which the progress in health and medicine has been unpredictable, to a new era in which healthcare has now become an information technology. And, as Ray has written about extensively, a key characteristic of information technology is very rapid exponential growth.

We've already started to reap the fruits of this new knowledge. We now have the means to dramatically reduce the risks of our biggest killers, heart disease and cancer, and to dramatically slow down the aging process itself.

Unfortunately, most conventional healthcare practitioners are caught up in the old paradigm and still don't practice medicine as an information technology. So, to make maximal use of the latest medical knowledge that is already available today, you'll need to take control of your own healthcare, become your own doctor in a sense. You can't rely on anyone else—not even your physicians, although they can still be a big help.

Our purpose in writing this book is to explain how you can take full advantage of the available information to help you eliminate your chance of disease and to drastically slow down the aging process—starting right now.

Reader: This sounds good to me, but tell me, what I'm really interested in is this: If I follow all of your ideas, how long can I live?

Ray: It is important to understand that we are talking about a moving frontier. There are substantial new knowledge and scientific advances every year. And these developments are coming at a faster and faster pace.

Reader: Okay, but if I follow your advice and implement exactly what's in this book, how many years will it give me?

Terry: A lot depends on your personal situation. Let's say, for instance, that, because of your genetics, you'd naturally be at high risk of a heart attack, but by following our recommendations, you reduce your risk by 95 percent—which we believe is feasible—then, to answer your question, it would give you 20 more years.

Reader: Wow, 20 years sounds great.

Ray: Ah, but wait. As we'll discuss later, those 2 extra decades will bring you to a future point in time when we'll have dramatic new technologies that can literally change your genes. There are already more than a thousand drugs in the development pipeline that use these methods. That will add a couple more decades again.

Reader: Now I'm up to 40 more years.

Ray: Correct, and that will bring you up to the middle of this century. That will be more than enough time for further developments such as microscopic devices that can travel in your bloodstream and keep you healthy at the cellular level from inside. That will give you a lot more decades.

Reader: Okay, I'm beginning to see the pattern here.

Ray: Right. The new developments in one period give you the time to get to the next. Because these new developments are accelerating, we believe that in less than 2 decades, we will be adding more than a year each year to life expectancy.

Reader: Of a newborn?

Terry: No, to your own remaining life expectancy.

Reader: So we're really at a tipping point, then.

Terry: Well said! It's not a guarantee, of course. You could be hit by lightning tomorrow, but we are close to the time when the sands of time will soon start to run in rather than run out of your personal hourglass.

Reader: But I don't want to live decades or centuries as the equivalent of a 95-year-old.

Ray: That's not what we're talking about. In this book, we'll show you how to slow down the aging process right now. This will keep you young until we get even more powerful knowledge and technology to completely stop—and even reverse—the aging process. So we're talking about staying young and vital indefinitely.

THE EXPONENTIAL GROWTH OF INFORMATION TECHNOLOGY

When Terry started college as a physics major at Brandeis University in Waltham, Massachusetts, in 1964, there was no computer available for student use. The situation was different a few miles down the road at the Massachusetts Institute of Technology in Cambridge. MIT was so advanced that the following year, when Ray started there as a freshman in 1965, it actually had its own computer. That was a key reason why Ray went to college there. The computer (an IBM 7094) cost $11 million (in today's dollars), took up a substantial portion of a building, and was shared by thousands of students and professors. Today, the computer in your cell phone is a million times smaller, a million times less expensive, and a thousand times *more* powerful. That's a billionfold increase in price-performance. As powerful and influential as information technology is already, we will experience another billionfold increase in capability for the same cost in the next 25 years (rather than the 40 years or so it took for the most recent billionfold increase) because the rate of exponential growth is itself getting faster.

The other important point to make is that this remarkable exponential growth is not just limited to computer and communication devices. It is now applicable to our own biology, and that is a very recent change. Consider, for example, the Human Genome Project. It was controversial when announced in 1990 because mainstream skeptics pointed out that with our best experts and most advanced equipment, we had only managed to complete one–ten thousandth of the genome in 1989. The skeptics were still going strong halfway through the 15-year project as they pointed out that with half of the time having gone by, only 1 percent of the genome had been completed!

But this was right on schedule for an exponential progression. As the first graph on the following page shows, if you double 1 percent seven more times—which is exactly what happened—you get 100 percent, and the project was completed not only on time but ahead of schedule. Similarly, as the second graph on the following page demonstrates, the cost for sequencing a single DNA base pair fell a millionfold over the same period, from $10 in 1990 to less than one-thousandth of a penny in 2008.

We have exactly doubled the amount of the genetic data collected each year since 1990, and this pace has continued since the completion of the Human Genome Project in 2003. The cost of sequencing a base pair of DNA—the building blocks of our genes—has dropped by half each year from $10 per base pair in 1990 to a small fraction of a penny today. Deciphering the first human genome cost a billion dollars. Today, anyone can have it done for $350,000. But, in case that's still out of your budget, just be patient for a little while longer. We are now only a few years away from a $1,000 human genome. Almost every other aspect of our ability to understand biology in information terms is similarly doubling every year.

Our genes are essentially little software programs, and they evolved when conditions were very different than they are today. Take, for example, the fat

GROWTH IN GENBANK DNA SEQUENCE DATA
LOGARITHMIC PLOT

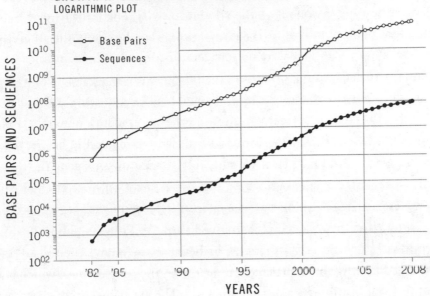

DNA SEQUENCING COST
LOGARITHMIC PLOT

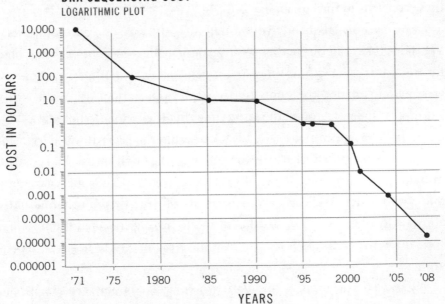

insulin receptor gene, which essentially says "hold on to every calorie because the next hunting season may not work out so well." That gene made a lot of sense tens of thousands of years ago, at a time when food was almost always in short supply and there were no refrigerators. In those days, famines were common and starvation was a real possibility, so it was a good idea to store as many as possible of the calories you could find in your body's fat cells.

Today, the fat insulin receptor gene underlies an epidemic of weight problems, with two of three American adults now overweight and one in three obese. What would happen if we suddenly turned off this gene in the fat cells? Scientists actually performed this experiment on mice at the Joslin Diabetes Center. The animals whose fat insulin receptor gene was turned off ate as much as they wanted yet remained slim. And it wasn't an unhealthy slimness. They didn't get diabetes or heart disease, and they lived and remained healthy about 20 percent longer than the control mice, which still had their fat insulin receptor gene working. The experimental mice experienced the health benefits of *caloric restriction*—the only laboratory-proven method of life extension—while doing just the opposite and eating as much as they wanted. Several pharmaceutical companies are now rushing to bring these concepts to the human market.

Consider how frequently you update the software on your computers. Yet the software in our bodies—our genetic code—has not been updated for millennia. The fat insulin receptor gene is just one of many genes that were better suited to the Stone Age than the Information Age. We're working on the update: Around the same time that the human genome was collected in 2003, a technique called RNA interference (RNAi) was perfected. The technique effectively turns off genes so they don't work in a mature individual; significantly, the work was recognized with a Nobel Prize only a few years later. We also have new forms of gene therapy that actually allow us to add new genes. So, in addition to having collected the software code upon which life is based, we now have the means to begin to reprogram it.

We also have the means of simulating biology on computers so that we can try out new drugs and interventions on simulators, a process dramatically faster than animal and human testing. These new simulators are doubling in power every year. We can also look inside the body and brain with unprecedented clarity and precision. The resolution of scanning technology is also doubling each year, providing dramatic new insights into how disease processes (such as heart disease) actually work.

Drug development used to be called drug discovery—basically just looking for substances that had an apparently positive effect. This is analogous to how primitive man and woman created tools by just finding them on the ground. *Oh, here's a useful stone. This will make a good hammer.* They lacked the means to actually shape tools for a task, whereas later on we learned to design and craft technology for precise missions. Now, for the first time, we can do that with our medical interventions. We are starting to develop new drugs in a rational manner rather than relying on luck, as in the case of Viagra. We can now design molecules to do very specific assignments, for example, to turn off an enzyme that promotes a disease process or to create proteins that directly fight a disease. We'll talk more about these new developments throughout this book.

The point is that health and medicine is now an information technology, and that represents a new frontier. As a result, our health technologies are subject to what Ray calls the law of accelerating returns, a doubling of capability each year. This means that the ability to understand, model, simulate, and reprogram the information processes underlying disease and aging processes will be a thousand times more powerful in 1 decade and a million times more powerful in 2 decades. According to Ray's models, we are also shrinking technology at an exponential pace—by about a hundredfold per decade. So these technologies will also be 10,000 times smaller just 20 years from now.

The most exciting development is that the first set of insights from this biotechnology revolution is now available to us. On the basis of the latest findings from the fields of genetics, medical scanning, and biological simulation, we have found that many of the ideas of how disease processes work have turned out to be dead wrong (and, as we will discuss, that is an apt phrase). Many of these insights, such as how heart attacks really develop, are very recent, and they form the basis for the recommendations we make in this book.

YOUR PERSONALIZED PATH TO DRAMATICALLY IMPROVED HEALTH AND LONGEVITY

Aging is not a single process. It consists of a dozen or so processes, each of which leads over time to the loss of our physical, sensory, and mental capabilities. We will show how to dramatically slow down these processes and in many cases how to stop or even reverse them. In this way, you can stay young until we have even more knowledge to become even younger.

Some of these aging processes are also disease processes. Consider atherosclerosis, which results in our arteries filling up with both soft and hard plaque (as we will explain in Chapter 2). This is a disease process that leads to both heart attacks and strokes, but it is also an aging process because not just the coronary and cerebral arteries are affected.

We'll mention one aging process that you can reverse right now. A vital substance with a complicated name—phosphatidylcholine (PC)—is a key constituent of the cell membranes of all 10 trillion cells that make up our bodies. PC is a remarkable substance that gives the cell its flexible structural integrity, allows nutrients into the cell, and facilitates the removal of toxins. In a 10-year-old child, about 90 percent of the cell membrane consists of PC. The body makes PC very slowly, so it is gradually depleted over our lifetimes. PC is typically down to about 10 percent of the cell membrane in the elderly. The cell membrane then gets filled in with hard fats and cholesterol, which do not work nearly as well. This is a key reason that the skin of elderly people is not supple and why their organs do not work as well.

You can reverse this aging process right now by supplementing with PC. For one thing, it is excellent for maintaining youthful skin, but the performance of your organs will also improve. We'll discuss this and many other ways to slow down, and in some cases to stop and reverse, the processes that underlie aging.

Key to our recommendations is an assessment of your own personal health profile. What are your key health issues? Do you have a high risk of heart disease? Are you at increased risk of cancer? Does your body process glucose efficiently? Do you have an overly active immune system? The answers to these questions have a profound effect on what your optimal program needs to be. We'll describe some simple tests that you can take, many of which are now available for home use, that will enable you to answer these questions. And we'll describe how you can immediately fashion a personalized program based on your assessment.

Keep in mind that we will not jump into the future world we describe in one big leap. Rather, it will come in countless small steps. Now that health technology is an information technology and subject to the exponential progress that underlies all information technology, these steps will come faster and faster.

As we have been putting the finishing touches on this manuscript, we've been fortunate enough to have avatars of our future selves visit us from time to time and share just how exciting the world of the future will be. **Ray2023** and **Terry2023** will tell us about *stem cell therapies, bionic replacement parts,* and *cloned organs,* which will all be realities in 2023. *Genomics* will allow us to see genetic structure with greater clarity by 2023, so that *finely tuned therapies* will enable us to treat and prevent disease with much greater precision on a personalized and individualized basis. These therapies exist in preliminary forms today, but because of the accelerating and exponential nature of technological change, they will mature rapidly in the near future. As we pointed out, because the power of information technology doubles every year, within a decade these technologies will be a thousand times more powerful than they are today and a million times more powerful in 2 decades.

By 2034, *nanotechnology* will have begun to bring fundamental changes to our lives, altering almost every aspect of how we live. **Ray2034** and **Terry2034** will help explain how nanotechnology will enable us to manipulate matter an atom at a time, allowing us to rapidly and inexpensively create almost any substance or structure we desire. Designs exist today—and we will soon have the technological expertise—to build submicroscopic nanobiotic robots or *nanobots*. There are already successful animal experiments of nanoengineered blood cell–sized devices that are performing therapeutic functions inside the animals' bloodstreams, such as a device that cures type 1 diabetes in rats. Within 2 decades, tiny nanobots will circulate throughout the bloodstream performing the same functions as our natural cells and tissues, only with far greater precision and reliability. Our future avatars will explain how nanobiotic red blood cells will deliver oxygen to our tissues more quickly and effectively than our native red blood cells; nanorobotic white blood cells will be able to destroy pathological invaders precisely and completely; and nanorobotic thrombovores will optimize blood clotting.

We often use the metaphor of three bridges to talk about the three steps toward radical life extension. This book will be your guide to Bridge One, which is what you can do right now to slow down and in many cases to stop the processes that lead to disease and aging. We will guide you on how to create a personalized approach based on your unique body and brain. Bridge One will take you over a moving frontier because our knowledge about biology and how to transcend its limitations is expanding—at an exponential pace.

Bridge One will take us to Bridge Two, which is the full flowering of the biotechnology revolution. In about 20 years, we will have the means to perfect our own biology by fully reprogramming its information processes. We will be able to actually change our genes in order to live—*well*—for decades longer than what we now consider a long life. This, in turn, will take us to Bridge Three—the full flowering of the nanotechnology revolution—where we can go beyond the limitations of biology and live indefinitely.

Ray2023 and Terry2023: We thought now would be a good time to introduce ourselves.

Reader: Wow, where did you guys come from?

Ray2023: We're avatars of our future selves.

Reader: Hmmm, so what advanced technology allows the two of you to talk to me from the future?

Ray2023: Actually, we're using a very old technology. It's called poetic license.

Reader: I see. You still look like yourselves. You really haven't aged much in the past 15 years.

Terry2023: That's right. We're in our mid-seventies now, but thanks to the dramatic advances in antiaging medicine that have taken place during the first quarter of the 21st century, we're still doing rather well.

Ray2034: And you'll be happy to see that we're still doing fine in 2034, when we're in our mid-nineties. If you take care of yourself the old-fashioned way for a while longer, you can be young in the 2030s also.

Reader: Wow, there's four of you now! But you older arrivals look even younger than you did in your seventies.

Terry2034: That's because slowing down the aging process was the best we could accomplish in the early 2020s. Now age reversal is a reality.

Reader: I like your clothes. Where'd you get them?

Ray2023: We used a custom online clothing shop, which is how almost everyone buys clothes these days. You have your body scanned, then you can dress your three-dimensional avatar in a full-immersion virtual environment, selecting the style, fabric, and colors you want. When you like the way your avatar looks, you can have the actual clothes quickly made for you by computerized custom tailors.

Reader: No disrespect, guys, but I meant your clothes in 2034.

Ray2034: Clothing—and almost everything else these days—is now made by home desktop molecular nanotech manufacturing devices. The nanofibers in our clothing make them stain resistant and bacteriostatic, meaning our clothes always remain clean and fresh.

Reader: Great, so no more laundry in the future.

Terry2023: True, but we came to tell you a little about how medicine is practiced in the future.

Reader: I'm sure early detection of disease will be much better than it is now.

Ray2023: Dramatically better, actually. Even in your day, many diseases could have been diagnosed far earlier in their course than they were, but with your antiquated diagnostic testing and imaging devices, doing so was cost prohibitive. The inefficient medical insurance bureaucracy in your day didn't help either.

Terry2023: Today we've perfected stem cell therapies and without the use of embryonic stem cells. We can turn your own skin cells into pluripotent stem cells.

Reader: Pluripotent?

Terry2023: Pluripotent stem cells are stem cells that can turn into any type of tissue. So, if you damage your heart, we can create new heart cells for you with your own DNA. If you have type 1 diabetes, we can create new pancreatic islet cells. We can rejuvenate all the organs in your body.

Ray2023: We've also identified the cancer stem cell that underlies the formation of cancer tumors. We now have effective therapies for most forms of cancer.

Terry2023: And, as far as cardiovascular disease goes, with the new generation of imaging devices, doctors can detect even the tiniest accumulations of cholesterol deposits or plaque within the arteries of the heart and brain. Tiny robots that travel through the arteries are fitted with laser tips able to vaporize these deposits before they have a chance to grow. Heart attacks, which were a leading cause of death in your day, are now a medical rarity.

Ray2023: We are now at the mature phase of the biotechnology revolution. Back in your day, medicine was basically hit-or-miss. Drug development was called drug discovery, basically just finding things that appeared to work, often with severe side effects. Several developments occurred around your time that turned medicine into an information technology. We finished mapping the human genome in 2003, so we had the software code of human life. Techniques to actually change our genes were also developed: RNA interference to turn genes off and new forms of gene therapy to add new genes. Rather than just finding drugs, interventions could be designed on computers and then tested on increasingly sophisticated biological simulators. One of my theses is that information technology doubles in power about every year for the same cost. So now, in 2023, these information-based medical technologies are about 30,000 times more powerful than they were in your day.

Reader: Sounds like you've really perfected biology.

Ray2023: We're getting there, but we're also realizing that even as we essentially reprogram our outdated human "software" that we've inherited from thousands of years ago, biology is inherently imperfect compared to what we can create through nanotechnology.

Terry2034: Indeed, we've now gone substantially beyond what even a reprogrammed biology can achieve. Today we have blood cell–sized nanotech devices called nanobots that patrol our blood vessels. They clear plaque from our arteries when the plaque is just a few molecules in size. Arteries of people in their nineties are now as clean as those of young children. These nanobots also find and destroy harmful viruses, bacteria, prions, and cancer cells before they can do any harm.

Reader: Sounds like these developments are really worth waiting for.

Ray and Terry: Indeed, that's well put. But we really need to do more than just wait around, because disease and aging processes are at work every minute we're alive. You'll want to be in good shape when the future arrives.

Reader: Yes, I'd certainly like to be alive at the very least.

Ray and Terry: That's what this book is all about. We'll tell you what you need to do so that you can live long enough to live forever!

FIRST THE FOREST, NOW THE TREES

In our previous book together, *Fantastic Voyage*, we painted the broad strokes for this concept of "living long enough to live forever." We felt that fantastic claims required fantastic evidence, so we needed to offer rigorous support. As a result, that book provided extensive scientific and technical data to support our claims. It was scientifically dense and contained 2,000 citations from the scientific and medical literature.

In this book, we translate our ideas into an easy-to-follow program. We offer specifics for a comprehensive exercise program, sample menus and recipes, precise dosages for supplements, when and where to obtain blood tests, and many other helpful details.

Living a long time is not in the natural order of things. It was not in the interest of the human species when our genes evolved for most people to live past child rearing, so the forces of natural selection filtered out those genes that would have conferred greater longevity to our ancestors. Unfortunately, we have inherited a genetic code programmed to provide us with optimal health for only the first couple of decades of our lives. Beginning at around age 25, we all begin to suffer the ravages of the process known as aging, which actively works to rob us of our health with only one goal in mind—to lead quickly and directly to our early demise.

THE PROBLEM AND THE PLAN

We have divided this book into two main parts. Part I outlines the Problem— why we are genetically programmed *not* to enjoy longevity and the main ways things go wrong. Our major killers today are chronic disease processes such as heart disease, cancer, stroke, diabetes, and Alzheimer's disease. Part I includes chapters focusing on the processes associated with each of these main killers, keeping in mind the overarching theme that we are in possession of an outdated genetic program that wants us to die young. For example, specific metabolic processes such as *glycation, methylation,* and *inflammation* (see Chapter 5) were developed millennia ago to improve survival in a primitive world quite unlike the world we live in today—a world of saber-toothed tigers but free of traffic jams; a world where all the food was organic but without refrigeration. Inflammation was needed to fight infections and heal injuries in a world without antiseptics or antibiotics, but today inflammation seems to create more problems than it solves. So, one of the topics we'll discuss is how you can ensure that your metabolic processes are functioning at an optimal level all the time.

After outlining the problem, Part II discusses the Plan, our program to overcome our outdated genetic code. Central to our thesis is the idea that we still don't possess the full technology and skills to enable us to do so, but it is very likely that we will have them in the near future. Therefore, our goal is to *live long enough* (and remain healthy long enough) to take full advantage of the biotech and nanotech advances that have already begun and will be occurring at an accelerating pace during the next few decades.

The two central pillars of the Plan are Prevention and Early Detection of disease. Key topics discussed include Talking with Your Doctor, Relaxation (stress management), Assessment, Nutrition, Supplements, Calorie Reduction, Exercise, New Technologies, and Detoxification. To help you remember the components of your comprehensive preventive healthcare program, we've arranged them into a mnemonic:

Talk with your doctor
Relaxation
Assessment
Nutrition
Supplements
Calorie reduction
Exercise
New technologies
Detoxification

These are the nine steps of our *TRANSCEND* program that you will use to achieve optimal health and longevity. A dictionary definition of *transcend* is "to go above or beyond what is expected or normal." To live long enough to live forever is to go above or beyond the original boundaries of our genetic legacy. So, we invite you to join us on this wonderful journey as we learn how we might be able to transcend the limits of our biology. If you stay on the cutting edge of our rapidly expanding knowledge, you can indeed *live long enough to live forever.* Let's get started on that path.

THE PROBLEM

1

BRAIN AND SLEEP

"All my life I have focused on being healthy, but I wanted to learn more about slowing the aging process. Your regimen is effective because I continually receive statements like: 'You look fantastic! What do you do to keep looking so young?' I know for a fact that people think I am considerably younger than my chronological age."

SANDY (55), NASHVILLE, TENNESSEE

Part I of this book describes the Problem—the fact that we are genetically programmed to age, to enjoy optimal health for a relatively short period of time, and then are forced to spend much of the rest of our lives dealing with the effects of *aging*, a process that has as its sole purpose the destruction of our health and our ultimate demise. Before we begin our discussion of the various processes associated with aging, it is important to realize that *growing older* (and wiser!) is not the same thing as aging. Everyone grows older all the time, but we aren't necessarily aging as we do so since, by definition, the aging process is one of deterioration.

You grew older today, but did you age as well? If you drank a few cups of green tea, had five servings of fruits and vegetables, exercised for at least 30 minutes at your target heart rate, took nutritional supplements optimized for your age and health situation, spent quality time with close friends and loved ones, consumed a glass of red wine, had a romantic (and sensual!) time with your spouse or significant other, and got 8 hours of quality sleep, then you probably aged very little if at all. If you were a coach potato, ate doughnuts for breakfast, skipped lunch, consumed an excessive amount of coffee, smoked cigarettes, and got into stressful arguments with friends, co-workers, and loved ones, then you probably aged a lot. People can look old in their thirties or young in their sixties, and the lifestyle choices you make every hour make all the difference.

Multiple processes cause us to age. Some are simple, such as the depletion of a vital substance called phosphatidylcholine in our cell membranes (which you can reverse by supplementing with that substance as we discuss below). Some are complex, such as keeping your most important organ—your *brain*—healthy. In this chapter, we'll discuss optimal brain health along with sleep since sleep is so vital to brain function. Then we'll move down to the heart, the digestive tract, and the sexual organs and hormones. We'll complete our overview of how the body works with a discussion of various metabolic processes, including inflammation, methylation, and glycation, and finally look at genomics, the new field that is unlocking the secrets of our genes, which control and regulate all bodily functions.

WE THINK, THEREFORE WE ARE

Your brain makes up only 2 percent of your weight yet receives 20 percent of the blood coming from the heart and uses 20 percent of your body's oxygen and glucose. It also represents 50 percent of your genetic complexity. In other words, half of your genes describe the design of your brain, with the other half describing the organization of the other 98 percent of your body. Moreover, your brain is the master puppeteer: It controls every beat of your heart, every blink of your eyes, the release of your hormones, not to mention all of your willful activities. It has long been regarded as the seat of consciousness, the true you. So it makes sense to consider what you can do to keep it healthy—and happy, too! As it turns out, there is a lot you can do. The ideas in this chapter can dramatically slow down brain aging and help you avoid the often catastrophic downsides of brain dysfunction.

Intelligence is arguably the most important phenomenon in the world because intelligence allows us to understand and shape our environments. The best example we have of an intelligent entity is the human brain itself. And the secret of its design is not hidden from us. Although there's a skull around it, we can see inside a living brain with increasingly precise scanning technologies. This is a wonderful example of Ray's law of accelerating returns: The spatial resolution of brain scanning is doubling every year, and the amount of data we are gathering on the brain is also doubling every year.

We now know the human brain is composed of about 100 billion neurons plus a trillion *glial* support cells. It was originally thought that the glial cells just provided physical support for the neurons, but recent studies have demonstrated that they play a role in influencing the synapses, which are the connections between neurons. We have about 100 trillion such connections, and that is indeed where most of the action takes place. So there is a lot of complexity.

We are gathering an exponentially expanding mountain of data on the brain, but can we understand it? A controversy going back thousands of years to the days of Plato is whether we are intelligent enough to understand our own intelligence. Computer scientist Douglas Hofstadter wrote that "it could be simply an accident of fate that our brains are too weak to understand themselves." Since Hofstadter wrote that line in 1979, we have shown that this is not the case. As we gather enough data on specific brain regions, we have been able to model these areas in precise, mathematical terms and actually simulate them on computers. For example, computer scientist Lloyd Watts and his colleagues have created a computer simulation of a dozen regions of the auditory cortex, the regions of the brain responsible for processing sound from the ears. Applying sophisticated psychoacoustic tests to Watts's simulation produces results very similar to applying these tests to human auditory perception. At MIT, there is a similar model and simulation of the visual cortex, which processes visual information.

At the University of Texas, there is a simulation of the cerebellum, an important region that makes up more than half of the brain's neurons and is responsible for skill formation, for example, catching a fly ball. We have always wondered how a 10-year-old accomplishes this feat. All she has to do is solve a dozen simultaneous differential equations in a few seconds, but most 10-year-olds have not yet taken calculus. We now understand how this works. Those equations are indeed solved by her cerebellum using a mathematical technique called basis functions. It takes place, of course, without conscious awareness, and we do have to train the cerebellum to learn specific tasks, which is why practicing a skill is important. Again, a variety of tests on this computer simulation of the cerebellum provides results similar to human skill formation using our biological cerebellum. This illustrates the oft-stated insight that although the brain is capable of some remarkable accomplishments, we perform these feats without much understanding of how our brains actually carry out these missions.

An ambitious project is underway at IBM to simulate the cerebral cortex, arguably the most important region of the brain and the one responsible for our abstract reasoning. As of the writing of this book, this simulation has successfully undergone its first set of tests.

As we continue the accelerating progress toward "reverse-engineering," understanding the methods of how the brain works, we'll gain far greater insight into our own human nature, which has been the goal of the arts and science since we first wrote symbols on stone tablets over 5,000 years ago. The results of this grand engineering project, which now includes over 50,000 scientists and engineers, will also provide us with methods for ever more intelligent computer software. But the benefit most relevant to this book is that we will gain far more powerful ways of fixing what goes wrong in our brains.

And there is a lot that does go wrong. As we pointed out before, evolution focused on our formative years and enough of our early adulthood to allow us to raise our children so that they became self-sufficient. As a result, keeping our brains healthy much past our twenties was not a trait selected by natural selection when our brains evolved. Our brains are subject to either sudden or gradual decline with age, to self-destructive addictive behaviors, to depression and anxiety disorders, and to many other limitations, not to mention potentially catastrophic lapses of judgment.

YOU CREATE YOUR BRAIN

Perhaps the most important insight relevant to brain health that has come from recent advances in information technology is the *plasticity* of the brain. Since the mid-19th century, it was thought that brain regions were hardwired for specific tasks and that neurons could not be replaced. In 1857, French neurosurgeon Paul Broca related specific cognitive deficits to particular regions of the brain affected by injury or surgery. For more than a century, it was believed that unlike other areas of the body that are capable of repairing themselves, the brain could not replace its neurons and connections that had been lost or damaged and that we are continually and irretrievably losing brain matter.

From recent brain imaging research, we now know the brain possesses *plasticity*, meaning it is perhaps the most dynamic and self-organizing organ of the body. Although there is some degree of specialization in the skills of different regions of the brain, stroke victims are often able to transfer skills from a damaged region to one that is undamaged. Moreover, we can see in recent brain scans how we actually grow new brain connections and even create new neurons from stem cells as a result of our thoughts.

In an experiment with monkeys at the University of California, brain scans obtained before and after the animals were trained to perform a specific task involving the nimbleness of one finger showed substantial growth in neural connections associated with controlling that finger. An experiment with humans who were taught how to play the violin showed substantial growth of connections associated with the fingers of the left hand responsible for controlling the notes. A brain scanning experiment at Rutgers and Stanford universities involved training dyslexic (reading-impaired) students how to distinguish between hard-to-resolve consonants such as "p" and "b." After the training, brain scans showed substantial growth and increased activity in the region of the brain responsible for this discrimination. Paula Tallal, one of the scientists who created this dyslexic training system, commented that "you create your brain from the input you get."

In the latest brain image studies, we can see real-time movies of individual interneuronal connections actually creating new synapses (connection points between neurons), so we can see our brain create our thoughts and in turn see our thoughts create our brain.

The true meaning of Descartes' famous dictum, "I think therefore I am," has been debated for centuries, but these findings provide a new interpretation: I do indeed create my mind from my own thoughts.

IN VIVO IMAGES OF NEURAL DENDRITES
SHOWING SPINE AND SYNAPSE FORMATION

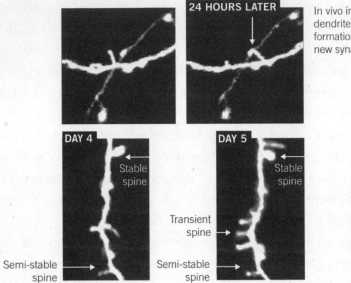

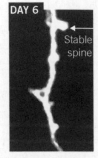

In vivo images of dendrites showing formation of possible new synapse

The lesson of these new insights is that our brain is entirely like any of our physical muscles: *Use it or lose it.* We all know what happens to your muscles if you are bedridden from illness or just living the couch potato life. The same thing happens to your brain. By failing to engage it in intellectually challenging activities, your brain will fail to grow new connections, and it will indeed become disorganized and ultimately dysfunctional. The converse is also true for both body and brain. If someone who has not been physically active for a sustained period starts a program of physical therapy and regular exercise, she can regain her muscle mass and tone within a matter of months. The same thing is true of your brain.

Many studies demonstrate that people who maintain their intellectual activities throughout life remain mentally sharp. A Canadian study called the Victoria Longitudinal Study has shown that older individuals who routinely engage in mentally challenging activities, including everyday activities such as reading, remain mentally alert, as compared with the substantial cognitive decline of those who do not engage in these activities.

Just as we have more than one muscle to keep fit, we have more than one region of the brain that we need to exercise. To keep the cerebellum—the region of the brain that controls voluntary movement—healthy, you should engage in physical activities, particularly those that involve the development of skills such as sports.

The concept that certain brain activities occur in the left half of the brain and others on the right is only partially true. A recently discovered type of neuron called the spindle cell crosses from one side of the brain to the other and appears to be heavily involved in higher-level emotions. In recent brain scanning experiments using new types of scanners that can image individual neurons, these cells "light up" (become especially active) when test subjects are shown a picture of a loved one or hear their child crying. The spindle cells are unusual in that they can be very long, spanning the entire length of the

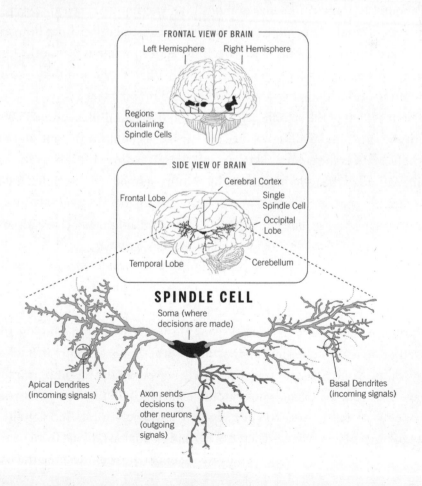

FRONTAL VIEW OF BRAIN

Left Hemisphere   Right Hemisphere

Regions
Containing
Spindle Cells

SIDE VIEW OF BRAIN

Cerebral Cortex

Frontal Lobe   Single
                           Spindle Cell

                           Occipital
                           Lobe

Temporal Lobe   Cerebellum

SPINDLE CELL

Soma (where
decisions are made)

Apical Dendrites   Basal Dendrites
(incoming signals)   (incoming signals)
              Axon sends
              decisions to
              other neurons
              (outgoing
              signals)

brain, and are deeply interconnected with other neurons. One spindle cell will typically have hundreds of thousands of connections to other cells. Unlike the highly organized cells of the cerebral cortex, the brain region responsible for rational thought, the spindle cells display unpredictable and fairly exotic structures and connection patterns.

They are connected to almost every other region, so they receive input from everything else going on in our brain. From these studies, it is apparent that the spindle cells are not doing rational problem solving, which is why we don't have rational control over our emotional responses.

Although each spindle cell is very complex, we don't have very many of them. Only about 80,000 of our 10 billion neurons are spindle cells. Only a few animal species have spindle cells at all. Gorillas have about 16,000, bonobos about 2,100, and chimpanzees about 1,800. Recently we have discovered that whales actually have more than humans. Interestingly, newborn humans don't have any spindle cells. They begin to appear at about 4 months and develop through 3 years of age, which exactly mirrors the ability of young children to deal with higher-level emotions and moral issues.

About 45,000 of the spindle cells are in the right hemisphere, and 35,000 are in the left. This small imbalance appears to account for the notion that the right brain is the emotional brain and the left brain is the more rational brain. Although the right brain does have more spindle cells, both halves of the brain are engaged in logical and emotional activities. Individuals with a rare disorder who use only half of their brain often appear to behave almost normally, engaging in both logical and emotional activities.

EXERCISING YOUR MIND

The concept that the right brain is responsible for creativity and emotion and the left brain is the center for rational and logical thought is more metaphor than reality. Nonetheless, in terms of exercising your brain, it is important to engage both your logical and your emotional faculties. To the extent that your job or educational activities do not engage your logical brain, find activities that require problem solving. There are myriad examples, ranging from board games such as chess to solving crossword or Sudoku puzzles. Keeping track of your finances or planning a trip will engage your logical mind. Reading and writing certainly engage both aspects of your brain. Express your creative and artistic urges by studying a musical instrument. Learn to create art using any modality, including computer graphics. Take up a hobby. Take an adult education course. Travel to new places. Engage in conversations with interesting and thoughtful people. Most important, emphasize interpersonal relationships. Strong connections to others deeply engage both types of mental activities and satisfy a basic human need, as discussed in Chapter 9.

So, here's a useful suggestion on what you can think about to keep your brain healthy: Contemplate how to keep your brain—and body—healthy. You can start by adapting the suggestions in this book into your personal plan!

Terry2023: We now have the capability of overcoming nerve damage, such as spinal cord injuries, using stem cells. Formerly disabled persons are now walking again.

Reader: That research must have already started today if it's working in humans in 2023.

Ray2023: It was not only started, but in your day, MIT scientist Robert Langer was already growing artificial nerves from stem cells in paraplegic mice, enabling these mice to walk again.

Reader: Completely normally?

Terry2023: It was not a completely normal gait but was pretty good. Any physically disabled human would be delighted to walk as well as these once-paraplegic mice could.

ADDICTION

One of the downsides of our brains that we alluded to above is its tendency to addictive behaviors. With our recent ability to model and simulate the information processes underlying our biology and mental activity, we are rapidly gaining an increasingly intricate understanding of the biochemical pathways to addiction. For example, dopamine is the neurotransmitter of pleasure. Dopamine is released when we achieve an accomplishment, win a contest, connect with someone in a loving and nurturing relationship, create a new idea, or even just appreciate someone else's new idea. When people have difficulty achieving regular releases of dopamine through these kinds of socially accepted activities, they will often seek a shortcut.

Gambling provides one such shortcut. Dopamine release is enhanced when a positive outcome is a surprise, so the thrill of winning at gambling can be especially effective in releasing dopamine. This reminds us of an episode of the old series *The Twilight Zone* in which a gambler dies and goes to heaven. Upon his arrival, he is delighted to find himself surrounded by gorgeous women, and, to his further delight, he discovers that he wins every turn of the roulette wheel. But his delight quickly wears thin, and he ultimately finds his persistent winning to be profoundly troubling. He tells the angel in charge that he would really rather go to the "other place." "But this is the other place," the angel replies. So the allure of gambling is dependent on its lack of predictability. And we all know that the odds in gambling are stacked against you; therefore, becoming dependent on this form of dopamine release can eventually become ruinous.

We see a similar self-defeating cycle in the dependency fostered by the initial pleasure generated from addictive drugs. Though addiction remains a troubling scourge, rapid progress is being made in understanding the genetic basis of such behavior. For example, mutations of the dopamine-receptor D2 gene have been associated with addictive abuse of substances, including illegal drugs such as cocaine and heroin, as well as smoking and abuse of food. These genetic mutations can produce unusually intense feelings of pleasure from early experiences with these addictive substances, but in a well-known pattern, the ability of these substances to continue to provide such pleasure becomes depleted. Other genetic mutations can also result in a generally diminished ability for dopamine release from everyday gratifications, leading people with these mutations to turn to other substances and activities to raise dopamine levels to normal.

Aside from the moral, ethical, and legal problems that may result from addictive behavior, the release of pleasurable neurotransmitters such as dopamine from substance abuse or another addictive behavior gradually depletes the brain's supply of dopamine and other pleasure neurotransmitters. This leads to an increasingly desperate dependence on the addictive substance or behavior. Over time, this leads to a catastrophic change in brain chemistry that often requires professional help to resolve.

Studies have shown that moderate use of alcohol is associated with increased longevity, and there is nothing wrong with occasional gambling as a recreational activity. The majority of the population is not genetically susceptible to addiction to either alcohol or gambling, but a substantial portion is; it's important to determine whether you have this susceptibility. If you find that you have this genetic tendency, you'll need to steer clear of activities that may cause you to fall into this pernicious form of mental quicksand.

A new generation of drugs in development promises to change the biochemistry of an addicted person closer to where it was before the addiction occurred. These drugs do not necessarily change the original susceptibility, so they appear to work best in conjunction with traditional drug treatment programs. Unfortunately, the rate of recidivism for addictions to drugs, gambling, and other addictive behaviors is very high, even with counseling programs. It is well known that drug addicts are considered never to be "cured" but to consider themselves "recovering addicts" indefinitely. The hope is that this new generation of drugs, which specifically targets the insidious neurotransmitter and hormonal cycles that reinforce addictive behavior, can be helpful in reducing recidivism.

HEALTHY LIFESTYLE, HEALTHY BRAIN

As we've discussed, in many ways you are what you think. But the old dictum that you are what you eat is also true. In addition to keeping your brain challenged, our dietary recommendations as outlined in Chapters 11 and 13 constitute your first line of defense for maintaining a healthy brain. Your brain is 60 percent fat, so consuming healthy fats is especially important for brain health. Both EPA and DHA, the principal components of the omega-3 fats found in fish, are important constituents in brain tissue.

Inflammation (overactivation of the immune system) is a major accelerator of brain aging, so our dietary recommendations aimed at reducing inflammation (such as avoiding high-glycemic-index carbohydrates such as sugary foods and starches) are also important for brain health.

The following brain nutrients have been shown in double-blind placebo-controlled studies to have significant benefits for brain health, as cited in leading medical journals such as *Nutrition*:

Vinpocetine, a natural supplement derived from the periwinkle plant, increases bloodflow to the brain as well as increases the production of adenosine triphosphate (ATP), the brain's energy source. It has been shown to enhance memory for people with normal memory as well as those with memory impairment.

Phosphatidylserine is a natural constituent of the cell membrane but is found in especially high concentrations in the brain. Supplementing with phosphatidylserine slows down memory loss and has been shown to reverse memory loss in some patients with age-related memory decline. It also lowers levels of cortisol, a principal hormone of aging.

Acetyl-L-carnitine is a natural substance that strengthens the mitochondria, the energy sources inside the cell. It also protects the brain from aging by slowing down inflammation of brain tissues.

Ginkgo biloba has been a staple of Chinese medicine for thousands of years. It increases bloodflow to the brain, and numerous studies show that it reduces short-term memory loss in the elderly. Ginkgo biloba is a prescription drug in Europe, where it is prescribed more than any other pharmaceutical substance for memory loss.

EPA and DHA are the principal components of omega-3 fats and are both found in high concentrations in brain tissue. Both help keep brain cell membranes flexible. As mentioned, the brain is 60 percent fat; when EPA and DHA levels are inadequate, the brain will substitute less desirable fats, such as omega-6 fats and even the dangerous trans fatty acids. When this happens, cell membranes lose their flexibility, and the transmission of signals between neurons becomes impaired. Many studies have shown improved mood and relief from symptoms such as depression and anxiety from supplementation with EPA/DHA.

Phosphatidylcholine (PC), as discussed in Chapter 2, is a key constituent of the cell membrane of all of our cells, including brain cells. Studies have shown that supplementing with PC can help with memory and learning in humans without mental impairment.

S-adenosyl-methionine (SAMe) is a natural derivative of an amino acid normally produced by the body, and it plays a role in methylation (see Chapter 5). Levels of SAMe in the body often become depleted by middle age.

Multiple clinical trials have shown that SAMe provides substantial benefit for patients with depression. This effect occurs relatively quickly, unlike the requirement to build up levels in the bloodstream that accompanies some prescription drugs for depression. It is, therefore, an effective, natural, and quick-acting treatment for mild depression. Human trials have also shown benefits for strengthening the liver and for relief from osteoarthritis.

NATURAL SUPPLEMENTS FOR BRAIN HEALTH

SUPPLEMENT	RECOMMENDED DOSAGE
Vinpocetine	10 milligrams twice a day
Phosphatidylserine	100 milligrams twice a day for 1 month, decreasing to 100 milligrams daily thereafter
Acetyl-L-carnitine	500 to 1,000 milligrams twice a day
Ginkgo biloba	80 to 120 milligrams twice a day
EPA/DHA	1,000 to 3,000 milligrams EPA daily 700 to 2,000 milligrams DHA daily
Phosphatidylcholine	900 milligrams two to four times a day
SAMe	200 to 400 milligrams twice a day

RayandTerry2034: Had we written this book today, we could have written it in a few days because of the nanobots in our brains.

Reader: Run that by me again.

Terry2034: We have many millions of blood cell–size robots in our bloodstreams. They keep us healthy from inside by destroying pathogens, repairing cells, removing toxins and debris, and correcting DNA errors. They also travel to our brains through the capillaries, basically providing computerized neural implants noninvasively, that is, without surgery. In our brains, they interact directly with our biological neurons to extend our memories and access to knowledge.

Reader: Okay, now that sounds awfully futuristic.

Ray2034: Yes, well, we are speaking to you from 2034.

Reader: But that's only about a quarter century from now.

Ray2034: Because of the doubling of the power of information technology in less than a year, computer technology today is millions of times more capable than it was 20 years ago, and it was pretty impressive then. It has also become more than 10,000 times smaller.

Reader: But we can't do anything like that today, so even with a millionfold increase in capability . . .

Ray2034: Actually, that's not the case. Back in your day, you could also have a computer put into your brain.

Reader: I can?

Terry2034: Well, if you're a patient with Parkinson's disease, you can. It was not blood cell size 20 years ago; it was more like pea size. But it did replace a region of the brain destroyed by the disease and communicated with the neurons in neighboring regions, namely the ventral posterior nucleus and the subthalamic nucleus, just like the original biological brain region did. And it was also possible to download new software to the computer in the brain from outside the patient.

Reader: This was an experiment?

Terry2034: No, it was an FDA-approved therapy. And there were other neural implants being developed back then, including retinal implants, chips that enable a stroke patient to control his computer from his brain, an artificial hippocampus for boosting short-term memory, and many others. If you apply the approximately 30 million–fold increase in capability and over 100,000-fold shrinking in size that has occurred in the past quarter century, we now have much more capable devices that are the size of blood cells.

Reader: Still, it's hard to imagine building something the size of a blood cell that can perform a useful function.

Terry2034: Actually, there was a first generation of blood cell–size devices back in your day. One scientist cured type 1 diabetes in rats with a blood cell–size device. It was an excellent example of nanotechnology from the first decade of this century. Its 7-nanometer pores let insulin out in a controlled fashion and blocked antibodies. At MIT, they developed a blood cell–size device that could scout out cancer cells and destroy them. So, if you just apply the law of accelerating returns to what was feasible 25 years ago, then having millions of blood cell–size devices in our bloodstreams today in 2034 to augment our physical and mental capabilities should not be so surprising.

Reader: Okay, so what do these nanobots in your brain allow you to do?

Ray2034: For one thing, they provide a search engine directly from our brains. The nanobots listen in on our thoughts, and if they see we're getting stuck on something, they provide useful information to keep the creative process going.

Reader: I can buy that. The fact that I have access to all human knowledge with a few keystrokes using a device that fits in my shirt pocket is already pretty amazing.

Ray2034: We now have direct brain-to-brain communication since the nanobots are on the Internet.

Reader: I guess no one wonders too much about telepathy in 2034?

Ray2034: That's true. The nanobots can also put us into full-immersion virtual reality environments.

Reader: How does that work?

Ray2034: If we want to go into a virtual reality environment, the nanobots shut down the signals coming from our real senses and replace them with the signals that our brains would be receiving if we were actually in the virtual environment. Then to our brain, it feels like we are in that virtual environment.

Terry2034: If you want to move your arm, it is your virtual arm that moves, so you can be an actor in these virtual reality environments. The environments are as imaginative as games were in your day—some are recreations of earthly environments like a beautiful beach, and some are fantastic environments that defy the laws of physics. You can go to these virtual environments yourself or meet with others. You can interact using all of the senses.

Reader: And you look just the way you do in real reality?

Ray2034: That's an option, but you can also change who you are. You don't have to be the same person all the time.

Reader: So you just forget about real reality?

Ray2034: At times. But most of the time, we live in augmented reality, basically a blend of real and virtual realities. Little avatars pop up explaining what's going on in the real world. It's hard to know where real reality stops and virtual reality starts; it's all kind of blended together.

Reader: Anything else?

Terry2034: The nanobots extend all of our mental capabilities, our memories, our ability to think logically, to perceive patterns. They're true brain extenders.

Reader: Well, I have a brain extender today in 2009. As I said, this computer in my pocket is already extending my mental reach.

Ray2034: Exactly. Extending our reach through our tools is really a very old idea.

TO SLEEP, PERCHANCE TO DREAM

Shakespeare's lines illuminated sleep and dreams as among life's precious pleasures. We all value a good night's rest, and yet research shows that one person in three is chronically sleep deprived. People often try to solve the problem by consuming a lot of caffeine in the morning, but this habit has cultivated a population of nervous—*and still tired*—individuals.

Sleep has also been shown to have many other important functions for health. The brain consumes 20 percent of the body's supply of glucose, and sleep improves glucose uptake into the brain. A baby's brain can use as much as 50 percent of the total glucose supply, which may help explain why babies need so much sleep. *Leptin* is a hormone that decreases appetite, and leptin levels rise during sleep. Many people today don't get enough sleep, which may help explain the dramatic rise in the number of people with weight problems. Sleep improves memory and the ability to learn and retain new material. Lack of sleep adversely affects mood and decreases energy. We feel that getting adequate sleep is just as important as diet and exercise in everyone's wellness program.

The most important phase of sleep, called REM (rapid eye movement), makes up about a quarter of our time spent sleeping. This is the phase during which you do the most dreaming and are most likely to remember your dreams. Your eyes move rapidly as if you were engaged in a drama from your waking life, although the body from the neck down is largely paralyzed.

Recent advances in brain scanning technology have begun to reveal why sleep is so important. We can actually see in brain scans of a living brain how our brain reorganizes itself during dreaming and processes the information that streamed into our brains during the day. In the journal *Science*, scientists reported studies of the brain during dreaming, especially during REM sleep, using a brain scanning technology called positron emission tomography (PET). They found that many regions of the brain are just as active during dreaming as when awake, some actually substantially more so. The brain continues to process visual images, though obviously not from the eyes; and the regions of the brain that process new visual information, including regions of the frontal lobe that combine processed information from the eyes with other sensory information, are quiet. But the regions of the brain involved in creating new memories and making sense of our emotions are even more active than when we are awake. The amygdala, a region responsible for intense emotions such as fear, as well as other regions responsible for

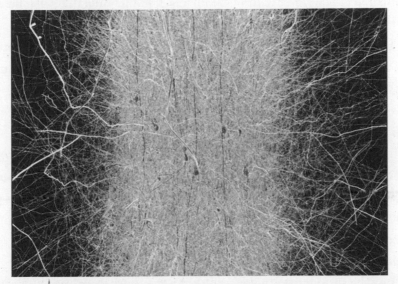

Working computer simulation of a slice of the cerebral cortex.

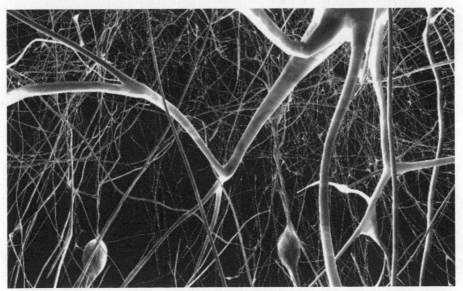

High Resolution Brain Scan

consolidating emotional memories, is especially active during REM sleep. Psychiatrist J. Allan Hobson of Harvard Medical School, a researcher on the PET brain imaging study, commented that "the PET results are consistent with Freud's idea that dreams have meaning."

Research at Harvard Medical School has demonstrated that sleep helps us absorb new information and process our experiences in a procedure called memory consolidation. In studies, people who slept adequately after studying a new task scored significantly higher on tests than people who did not sleep well. With the most recent brain scanning technology, we can now actually see the brain make new connections as it creates new memories and new insights by processing information gathered through the day.

An agreement on the exact role of dreaming will require further progress on reverse-engineering the brain in general, but there is a growing consensus that dreaming is not just a random process of neural firing and that it is vital for our mental and even our physical health.

Each of us will experience occasional periods of poor sleep resulting from a wide range of distractions, such as gastrointestinal upset and daytime worries, but the serious health concern is chronic sleep deprivation. Harvard Medical School reports a wide range of negative effects from a consistent failure to get a good night's sleep. It can cause weight gain by affecting leptin levels that control appetite as well as our ability to process carbohydrates efficiently. Sleep deprivation has been linked to hypertension and increased levels of stress hormones, which heighten the risk of heart disease. It can suppress the immune system, which can increase the risk of cancer and other diseases. Not sleeping can make you accident prone. And not sleeping adequately can wreck your mood and ability to concentrate.

In our own experience, we have found that if we get a good night's sleep (which is most of the time), we have a positive attitude toward life and have the energy and optimism to deal with the challenges that each day brings. Conversely, if we've slept poorly, even small problems can stick and become upsetting.

Our first recommendation with regard to sleep is to recognize its importance and to give it a high priority. Pulling an all-nighter to make a deadline is invariably self-defeating. Following the recommendations in this book will put you in close touch with your body and your needs so you will be able to identify just how much sleep you need. Although sleep requirements do vary from person to person, most people need at least 7 to 8 hours per night.

Here's our seven-point program for getting a good night's sleep every night:

1. Eat right. By following the nutrition recommendations in this book, you'll feel better overall, your gastrointestinal system will be happier, and you'll sleep better.

2. Remember that exercise promotes a natural cycle of sleep. If you have difficulty sleeping, increase your aerobic exercise, although you shouldn't exercise right before retiring for the night. Aerobic exercise releases endorphins, which are natural chemicals that reduce stress.

3. Follow our guidelines for reducing stress, as outlined in Chapter 9.

4. Practice good sleep hygiene before retiring. This means slowing down and engaging in relaxing activities such as reading before going to bed. Working on a stressful project or listening to stimulating music is not the best way to wind down. Having a regular routine at the end of the day is conducive to sleep.

5. If you have difficulty sleeping, cut down on caffeine or avoid it altogether. Don't consume caffeine in the afternoon or evening.

6. Assess whether you suffer from sleep apnea, a common condition in which the mouth opens widely during sleep, causing a temporary blockage of air and decline in available oxygen. This is a very common reason that people do not sleep well. The periods of apnea (literally, "no air") during the night increase risk of both high blood pressure and heart disease. A person undergoing a sleep apnea event may appear to be gagging, although many apnea events produce no visible symptoms. People with moderate to severe sleep apnea may have dozens to hundreds of such events each night. This condition, found in many patients who snore a lot, can be diagnosed in a sleep clinic.

There are also home tests that use a finger-mounted electronic probe to monitor blood-oxygen levels. Excess body weight contributes to sleep apnea, so achieving optimal weight is one approach to solving the problem. One popular treatment is CPAP (continuous positive airway pressure), in which the patient wears a mask connected to a device that maintains positive airway pressure, reducing sleep apnea events. This cumbersome device is indeed intrusive, but some people who suffer from severe sleep apnea find that the inconvenience is worth it to be able to sleep well. Another approach is a prescription dental appliance that is form-fitted to your teeth. Worn at night, this appliance forces your lower jaw forward and thereby prevents sleep apnea events. We recommend you look into this dental appliance before considering CPAP, as it is much less intrusive and for many people works just as well. For mild apnea, try not to sleep on your back because apnea is much more likely in this position. If necessary, sew a tennis ball to the back of your pajama top to keep you from sleeping on your back.

7. Consider the following natural supplements, which are helpful for ensuring a good night's sleep (the authors take some of these):

   o L-theanine is a substance found in tea and promotes relaxation.

   o GABA is a neurotransmitter and a natural, mild tranquilizer. We recommend 500 to 1,000 milligrams before retiring.

   o Melatonin is a natural hormone that controls the body's sleep clock. Normally, the body's level of melatonin dramatically increases when it is time to go to sleep. This in turn triggers a cascade of other hormonal changes to prepare the body for sleep. Melatonin levels decline with age, which is one reason people have more difficulty falling and staying asleep as they get older. You should take melatonin only when it is time to start your night's sleep. Do not take it in the middle of the night if you wake up because that will confuse your body's sleep clock. If you have trouble falling asleep, we recommend the sublingual form, which will go directly into the bloodstream. Standard oral preparations or timed-release preparations are better if you have trouble staying asleep. A wide range of doses, from 0.2 to 10 milligrams, are usually effective. The sublingual form is also available in doses of 2.5 to 3 milligrams. Usually a 1-milligram sublingual dose is sufficient, and many people find that larger doses leave them groggy the next day. Melatonin is also useful for resetting your sleep clock when you change time zones. In this case, take melatonin when it is time to go to sleep in the new time zone.

Terry2023: Sleep scientists have developed methods so that people don't need to sleep much at all.

Reader: But I love to sleep, and dreaming is fun—how else can you fly through the clouds or dance with the elephants?

Ray2023: We now have virtual reality technologies that allow us to do these things while we're awake, but most people still get normal sleep most of the time. A good night's sleep and dreams can certainly be enjoyable and comforting.

Terry2034: Sleeping is optional now. These days, most people still choose to get normal sleep, but we now have devices and medications that produce many of the same benefits of sleep without your needing to lie down. But, like most people, I still find that I feel better with a good night's sleep than with these new versions.

Ray: I agree. I find I'm much more productive if I've slept well.

Reader: I know, but every once in a while it would be great to not have to spend 7 or 8 hours in bed—like when I have an important deadline.

Terry: Ray, why don't you tell the readers the technique you've developed for solving problems while you sleep—your *lucid dreaming* technique?

Ray: I've developed my problem-solving method over several decades and have learned the subtle means by which it is likely to work better.

I start out by assigning myself a problem when I get into bed. This can be any kind of problem. It could be a math problem, an issue with one of my inventions, a business strategy question, or even an interpersonal problem.

I'll think about the problem for a few minutes, but I try not to solve it. That would just cut off the creative process to come. I do try to think about it. What do I know about this? What form could a solution take? And then I go to sleep. Doing this primes my subconscious mind to work on the problem.

Terry: Sigmund Freud pointed out that when we dream, many of the censors in our brain are relaxed, so that we might dream about things that are socially, culturally, or even sexually taboo. We can dream about weird things that we wouldn't allow ourselves to think about during the day. That's at least one reason why dreams are strange.

Ray: There are also professional blinders that prevent people from thinking creatively, many of which come from our professional training, mental blocks such as "you can't solve a signal processing problem that way" or "linguistics is not supposed to use those rules." These mental assumptions are also relaxed in our dream state, so I'll dream about new ways of solving problems without being burdened by these daytime constraints.

Terry: There's another part of our brain not working when we dream, our rational faculties to evaluate whether an idea is reasonable. So that's another reason that weird or fantastic things happen in our dreams. When the elephant walks through the wall, we aren't shocked as to how the elephant could do this. We just say to our dream selves, "Okay, an elephant walked through the wall, no big deal." Indeed, if I wake up in the middle of the night, I often find that I've been dreaming in strange and oblique ways about the problem that I assigned myself.

Ray: The next step occurs in the morning in the halfway state between dreaming and being awake, which is often called *lucid dreaming*. In this state, I still have the feelings and imagery from my dreams, but now I do have my rational faculties. I realize, for example, that I am in a bed. And I could formulate the rational thought that I have a lot to do, so I had better get out of bed. But that would be a mistake. Whenever I can, I will stay in bed and continue in this lucid dream state because that is key to this creative problem-solving method. By the way, this doesn't work if the alarm rings.

Reader: Sounds like the best of both worlds.

Ray: Exactly. I still have access to the dream thoughts about the problem I assigned myself the night before. But now I'm sufficiently conscious and rational to evaluate the new creative ideas that came to me during the night. I can determine which ones make sense. After perhaps 20 minutes of this, I invariably will have keen new insights into the problem.

I've come up with inventions this way (and spent the rest of the day writing a patent application), figured out how to organize material for a book such as this, and come up with useful ideas for a diverse set of problems. If I have a key decision to make, I will always go through this process, after which I am likely to have real confidence in my decision.

The key to the process is to let your mind go, to be nonjudgmental, and not to worry about how well the method is working. It is the opposite of a mental discipline. Think about the problem, but then let ideas wash over you as you fall asleep. Then in the morning, let your mind go again as you review the strange ideas that your dreams generated. I have found this to be an invaluable method for harnessing the natural creativity of my dreams.

Reader: Well, for the workaholics among us, we can now work in our dreams. Not sure my spouse is going to appreciate this.

Ray: Actually, you can think of it as getting your dreams to do your work for you.

THE POWER OF IDEAS

When Ray was 7 years old, he started reading the Tom Swift Jr. series of books. The plots were all very similar: Tom and his friends would get into a terrible jam (often with the fate of the human race hanging in the balance). Tom would retreat to his basement lab and would come up with an idea—usually an invention—that saved the day. These books were one of the inspirations that led Ray to become an inventor. Both Ray and Terry are committed to the basic idea articulated in this children's series: No matter what difficulties we face, there exists an idea that can overcome the problem. So when you confront life's challenges, don't despair. Retreat to your "basement lab" knowing that the solution exists and you can find it and implement it. Both of us have in fact done this by overcoming the health challenges in our lives, the most recent of which has been "middle age" and the attendant challenges that come with it.

Often the idea that saves the day is a set of ideas, none of which by itself would be sufficient. We can reduce our risk of heart disease by as much as 95 percent, but this is not achieved by a single supplement or a single dietary rule. By aggressively reducing each and every risk factor, you can almost eliminate the risk of the number one killer in the developed world today. You can find these ideas in Chapter 2.

Here is a suggestion for an idea that you can adopt today that will change your life: You alone are responsible for your health—not your doctor, not your relatives, not your friends. You are not only the pilot, you're the only one on the plane. The ideas you need to get started on the right path are in this book. Once you're on the road to slowing down aging and dramatically reducing your risk of disease, you'll find that you will discover new ideas on a regular basis as our knowledge of how biology works continues to grow exponentially.

We are the only species that uses its brains to extend its horizons. We'll eventually have powerful new technologies to improve our brains in dramatic ways, but you can apply your own mental powers to start enhancing your life today.

2

HOW TO KEEP YOUR HEART BEATING . . .

*until we come up with better ways
to move your blood along!*

"I feel better today at 50 than I did 10 years ago at 40. Your diet, supplements, and advice are helping me feel younger and more confident as the days go by. I have restored my cholesterol levels to normal after having them extremely high."

OSVALDO (50), SPAIN

Your heart is a seemingly tireless organ that has beat about a billion times by the time you're 30 years old. When healthy, its rhythm is more like a delicate dance than just a repetitive mechanical pump. It responds, of course, to our physical need for greater bloodflow when we exert ourselves, but its patterns are also affected by our moods and emotions, hence its reputation as the organ of love and affection.

As we discussed in the Introduction, our bodies evolved in an era when it was not in the interest of our species for people to live much past their twenties. In addition, our modern diets, which are high in saturated and other unhealthy fats, sugars, and starches, and our often sedentary lifestyles exacerbate the processes that lead to heart disease. Both of the authors have intensely studied the process of heart disease—and how to thwart it— for most of our lives. Ray was 15 when his father had his first heart attack at only 51; his father then died of heart disease at the age of 58. Ray's paternal grandfather also died from heart disease in his fifties. Because of this strong family history of premature heart disease, Ray has taken aggressive steps to overcome his own genetic legacy and has counseled hundreds of others to do the same. Terry has treated thousands of patients for heart disease and, being a male baby boomer over 45 himself, is also personally concerned about this disease—just being male is a risk factor, and so is being over 45 for males and over 55 for females. We should all be concerned about heart disease—it is the number one killer of both men and women, taking 600,000 Americans each year, and is the leading cause of death in the developed world.

Yet, the authors are convinced that almost no one needs to die of a heart attack. Every time we hear of someone who has died of a heart attack, we are filled with regret that the message of how to avoid this circumstance did not reach that person in time. One-third of first attacks are fatal, and another third result in permanent damage to the heart. However, the good news is that if you follow the simple guidelines in this chapter, and throughout this book, you can gain the comfort and security of substantially protecting yourself from this devastating disease.

Like many goals in life, our strategy doesn't rely on a single magic bullet. Rather, by persistently and aggressively chipping away at the risk factors underlying heart disease from multiple directions, we show how you can reduce your likelihood of a destructive event to extremely low levels. This is possible because we now have the knowledge to reduce heart attack rates by more than 90 percent. Although it may be difficult to move the entire society to healthier patterns of nutrition and lifestyle, you can drastically reduce your own risk in just a few weeks.

THE NEW UNDERSTANDING OF HEART ATTACKS

Before discussing how you can dramatically reduce your risk of a heart attack, it is important to understand the process of heart disease. Recent large-scale follow-up studies of patients, as well as new scanning technologies that provide an unprecedented clear view of what is actually going on in the coronary arteries, have completely changed our understanding of the disease. Recent studies of the most popular form of heart surgery have shown that the long-held model of heart disease as basically a plumbing problem in which the coronary arteries that supply blood to the heart become increasingly filled with cholesterol-laden sludge is fatally flawed. A major study of 2,300 heart patients, both men and women, published in 2007 in the prestigious *New England Journal of Medicine*, examined the effectiveness of angioplasty, the most common form of heart surgery. This surgery involves smashing the deposits blocking the coronary arteries—the arteries that provide blood to the heart itself—up against the arterial walls and inserting a "stent" (a wire mesh tube to keep the artery open). All of the patients in this study were considered candidates for angioplasty according to the standard surgical criteria and were divided into two groups. One group underwent angioplasty surgery plus standard medical care, which included lifestyle recommendations and standard-of-care cardiac medications such as aspirin (to reduce blood clots), beta-blockers (to reduce strain on the heart), and statin drugs (to lower "bad" cholesterol levels and inflammation). The control group received the same standard medical care but no surgery. After 4½ years, no benefit was seen from the surgery in reducing heart attacks or deaths.

According to the numerous studies that have been done, the primary circumstance in which angioplasty aids survival is immediate administration after a heart attack. Proponents of angioplasty and stenting countered that even though the patients who underwent these procedures had no reduction in heart attacks or deaths, the surgery was still worth doing because these patients would have less angina or chest pain. Some studies have shown a reduction in angina, but this new study also found that patients who underwent angioplasty did not have less chest pain either. Angioplasty procedures are still done more than 1.2 million times a year. At an average cost of about $44,000, the American public is spending more than $50 billion a year on a procedure that has never been shown to prolong life.

The second most common type of heart surgery is coronary artery bypass grafting (CABG). With this very invasive surgery, occluded arteries are bypassed with grafted veins or mammary arteries. The surgery involves stopping the heart, maintaining the patient on a heart–lung machine during surgery, and then restarting the heart when the bypass surgery is completed.

Studies of the effectiveness of bypass surgery show it to be more effective than angioplasty. Unlike angioplasty, CABG does bypass both the hard, calcified plaque and the soft plaque in the treated occluded arteries. If postoperative patients are aggressive in preventing the bypassed arteries from becoming diseased again, the surgery can be successful in "resetting" these arteries from both types of plaque.

If the primary objective is reducing angina pain, less expensive and safer, noninvasive ways are available to accomplish the same thing, such as judicious use of cardiac medications and a noninvasive procedure known as enhanced external counterpulsation (EECP), discussed below.

The old scientifically discredited theory of heart disease—that it is a plumbing problem that can be fixed by unclogging a stopped-up pipe—explains why angioplasty is relatively ineffective at preventing subsequent heart attacks. Angioplasty burrows through calcified plaque but does not eliminate the soft vulnerable plaque that causes most heart attacks. Bypass surgery does "bypass" both forms of plaque and, under the right circumstances, can reduce subsequent heart attacks and death. Bypass is a major operation and requires over a month of recovery and obviously should be considered a last resort.

Let's compare the old plumbing model with our new understanding, because all of our recommendations stem from a proper understanding of the real causes of heart attacks. The old model works like this: Hard, calcified plaque builds up in your arteries, gradually occluding them. Then, when an artery becomes sufficiently blocked—75 percent or more—there is a risk that a clot will get stuck in the narrowed opening. When that happens, the artery becomes completely blocked, no blood can get through to the heart muscle—and that's a heart attack.

We now know that most heart attacks do not result from arteries blocked with the hard calcified deposits, or *calcified plaque*, that patients are shown by their surgeons. In fact, this type of hard plaque is rarely the cause of heart attacks; rather, it appears to be the *result* of the body's attempt to wall off the real culprit, which is soft, noncalcified, or *vulnerable*, plaque. Soft or vulnerable plaque is flexible and dynamic. It rarely produces symptoms, does not appreciably block arteries, and is difficult to see on angiograms. Yet, vulnerable plaque is the real villain in the story.

This new understanding replaces the old model of heart disease and looks at it instead as a dynamic multistep process in which *inflammation* (the overactivation of the immune system) works first to create vulnerable plaque and then to lead it through an intricate and insidious cascade of events that ultimately ends in a heart attack. It is worthwhile to review the steps in the process that leads up to a heart attack because it guides our thinking on how to thwart this process at every stage.

The story begins with LDL (low-density lipoprotein) cholesterol particles—the aptly named "bad" cholesterol. We should note that LDL is not all bad; indeed we could not survive without it. LDL transports cholesterol from the liver to the body's tissues, where it is needed to keep cell membranes healthy. It is also a precursor of our sex hormones. But when levels of LDL are higher than we need for these vital life processes, it accumulates inside the artery walls, where it can undergo pathological changes. LDL can react with oxygen to become oxidized and with excess glucose in a process called glycation (binding with sugar molecules). Once modified in this way, the LDL particles take on a different appearance. They no longer look friendly to the immune system and are easily mistaken for foreign invaders. The immune system responds by sending in different types of white blood cells, including monocytes and T lymphocytes, in an attempt to destroy the pathological LDL molecules.

After the monocytes encounter the LDL deposits, they become *macrophages* and begin to gobble up these deposits. These macrophages (from the Latin *macro* for big and *phage* for eater) have such big appetites, they eventually become stuffed with the LDL particles and become "foam cells," so named because they look like bubbles of foam. This is the beginning of vulnerable plaque, which at this stage is called a fatty streak. Autopsies of soldiers killed in battle have shown that this early form of vulnerable plaque is quite common in 20-year-olds, and can even be found in children.

Note that the entire process above is associated with inflammation, basically an overactivation of the immune system. Inflammation, in fact, underlies every stage of this process. In the next step, inflammation causes the blood vessel's smooth muscle cells to grow over the foam cells and form a fibrous cap. This is now a mature vulnerable plaque, which typically does not restrict bloodflow but just appears as a slight bulge in the outer diameter of the blood vessel. Vulnerable plaque has been notoriously difficult to visualize, but we have recently begun to be able to see images of it in the arteries of a beating heart using a new generation of noninvasive scanners, which are emerging as promising diagnostic tools.

The stage is now set for the coup de grâce event of a heart attack, and is again fueled by inflammation. Prompted by substances produced by an over-active immune system, the fibrous cap can rupture, spilling the contents of the foam cells and other dangerous chemicals that they have produced. Specific elements in the bloodstream respond by forming a blood clot or thrombus to keep the contents of the foam cells from entering the bloodstream. If the thrombus that forms is large enough to completely block the coronary artery, that's a heart attack. The region of the heart normally supplied by this artery is now deprived of oxygen and other nutrients and will die if the blockage is not quickly reversed. It is important to note that, in most cases, until just moments before the heart attack, the artery was *not* significantly blocked by the vulnerable plaque. The thrombus formed suddenly after the rupture of the fibrous cap, with devastating consequences.

This new understanding motivates all of our recommendations for heart attack prevention. Since the process starts with excess LDL particles, keeping LDL at healthy low levels is our first recommendation. In addition to LDL, there is a form of cholesterol called HDL (high-density lipoprotein), the "good cholesterol," which clears LDL particles from the bloodstream and carries them back to the liver. So, keeping HDL levels high is another important approach.

Keeping in mind that every stage leading to a heart attack is fueled by inflammation, we see once again another way in which our evolutionary Stone Age heritage is not on our side when we get to middle age. Infections were the most common form of death tens of thousands of years ago, so having a strong and highly reactive immune system was critical to the survival of the human species. At this earlier time in our evolution, very few people lived long enough to die from heart attacks, so there was little need to worry about the downsides of an overly active immune system that might cause a heart attack later in life. In addition, many aspects of our modern lifestyle, such as the wrong diet and excessive stress, increase the activation of the immune system and increase inflammation. So, our next and perhaps most important strategy in preventing heart attacks is keeping our immune system robust enough to combat infections but avoiding its overactivation and subsequent inflammation. Each of our recommendations below fits into one of these overarching themes.

A MULTIPRONGED STRATEGY TO AVOID HEART ATTACKS: COMBAT EVERY RISK FACTOR

Several other risk factors have been associated with higher levels of heart disease, and removing each such factor has been shown in extensive studies to lower heart attack risk. We will discuss how each risk factor fits into the new understanding of how heart attacks arise. In addition, a stroke results from the same set of steps except that it takes place in the arteries feeding the brain rather than the heart. So, an added benefit is that by reducing your risk of a heart attack, you will also be reducing your risk of strokes.

To start, we recommend that you get a basic set of heart-related blood tests, which should include:

- **A lipid panel,** which includes total cholesterol as well as LDL, HDL, and triglycerides (a measure of fat in the blood)

- **High-sensitivity C-reactive protein** (CRP, a measure of inflammation in the body)

- **Homocysteine** (a measure of an independent risk factor)

Then count the number of major risk factors you have based on the 11 we list below. If you have three or more major risk factors, we recommend that you also get an exercise stress test and ultrafast computed tomographic scan of the heart, which will provide additional information on your risk of having a heart attack in the next several years.

Major Risk Factors for Heart Disease

1. **Genetic inheritance:** Did your father have a heart attack before the age of 55 and/or did your mother have a heart attack before the age of 65?   Y/N

2. **Age:** If you are male, are you 45 or older? If female, 55 or older?   Y/N

3. **Smoking:** Do you smoke cigarettes and/or have you been a smoker any time in the last 10 years?   Y/N

4. **Weight:** Are you 20 percent or more over your optimal weight? (See Tables 13-2 and 13-3)   Y/N

5. **Cholesterol and triglycerides:** Do you have any of the following:

   o Total cholesterol over 200
   o LDL over 130
   o HDL over 130 (in men over 40 and women over 50)
   o Ratio of total cholesterol to HDL over 4?   Y/N

6. **Homocysteine:** Is your homocysteine more than 10.0? (See Chapter 5, page 118)   Y/N

7. **High-sensitivity CRP:** Is your high-sensitivity CRP more than 5.0? (See Chapter 5, page 113)   Y/N

8. **Fasting glucose:** Is your fasting glucose (blood sugar) under 110? (Fasting glucose >110 is a risk factor for metabolic syndrome, see Chapter 11, pages 211–13)   Y/N

9. **Blood pressure:** Is your systolic 140 or higher and/or is your diastolic 90 or higher?   Y/N

10. **Stress:** Are you a type A personality with a high level of anger and/or lack of social connectedness (type D)? (See Chapter 2, page 45)   Y/N

11. **Exercise:** Are you sedentary?   Y/N

Let's discuss some of the most important risk factors and what you can do to minimize them.

Genetic Inheritance

Your genetic profile affects your predisposition to many of the other risk factors, and many studies have shown that overall heart disease risk is inherited. However, it is our strongly held view that your genetic inheritance is not destiny. The conventional wisdom used to be that your risk of diseases such as heart disease was 80 percent genetic and 20 percent determined by your lifestyle. New research from the field of epigenetics, however, suggests that this thinking is completely backwards! It now appears that only 20 percent of risk comes from your genes and 80 percent from the lifestyle choices you make every day. Thinking that the opposite was the case was perhaps a reasonable perspective given how watered down public health recommendations for prevention of heart disease were up until fairly recently. If you are really diligent, we believe that you can reduce your risk of heart attacks significantly. New public health guidelines (such as keeping LDL below 70 if you are in a high-risk category) have been positively influenced by recent research. The bottom line is that we now have the knowledge to largely overcome most of the risks associated with our genetic heritage.

Ray: There is a new technique called RNA interference that essentially allows us to turn off genes in a mature human. This method is only a few years old, but has already been recognized with the Nobel Prize, another indication of the acceleration of progress. We also have new forms of gene therapy that allow new genes to be added. For example, I am involved with a company that takes lung cells out of the body, adds a new gene in a Petri dish, ensures that it has been inserted properly, replicates the cell a millionfold (using another brand-new technology), and then injects these million cells—with the added gene—back into the body, where they end up in the lungs. This has already been shown to cure a fatal disease called pulmonary hypertension in animals and is now undergoing human trials. There are now over 1,000 drugs and procedures in various stages of the development pipeline to either turn off or add genes.

Terry2023: Today in 2023, the first batch of drugs to turn off genes and methods to add new genes are now approved therapies. We now have a direct and elegant way to remove your genetic disposition to heart disease.

Reader: So I will be able to go back in time and pick new parents?

Ray2023: Well, as far as the "nature" side of the equation is concerned, that is exactly what you can do in 2023. As for changing the results of the "nurture" experiences you've had with your parents, I'm afraid you'll have to wait a little longer.

Reader: Well, changing my genes is a good place to start. I know that the concept of "designer babies" has been somewhat controversial, but I kind of like the idea of being a "designer baby boomer."

Gender and Age

The common wisdom used to be that only men need to be concerned about heart disease. A 2002 survey by the society for Women's Health Research showed that 60 percent of women fear cancer the most, compared with only 5 percent who were afraid of heart disease. So it may come as a surprise that heart disease is the number one killer of *both* men and women. Of the 1.1 million heart attacks each year, almost half occur in women.

It is true that women have some protection from heart disease while they are menstruating, but after menopause all bets are off. The statistics show that women's risk is delayed by about 10 years.

Iron in the blood can act as a catalyst for the process of oxidizing LDL, one of the first steps in the formation of vulnerable plaque. This is one reason that premenopausal women have some level of protection since menstruation helps keep iron levels low. Women also receive protection before menopause thanks to hormone levels that may inhibit these LDL changes.

If you are a man 45 years or older or a woman over 55, then you already have one major risk factor. If you have two additional major risk factors, then you should give a high priority to adopting all of the recommendations in this chapter.

Smoking

The risk of heart attack for smokers is 200 to 400 percent greater than that of nonsmokers. There are 4,000 poisons contained in tobacco and tobacco smoke, many of which greatly accelerate the processes that lead to a heart attack. Cigarette smoke significantly increases the overall level of inflammation in the body and dramatically increases free-radical activity, which accelerates the oxidation of LDL. Smoking also increases heart rate and blood pressure, which accelerate damage to the arteries. We could go on, but the recommendation is obvious: Don't smoke, and avoid secondhand smoke.

Weight

Being overweight contributes to a wide range of diseases and to several of the other risk factors. It is a major contributor to development of metabolic syndrome, type 2 diabetes, and hypertension. The Framingham study, a major study that has followed tens of thousands of individuals for several decades, found that obesity significantly increased risk of heart disease in both men and women. Excess weight is also a major risk factor for increasing the level of inflammation in the body.

Maintaining your optimal weight, as discussed in Chapter 13, is critical to heart disease avoidance, but even losing as little as 10 pounds can significantly decrease heart attack risk.

Cholesterol and Triglycerides

Cholesterol and its LDL and HDL components continue to play major roles in the new inflammation-based understanding of heart disease. We know that the inflammation process starts with excess LDL particles, which enter the coronary artery lining and become oxidized. HDL (good cholesterol) particles reduce heart disease risk by transporting excess LDL cholesterol back to the liver and also by keeping it from becoming inflamed and oxidized.

Based on the statistics for the general population, total cholesterol less than 200 is considered optimal. However, we feel the optimal range for total cholesterol is 160 to 180. Ideally, LDL should be 80 or less, and, depending on the number of your risk factors, HDL should be 60 or higher. An ideal ratio of total cholesterol to HDL is under 2.5.

Recent research has confirmed that reducing LDL cholesterol to much lower levels than the standard recommendation (below 100) substantially reduces the risk of heart disease. A 2004 study by researchers at Harvard Medical School also published in the *New England Journal of Medicine* examined whether reducing LDL levels well below 100 would substantially reduce heart disease risk. The group that took the more aggressive LDL-lowering therapy had a median LDL level of 62, compared with 95 for the control group, who took a more moderate course of statin drug therapy. The group with lower LDL had substantially fewer heart attacks as well as fewer recommendations for bypass or angioplasty surgery. "This is really a big deal," commented Dr. David Waters, professor of medicine at the University of California, San Francisco. Dr. Waters, who was not involved in the research, added, "We have in our hands the power to reduce the risk of heart disease by a lot." On the basis of this and other corroborating research, we recommend that you keep your LDL levels at approximately 80 (if you have fewer than three major risk factors) or 70 or less (if you have three or more major risk factors). Another independent risk factor for heart disease is the level of triglycerides (free-floating fat) in the blood. Conventional recommendations are for the triglyceride level to be less than 150, but we feel that less than 100 is optimal. Excessive consumption of high-glycemic carbohydrates and alcohol are common causes of elevated triglycerides.

The first step toward improving cholesterol and triglyceride levels is to adopt a healthy diet by following the nutritional recommendations in Chapters 11 and 13. Most important, you should sharply reduce saturated fat, which is the most significant dietary influence. No other major dietary nutrient increases LDL levels more than does saturated fat.

There is some controversy regarding dietary cholesterol. Cholesterol levels in the blood are regulated by the liver, so a healthy system is able to maintain healthy levels of cholesterol in the blood despite consumption of dietary cholesterol. However, if you have unhealthy lipid levels, these cholesterol-regulation mechanisms are probably not working optimally. If your blood cholesterol levels are not optimal, we recommend reducing dietary cholesterol to no more than 100 milligrams per day. One egg yolk has about 220 milligrams of cholesterol.

The most powerful method of lowering cholesterol levels is with the use of statin medications such as Zocor (now available as inexpensive generic simvastatin), Lipitor, and Crestor. Before you resort to statins, however, consider the many effective nonprescription supplements that can significantly improve cholesterol, LDL, HDL, and triglyceride levels. We recommend that you try these over-the-counter supplements first and then turn to prescription statin drugs as your second line of therapy if these prove insufficient. The supplements described here have mechanisms that are independent from the statins, so they can be used together with the drugs.

The most effective cholesterol-lowering over-the-counter supplements include the following:

• **Red yeast rice** is a supplement that naturally contains small amounts of lovastatin, the active ingredient in Mevacor, a prescription drug used to lower cholesterol. In a paper published in the July 2008 issue of *Mayo Clinic Proceedings*, researchers compared a group of patients who took red yeast rice and fish oil and followed a healthy diet with a group that took 40 milligrams (a large dose) of prescription-strength Zocor (simvastatin). Cholesterol fell in the red yeast rice group 42.4 percent compared with 39.6 percent in the group that took the drug. In addition to lowering cholesterol, red yeast rice also possesses other properties that appear to protect the heart. In a study reported in the June 15, 2008, issue of the *American Journal of Cardiology*, red yeast rice was found to lower the risk of subsequent heart attack in 5,000 patients with a history of heart attack by half and risk of death from any cause by one-third.

• **Plant sterols** can lower cholesterol levels significantly. They have been marketed in cholesterol-reducing margarines, but these products contain unhealthy fats, so we recommend taking plant sterols as a supplement in pill form.

• **Policosanol** is an effective supplement for improving lipid levels, with results similar to those seen with statin drugs. Studies have also demonstrated that combining policosanol with statins provides even greater effects. A study published in the *American Heart Journal* showed that at dosages of 10 to 20 milligrams per day, policosanol "lowers total cholesterol by 17 percent to 21 percent and LDL cholesterol by 21 to 29 percent. It also raises high-density lipoprotein cholesterol by 8 to 15 percent." Similar to lipid drugs, policosanol also inhibits the oxidation of LDL, a critical first step in the creation of deadly foam cells.

• **Vitamin E** may also be effective both in lowering cholesterol and dramatically reducing overall heart disease risk. In the 1996 Cambridge Heart Anti-Oxidant Study (CHAOS), 1,000 heart patients were given 400 or 800 international units of vitamin E, while a control group of another 1,000 patients (with the same health profile) was given a placebo. Eighteen months later, the vitamin E groups had 75 percent fewer heart attacks.

• **Phosphatidylcholine** (PC) is a major component of your cell membranes. As you age, the level of PC in the cell wall diminishes, which is an important aging process. By supplementing with PC, you can stop and even reverse this process. Research indicates that PC can stimulate reverse cholesterol transport—that is, removal of cholesterol from artery plaque— essentially the same process that HDL promotes. PC, both as an oral supplement and as an intravenous therapy, is widely used in Germany and approved by the German equivalent of the FDA. When taking oral PC, it is important to use one that is at least 50 percent pure. Many supplements labeled as phosphatidylcholine are actually only about 30 percent PC. Food-grade lecithin contains PC, but only about 20 to 25 percent is PC.

The following table lists doses of the supplements mentioned above as well as a few additional supplements that have been found of value in helping lower cholesterol to optimal levels. We recommend that you start with one or more of the supplements in the table below and measure your results 2 months later. A common regimen is to begin by taking red yeast rice and plant sterols as separate supplements and vitamin E as part of your daily multiple. If your heart lipid levels still need improvement, you can consider adding additional supplements from the list below or a statin drug in consultation with your physician.

TABLE 2-1: NATURAL SUPPLEMENTS TO IMPROVE HEART LIPID LEVELS

SUPPLEMENT	AMOUNT PER DOSE	TIMES PER DAY	TOTAL DOSE PER DAY
Red yeast rice	600–900 millligrams	2	1,200–1,800 millligrams
Plant sterols	1,800 millligrams	2	3,600 millligrams
Policosanol	10 millligrams	2	20 millligrams
Vitamin E (mixed tocopherols)	200 international units	2	400 international units
Garlic	900 millligrams	3	2,700 millligrams
Curcumin	900 millligrams	1–2	900–1,800 millligrams
Niacin*	100–500 millligrams	2	200–1,000 millligrams
Phosphatidylcholine	900–1,800 millligrams	2	1,800–3,600 millligrams
Soluble fiber**	4–6 grams	2–3	8–18 grams

*Dosages of up to 3,000 milligrams per day are often used, although we recommend starting with closer to 200 milligrams per day. Periodic monitoring of liver function is recommended when taking niacin.
**Soluble fiber, such as pectin, guar gum, or psyllium, is recommended, especially before meals high in fat. If you take the prescription drugs nitrofurantoin or digitalis, do not take soluble fiber.

Statin Drugs

If natural supplements fail to move your cholesterol, LDL, HDL, and triglyceride levels to an ideal range, you and your physician may wish to consider one of the HMG-CoA reductase enzyme inhibitors, also known as statin drugs. Statins slow down the creation of cholesterol by the liver and increase the rate at which LDL is cleared from the blood.

They also appear to inhibit the oxidation of LDL, thereby slowing down the first step of vulnerable plaque formation. Perhaps most important of all, statins reduce the likelihood that cholesterol in plaques will become inflamed.

Like all prescription medications, statin drugs are associated with side effects. They may have toxic effects on the liver, so your physician will want to monitor your liver enzymes periodically. The same enzyme that the body uses to make cholesterol, HMG-CoA reductase, is also used in the manufacture of coenzyme Q_{10}. Since taking statins depletes the body of coenzyme Q_{10}, which is needed to maintain the health of the mitochondria (the energy furnaces in every cell), *it is vital to take supplemental coenzyme Q_{10} when taking statin drugs.* You should take 50 to 150 milligrams of coenzyme Q_{10} twice a day or 50 milligrams of the activated form of coenzyme Q_{10} known as ubiquinol twice a day if you are receiving a statin drug. These are available over the counter without prescription.

A particularly effective statin drug is atorvastatin, also known as Lipitor. Unlike other lipid drugs, Lipitor is approved as a treatment to reduce triglycerides in addition to lowering cholesterol levels. Lipitor can reduce LDL by 40 to 60 percent and triglycerides by 20 to 40 percent. It also boosts HDL by 5 to 10 percent. Lipitor has been shown to significantly reduce heart attacks and deaths in people at high risk of heart disease.

There is clear research showing that statins lead to a reduction in heart attacks and deaths from heart attacks in men who are at high risk. Use of statins in other groups has become controversial within the medical community. The same types of beneficial results have not yet been shown in women. In addition, even though statin therapy has been shown to decrease heart attack risk, no studies have shown that statin therapy increases life expectancy for any group other than men with a history of heart attack. It should be noted, however, that studies demonstrating significant increases in life expectancy are hard to conduct because of the significant time periods required to see an effect.

The bottom line is that changing to a heart-healthy diet has been shown to be much more effective than taking statin drugs for preventing heart attacks.

In a French study conducted in the 1990s, people with prior heart attacks who ate a diet high in fruits and vegetables, replaced simple starches with whole grains, consumed more olive oil and fish, and avoided red meat, butter, cheese, and egg yolks—followed a prudent diet not unlike the one we outlined in *TRANSCEND*, in other words—had a substantially reduced risk of subsequent heart attacks and deaths. The benefit was two to three times greater than what was possible with taking statin drugs.

Thus, the most important first steps are to follow our recommendations for a healthy diet (see Chapters 11 and 13), regular exercise (see Chapter 14), and stress reduction (see Chapter 9). If your cholesterol and other lipid levels still remain above the optimal ranges, add one or more of the natural supplements shown in Table 2-1. If you have three or more major risk factors and your levels are still too high, you can discuss addition of a statin drug with your physician. Be sure to take coenzyme Q_{10} or ubiquinol if you are taking statin drugs—it is a valuable health supplement in any event.

Blood Pressure

Even under normal circumstances, blood pressure in the coronary arteries is quite high, which increases the inflammation that begins the process of plaque formation. Inflammation in the coronary arteries is worsened by elevated blood pressure. A study of 10,874 men reported in the *Archives of Internal Medicine* showed that people with mild hypertension—blood pressure of 140/90 to 160/105 mm Hg—had a 50 percent higher risk of dying of coronary heart disease. Even those with high-normal blood pressure (also known as *prehypertension* and defined as readings between 120/80 and 140/85 mm Hg) had a 34 percent higher risk of heart attack. Many other studies have demonstrated how high blood pressure can accelerate the build-up of plaque in the arteries and increase the likelihood of a heart attack. Hypertension is also a symptom of metabolic syndrome.

Optimal blood pressure is less than 120/80 mm Hg. If your blood pressure is higher than this, we recommend following a lifestyle and supplement program to get as close to this level as possible. The first step is to adopt our nutritional recommendations in Chapter 11 and attain your optimal weight.

Determine whether you have metabolic syndrome or type 2 diabetes and follow our program in Chapter 5. These steps, particularly adopting a low-carbohydrate, very-low-glycemic-index diet, are often adequate by themselves to resolve hypertension. If blood pressure remains elevated despite these measures, we will often recommend a traditional Chinese medicine formulation of six herbs known as Uncaria-6 (also called Gou Teng Jiang Ya Pian). This formulation will frequently lower blood pressure without the side effects of many blood pressure medications. Uncaria-6 is available from acupuncturists and practitioners of traditional Chinese medicine.

There are many other supplements that can help lower blood pressure. In addition to or in place of Uncaria-6, you might try a combination of magnesium, garlic, and arginine as shown in Table 2-2. If your blood pressure is still suboptimal, consider some of the other supplements listed below.

TABLE 2-2: NATURAL SUPPLEMENTS TO IMPROVE BLOOD PRESSURE

SUPPLEMENT	AMOUNT PER DOSE	TIMES PER DAY	TOTAL DOSE PER DAY
Magnesium	200 milligrams	2	400 milligrams
Garlic	900 milligrams	3	2,700 milligrams
L-arginine *	1–2 grams	3	3–6 grams
Coenzyme Q_{10}	100 milligrams	3	300 milligrams
EPA/DHA (fish oil)	EPA (500–1,500 milligrams) DHA (350–1,000 milligrams)	2	EPA (1,000–3,000 milligrams) DHA (700–2,000 milligrams)
Vitamin C	1,000 milligrams	2	2,000 milligrams
Vitamin E	200 international units	2	400 international units
Calcium	500 milligrams	1–2	500–1,000 milligrams
Alpha-lipoic acid (ALA)**	250 milligrams	2	500 milligrams
Potassium	200 milligrams	1	400 milligrams
Green tea extract	500–1,000 milligrams	2	1,000–2,000 milligrams
Hawthorn	250 milligrams	2–3	500–750 milligrams

*L-arginine has additional benefits in improving vessel health.
**ALA is an important supplement for preventing and treating metabolic syndrome.

If these recommendations prove insufficient and prescription drugs are considered, angiotensin II antagonists such as Cozaar or Hyzaar appear to be safer and more effective than other classes of blood pressure medications, such as calcium-channel blockers. Diuretics and beta-blockers appear to increase insulin resistance, which is counterproductive because it increases the risk of developing metabolic syndrome and type 2 diabetes.

Stress

Given the prominent role of inflammation at every step of the process leading up to a heart attack, it is not hard to understand why stress is a risk factor. Studies have demonstrated that feelings of aggression and rage increase levels of homocysteine. The continual self-imposed stress associated with the type A personality results in higher levels of adrenaline, which worsens inflammation. As we discuss in Chapter 9, not everyone with a type A personality is at risk. People with short tempers who are continually getting angry have the personality type with higher risk. The type D personality, characterized by a lack of social connectedness and inability to express emotion, also has increased heart disease risk.

Exercise

To put this in a positive context, adequate levels of exercise reduce all of the controllable risk factors, including improving insulin sensitivity, which contributes to weight loss and reduces blood pressure, stress, and inflammation. We discuss this key issue in Chapter 14.

SECONDARY RISK FACTORS FOR HEART DISEASE

Several other factors can contribute to heart attack risk. Let's look at a few of these secondary cardiac risk factors, as well as the tests you can do to assess whether you have these.

- **Obstructive sleep apnea** is a common condition in which the mouth opens widely during sleep, causing a blockage of air. Most people who have it are unaware of the condition, and it has been shown to be a risk factor for heart disease. See our discussion of sleep in Chapter 1 for a description of how to diagnose and treat sleep apnea.

• **High levels of iron in the blood** (a hereditary condition called *hemochromatosis*), particularly combined with elevated LDL levels, promote the oxidation of LDL, which is the critical first step in creating deadly foam cells. The easiest way to test for the amount of iron in the blood is with two blood tests: the *serum ferritin* and the *iron binding capacity*. If you have elevations of either of these, the simplest treatment is regular phlebotomies, or donations of blood. Giving blood a few times a year can help lower your iron level, as well as help many patients in need of blood transfusions at the same time. Supplements that reduce iron levels include fiber, calcium, magnesium, garlic, vitamin E, green tea, and red wine. Unless you are anemic, you should not take supplements (particularly mineral supplements) that include iron, and you should avoid iron cookware.

• **Periodontal disease,** such as gingivitis, is characterized by chronic inflammation of the gums and has been linked to increased risk of heart disease. We do not yet know whether the existence of gum disease itself contributes to heart disease, or whether underlying inflammatory and infectious processes contribute to both gum disease and heart disease. It is also possible that the varied bacteria involved in gum disease may contribute to the process of atherosclerosis. Proper dental hygiene, including daily flossing and regular dental visits, can reduce the likelihood of gum disease, and, in turn, reduce the likelihood of coronary heart disease.

• **Hypothyroidism (low thyroid function)** has been linked to elevated cholesterol levels and increased heart disease risk. Half of hypothyroid patients have high levels of homocysteine, compared with 18 percent of the overall population. In addition, more than 90 percent of hypothyroid patients have excessive levels of cholesterol or homocysteine, compared with only about a third of the general population. Tests to check thyroid function (free T_3 [triiodothyronine], free T_4 [thyroxine], and TSH [thyroid-stimulating hormone] levels) should be a routine part of your annual examination, and impaired thyroid function should be treated.

Ray2034: You can now replace a portion of your biological red blood with nanobots called respirocytes that perform the same function. These robotic red blood cells were designed more than 20 years ago by nanomedicine pioneer Robert Freitas, and are now approved methods to enhance the performance of your blood. Like most of our biological systems, red blood cells perform their oxygenating function very inefficiently, but these tiny respirocyte robots are a thousand times more capable. By replacing a portion of your blood with these devices, you can now do an Olympic sprint for 15 minutes without needing to take a breath.

Reader: So will I be able to sit at the bottom of my pool without oxygen or go underwater diving without scuba gear?

Ray2034: Yes, for about 4 hours.

Reader: "Honey, I'm in the pool" will take on a whole new meaning.

Terry2034: Indeed it will.

Reader: But what about our athletic competitions? Today we have controversies with injections of steroids and human growth hormone, but robotic red blood cells are going to blow that out of the water, so to speak.

Ray2034: Well, there will always be specific rules in athletic contests. For example, it was quite feasible back in 2008 to develop cars that go much faster than the winning cars in NASCAR competitions, but there were very detailed rules as to how you can soup up your car. We'll have to determine rules for how you can soup up your body.

Reader: Assuming these things can be detected.

Ray2034: Actually, they'll readily show up in a blood sample, but we do want to point out that there is a good reason not to ban them. Anabolic steroids and human growth hormone in the absence of specific medical conditions requiring their use (such as the condition of low levels of human growth hormone or other medical reasons) *should* be illegal because they are bad for your health. If we did not ban them, then athletes would be forced to harm their health in order to be competitive. Respirocytes, on the other hand, are good for your health. They provide better oxygenation of your tissues and superior removal of carbon dioxide and toxins.

Reader: How about if I have a heart attack? Are these things going to get in the way?

Terry2034: Here in 2034, the incidence of heart attacks has been reduced by more than 95 percent because of the widespread use of effective medications and procedures that change your genes to be heart protective. But in the rare event that you do have a heart attack, you'll be glad you have these little robots. They'll keep your heart and brain and all your vital organs supplied with oxygen for at least 4 hours. You can walk into your doctor's office and calmly explain that you're having a heart attack. She'll inject you with more respirocytes and then deal with removing the clot and fixing the problem.

Reader: I suppose you have more nanobots for that too?

RayandTerry2034: Yes, we do. They travel through the bloodstream and destroy the clot that caused the heart attack.

Reader: What about the other parts of the blood?

Terry2034: We also have micron-size artificial platelets that are capable of achieving homeostasis (bleeding control) up to 1,000 times faster than biological platelets. Right now nanorobotic microbivores (white blood cell replacements), which can destroy specific infections and are effective against all bacterial, viral, and fungal infections, even cancer cells, and with no limitations of drug resistance, are being tested for human use. These robotic microbivores can destroy a pathological organism like a harmful bacterium or virus in 30 seconds. The pathogens are broken down into harmless amino acids and other nutrients rather than the often-toxic result from the action of our biological immune system.

Reader: That doesn't sound so impressive. I studied microbiology in college and I know that my own white blood cells can quickly destroy a pathogen right now.

Ray2034: Actually, back in your day I observed my own white blood cells destroy a bacterium through a special microscope at Terry's clinic. The white blood cells were indeed very clever at blocking the bacterium's escape, but they were very slow. It took over an hour. Our new nanorobotic microbivores can do that in seconds—they can also download software from the Internet so they'll know what germs are in the community at the moment, as well as be able to treat any engineered biological agents.

Enhanced External Counterpulsation

In addition to the noninvasive remedial procedures involving diet and supplements described above, an ingenious method for reducing angina pain and improving cardiac function in patients with heart failure is enhanced external counterpulsation (EECP). This completely noninvasive treatment involves placing air-filled cuffs around the patient's calves, thighs, and buttocks. While the patient lies on a table, the cuffs are compressed with air in a specific rhythm controlled by a computer that receives input from the patient's real-time electrocardiogram. The inflation of the cuffs is timed to occur precisely during the resting phase of the heart rhythm, called *diastole*. As the computer inflates the cuffs, blood is propelled from the lower body back into the heart. This treatment, which is approved by the FDA for some cases of angina pectoris and heart failure, rapidly promotes the development of collateral coronary blood vessels (very small coronary arteries that augment the main coronary arteries). In other words, EECP causes the heart to grow its own natural bypasses.

EECP greatly accelerates the natural process of growing collateral bypass circulation. It also appears to provide the heart with a profound form of exercise. It is well known that elderly heart patients, who have had more time to grow collateral circulation, have a lower risk of dying from a heart attack for this reason. With EECP, however, people can grow effective collateral circulation at any age. It dramatically improves blood circulation and has been shown to improve a variety of conditions that benefit from improved circulation, such as Parkinson's disease. A typical course of EECP treatment is 1 hour per day, 5 days a week for 7 weeks. Although this involves a significant commitment of time and inconvenience, it is far preferable to invasive surgery and involves a healthy, healing process, rather than the risks and complications of surgery. EECP is both FDA- and Medicare-approved under certain circumstances, such as forms of congestive heart failure. It is the leading form of heart therapy in China.

Terry2034: Now that robotic red blood cells are in daily use, research is gearing up to replace the heart altogether. We expect this to happen later in the 2030s. The heart is a remarkable machine, but it has a number of severe problems. It is subject to a myriad of failure modes—as discussed at length in this chapter—and it represents a fundamental weakness in our potential longevity. The heart usually breaks down long before the rest of the body, and often very prematurely.

Ray2034: Although artificial hearts have come a long way in the past 30 years and the new models work quite well, a more effective approach is to get rid of the heart altogether. We can do this by using robotic blood cells that provide their own mobility. If the blood system moves on its own, the engineering issues of the extreme pressures required for centralized pumping by the heart can be eliminated. With the self-propelled respirocytes providing greatly enhanced access to oxygenation, we will be in a position to eliminate the lungs, too, since the nanobots can also provide oxygen and remove carbon dioxide.

Reader: Okay, now hold your horses, I kind of like breathing. Going into the great outdoors and taking a deep breath is one of the great pleasures of life. And for that matter, I like the feeling of my heart beating also.

Ray2034: The therapies that are now being developed are intended to augment our heart and lungs, so we'll have the best of both worlds.

Reader: Yes, but from the way you are talking, it sounds like the heart and lungs, eventually, won't be needed at all.

Ray2034: If you like breathing that much, we are also developing virtual ways of having this sensual experience.

Reader: Well, for some things, I kind of like real reality.

Terry2034: You can keep your biological heart and lungs as long as you like. But, I hope you'll come talk to us about this in a quarter century. It will be comforting to know that you have a backup if something goes wrong.

EARLY DETECTION: CARDIOVASCULAR DISEASE

Finding out that you have some type of vascular disease before a catastrophe occurs—early detection, in other words, can be lifesaving. More people die of cardiovascular disease than from any other cause, and, in 2005, one American died of cardiovascular (heart and blood vessel) disease on average every 96 seconds, for a total of 151,671 fatal heart attacks and 143,948 fatal strokes.

In more than half of these cases, the people who died had no prior warning symptoms. There was neither chest pain nor skipped heartbeats to let them know something was wrong with their hearts—right up to the very day they suffered the heart attack that killed them. Most didn't have any strokelike symptoms before the day of their fatal stroke. Sadly, in the majority of cases, there was nothing to suggest that there was anything wrong with these folks, which would have brought them to the doctor so that something could be done in time.

This is an unnecessary tragedy since several simple, safe, and inexpensive screening tests can easily detect cardiovascular disease long before a heart attack or stroke occurs. Yet, the overwhelming majority of practicing physicians still do not routinely order these tests on their patients, leading to hundreds of thousands of unnecessary deaths.

Part of the reason is too much emphasis on measuring and aggressively treating blood lipids, such as cholesterol, while not directing enough attention toward other critically important risk factors, such as homocysteine and CRP. In addition, only a small fraction of the population has had their coronary calcium score measured or undergone carotid intima-media thickness measurement, two simple screening tests discussed below that can detect the presence of cholesterol build-ups in the arteries and alert people to potential problems that can be corrected long before heart attacks or strokes occur.

An effective program for the early detection of cardiovascular disease relies on a combination of blood tests and imaging studies, such as the coronary calcium score and ultrasound examination of the arteries. We'll outline for you a very effective program that will enable you to discover whether you have any problems early on—at a time when almost all of the damage can be avoided. Luckily, you don't need to be independently wealthy or have a doctor's orders or permission to have any of the testing we recommend done. Hopefully, you have a forward-thinking physician who is already using these tests. If not, talk with your doctor (one of the TRANSCEND principles) to see whether you can arrange to have them done on your own.

Blood Tests for Early Detection of Cardiovascular Disease

Lipid Panel

The lipid panel tests for four major cardiovascular risk factors: **total cholesterol, LDL cholesterol, HDL cholesterol, and triglycerides**. This is one test for the detection of cardiovascular disease that is routinely done by all practicing physicians. We agree with the National Cholesterol Education Program (NCEP) and the American Heart Association recommendations that everyone get a lipid panel beginning at age 20, and, if results are normal, every 5 years thereafter.

This test should be done in the morning after you have been fasting for at least 10 hours. If your lipid levels are not within the goal ranges, you should have testing done more often—say, every 4 to 6 months—until you get your numbers into the desirable ranges. In addition, if you do the imaging studies recommended below and find that you have either early heart disease because of a positive coronary calcium score or early cerebrovascular disease as a result of ultrasound testing of the carotid arteries in your neck, then the optimal values below are not merely desirable—they are mandatory.

TOTAL CHOLESTEROL

NCEP recommendations are for total cholesterol to be less than 200 mg/dL. Readings between 200 and 239 are considered borderline high, while levels above 240 are associated with twice the risk of heart attack compared with levels less than 200. On the basis of the most recent evidence, we believe that the optimal range for total cholesterol should be 160 to 180. This range is supported by research suggesting that these lower levels can reduce risk of cardiovascular disease even further. People who have positive results on their coronary artery calcium scans or carotid ultrasounds need to achieve this optimal level.

LDL CHOLESTEROL

To determine your optimal goal for LDL cholesterol, you need to count how many of the following types of risk factors you have:

Severe Risk Factors:

- Established coronary heart disease

- Diabetes

- Metabolic syndrome

- Coronary artery calcium score is below 75th percentile for your age (our recommendation—not included in the NCEP list)

If you have one or more of these severe risk factors, you are considered very high risk and need to lower your LDL cholesterol very aggressively. NCEP suggests 100 as the upper limit for LDL in this group.

Major Risk Factors:

- Age—male older than 45 years of age or female older than 55 years of age

- Cigarette smoking

- Family history of premature heart or blood vessel disease (age older than 55 in a first-degree male relative or older than 65 in a first-degree female relative; a first-degree relative is a parent, sibling, or child)

- High blood pressure (140/90 or higher, or on blood pressure medication)

- HDL cholesterol < 40

- Coronary artery calcium score > 25th percentile (our recommendation— not included in the NCEP list)

According to NCEP, if you have zero or one major risk factor, your LDL goal is less than 160. If you have two or more major risk factors, NCEP suggests an LDL goal of less than 130, with an optimal goal of less than 100. If you are in the severe risk group for heart attack or stroke, NCEP feels your LDL goal is less than 100. Although we agree with the NCEP categories, we feel LDL should be treated more aggressively in people in the high- or severe-risk categories. For the severe-risk group, we recommend that the LDL level be less than 70 (rather than 100); for the high-risk group, we feel it should be less than 100 (rather than 130). Terry's clinic has set a goal of LDL less than 80 if there is any evidence of early coronary artery disease based on a positive calcium heart score (any calcium whatsoever) or of carotid artery disease because of ultrasound abnormalities on the intima-media thickness test. Reaching this goal should help prevent further disease progression even in the absence of any other risk factors.

HDL CHOLESTEROL

HDL cholesterol is the "good" cholesterol because it works to remove plaque from arteries. HDL levels less than 40 mg/dL are a major risk factor for cardiovascular disease. Levels above 60 mg/dL are protective, while levels below 40 mg/dL in men and below 50 mg/dL in women are a symptom of metabolic syndrome, which is another major risk factor for cardiovascular disease.

TRIGLYCERIDES

Triglycerides are a measure of fat in the blood. A high triglyceride level, along with a low HDL level, is classic for metabolic syndrome. High triglycerides are commonly the result of a diet high in sugary and high-glycemic-index foods. Triglyceride levels less than 150 mg/dL are considered normal, and the optimal values are less than 100.

The VAP Lipid Panel

The VAP, or Vertical Auto Profile, test provides more detailed information than a conventional lipid panel. In one study of families with premature coronary artery disease in Utah, Roger Williams, MD, found that only 25 percent of the cases of heart disease could be accounted for by elevated total cholesterol and LDL alone. At least 60 percent were the result of other lipid abnormalities, such as low HDL, elevated lipoprotein(a), intermediate-density lipoprotein (IDL), or very-low-density lipoprotein (VLDL).

When a standard lipid panel is run, the total cholesterol, HDL, and triglycerides are measured, while the LDL is calculated from the other results.

In the VAP test, LDL is measured directly, providing much more accurate information than a calculated value. In addition, the VAP provides information about the size and actual number of LDL particles, and tells how many of the less dangerous, larger, fluffy "A" LDL particles you have compared with the number of more dangerous, small, dense "B" particles. Because they are light and fluffy, "A" particles tend to bounce off the walls of the arteries. In contrast, the small, dense "B" particles act more like little bullets and penetrate the arterial wall more easily, where they deposit the cholesterol they contain. Higher numbers of small "B" particles are more common in patients with diabetes or metabolic syndrome.

VAP also measures lipoprotein(a), IDL, and VLDL, risk factors found to be more important than total cholesterol and LDL. Finally, VAP fractionates or splits HDL, the beneficial form of cholesterol, into HDL-2 and HDL-3 subunits. HDL-2 is much more protective than HDL-3. Low HDL-2 is a significant risk factor for cardiovascular disease.

IDL is a genetic risk factor and tends to be elevated in patients whose family tree includes diabetes. VLDL carries triglycerides in the blood and, if elevated, suggests the need to eat less sugar and high-glycemic-index carbohydrates. Elevated lipoprotein(a) is another hereditary factor that is associated with high risk and rarely responds to conventional cholesterol-lowering drugs. Elevated lipoprotein(a) levels can be lowered by taking 1 gram of vitamin C, 1 gram of lysine, and 1 gram of proline twice daily.

The Metabolic Factors—Glycation, Inflammation, and Methylation

In Chapter 5, we discuss how the metabolic factors—glycation, inflammation, and methylation (GIM)—can contribute to cardiovascular as well as other types of health risks. Three simple laboratory tests can provide a wealth of information on how well your body is performing these critical metabolic functions. You can assess your own GIM factors with three simple blood tests: *Hemoglobin A_{1c}* directly measures glycation, *CRP* measures inflammation, and *homocysteine* measures methylation status. These tests are also discussed at further length in Chapter 5.

Fasting Glucose and Insulin

Metabolic syndrome and its more serious cousin, type 2 diabetes, have far-ranging implications for heart disease risk. Patients with these conditions have insulin resistance, which results in high blood levels of insulin. Insulin is a growth promoter and accelerates coronary plaque formation. It also is associated with hypertension (high blood pressure), another risk factor for heart disease. High levels of glucose in the blood increase the glycation of LDL, a key step in turning macrophages and LDL into pathological foam cells. Fat metabolism is also likely to be disrupted by insulin resistance, causing excessive levels of triglycerides, another coronary risk factor.

We recommend having your fasting glucose and insulin levels checked and suggest that you follow the guidelines in Chapter 5 for fasting glucose less than 90 and fasting insulin less than 5.

TABLE 2-3: SUMMARY OF REFERENCE VERSUS OPTIMAL BLOOD LEVELS FOR HEART HEALTH

BLOOD TEST	STANDARD REFERENCE RANGE	OPTIMAL LEVEL
C-reactive protein (mg/L)	< 5	< 1.3
Homocysteine (μmol/L)	< 15	< 7.5
Cholesterol (mg/dL)	100–199	160–180
LDL (mg/dL)	0–129	< 80 (if you have < 3 major risk factors) < 70 (if you have > 3 major risk factors)
HDL (mg/dL)	40–59	> 60
Cholesterol-to-HDL ratio	2.5–4.0	< 2.5
Triglycerides	0–149	< 100

Imaging Tests

Coronary Artery Calcium Score

Even though most heart attacks are caused by soft or *vulnerable* plaque, the amount of hard, calcified plaque in your coronary arteries is important because there is a direct correlation between the levels of hard and soft plaque in your arteries. Soft plaque still cannot be readily measured, while hard plaque is easier to detect. As we discussed earlier, our current understanding is that soft or vulnerable plaque in coronary arteries is the cause of most heart attacks. As noted, when inflammation is present, soft plaque might rupture, thereby stimulating local clot formation and downstream obstruction of bloodflow—also known as a heart attack.

The primary reason for the direct correlation between hard and soft plaque is that the body walls off vulnerable plaque with calcified deposits, so the rate at which calcified plaque is created also correlates to the amount of soft plaque. You can get an indirect measurement of how much of the more dangerous soft plaque you have by measuring your hard plaque with ultrafast or electron-beam computed tomography (EBCT), also known as the coronary artery calcium (CAC) score. The technique is fast, noninvasive, reasonably priced, and widely available.

Higher CAC scores relate to higher risks of a heart attack, but physicians differ about the usefulness of this method of risk assessment. In 2007, the American Heart Association's Consensus Document concluded that it "may be reasonable" to measure CAC in asymptomatic patients with intermediate risk (two or more major risk factors) of coronary disease, but not in low-risk patients (zero to one major risk factor) or in the general population. The American Heart Association document also advised against measuring CAC scores in asymptomatic high-risk patients because they already qualify for "intensive risk reducing therapies" regardless of the CAC results. We believe the CAC test is a useful measurement if you understand the different roles of calcified and soft plaque and if the results are properly interpreted. One caveat is that CAC cannot be used in patients who have undergone previous invasive cardiac procedures, such as bypass surgery, angioplasty, or stenting. The EBCT has been available for over 15 years, and it is unfortunate that it has taken this long for conventional medicine to begin to recognize its value.

Optimally, your CAC score will be zero, meaning you have no detectable plaque, but for any non-zero score, the higher the level, the greater your risk of a heart attack. Your score should be compared to the range of scores observed in people of your age and gender, expressed as a percentile rating (see Table 2-4). If your score places you in the 75th percentile or higher (meaning that 75 percent of people of your age and gender have scores lower than yours), we recommend that you address coronary plaque reduction urgently. It is actually the rate of increase in calcified plaque that indicates your level of vulnerable plaque. Without treatment, CAC

TABLE 2-4: CORONARY ARTERY CALCIUM SCORES (AVERAGE AND 75TH PERCENTILE)

MEN

AGE	AVERAGE (50TH PERCENTILE)	75TH PERCENTILE
40–45	2	11
46–50	3	36
51–55	15	110
56–60	54	229
61–65	117	386
66–70	166	538
70+	350	844

WOMEN

AGE	AVERAGE (50TH PERCENTILE)	75TH PERCENTILE
40–45	0.1	1
46–50	0.1	2
51–55	1	6
56–60	1	22
61–65	3	68
66–70	25	148
70+	51	231

scores can progress by 40 percent per year or more. With aggressive maneuvers, that rate of progression can be lowered to 10 percent or less and, as first demonstrated by Dr. Dean Ornish, even reversed by aggressive dietary manipulations.

Because the heart is imaged multiple times during this procedure, this test results in radiation exposure. In men, a single EBCT screening administers the equivalent amount of radiation as eight standard two-view chest x-rays. For women, because of the greater amount of radiation to the breast tissue, it is equivalent to about 15 chest x-rays (or five mammograms). Therefore, this procedure should not be repeated too frequently and should be done less often in women than men.

Intima-Media Thickness Measurement

Arteries have three layers; the intima and media are the two innermost layers. Increased thickness of these two layers is a sign of plaque build-up in the arteries. Measurement of intima-media thickness (IMT) is usually done on the carotid arteries, which are in the neck and carry blood to the brain. Carotid IMT is a diagnostic test that uses ultrasound waves and is a safe, noninvasive, inexpensive, and rapid method for determining carotid wall thickness and plaque. Carotid plaque appears on the ultrasound as an abnormal thickening between the intimal (innermost) and medial (midlevel) layers of the artery. Carotid IMT provides an estimate of a person's risk of stroke, the third leading cause of death.

Despite its many advantages and the amount of information it provides, use of this test is still not routine. In early 2007, a report in the journal *Circulation* found that carotid IMT is a strong predictor of both stroke and heart attack. Even so, a few months later, the U.S. Preventive Services Task Force recommended that asymptomatic adults *not* get routine carotid IMT screening.

Terry's clinic has been using the CAC score since 1996 and has recently added carotid IMT imaging to its testing panels. We feel these two tests provide critical information about a patient's risk of some of the leading causes of death and believe they should be included in the comprehensive health evaluations of any patient with two or more cardiovascular risk factors.

Peripheral Arterial Disease Screening

Peripheral artery disease (PAD) refers to plaque build-up in the peripheral arteries (the arteries in the arms and legs). The prevalence of asymptomatic PAD has been steadily increasing among American adults and is found in about 5 percent of adults 40 years and older, as reported at the 2007 Scientific Sessions of the American Heart Association. PAD occurs when plaque accumulates in the walls of arteries supplying blood to the limbs, especially the legs and feet. When PAD becomes more severe, individuals develop pain when they try to do routine tasks that involve use of the lower extremities, such as walking or climbing up stairs. Eventually, the pain becomes so severe that some type of bypass surgery or stenting must be done to restore blood circulation to the legs, although in some cases, amputation becomes necessary to prevent gangrene. PAD is also associated with increased risk of heart attack and stroke.

PAD can be easily diagnosed before symptoms develop by measuring the blood pressure at the ankles and comparing it to the blood pressure in the arms. The ratio between the two is referred to as the ankle–brachial index or ABI (*brachium* is Latin for "arm"). The blood pressure in the legs is normally higher than in the arms, so the ABI should always be greater than 1.0. When the ratio drops below 0.8, people often experience pain in their legs when walking. A person with an ABI less than 0.4 will typically have pain even at rest and often needs surgery.

A convenient and inexpensive way to get both carotid IMT and PAD screening is through mobile screening clinics, which are available several times each year in most cities. These tests are typically done in a van that travels to different locations each day.

Graded Exercise Testing

Graded exercise testing (GXT) is commonly known as a stress test or exercise tolerance test and evaluates how well an individual can tolerate the stress of exercise. During a GXT, the patient performs gradually increasing levels of exertion while the physician continuously monitors the electrocardiogram tracing, frequently checks the heart rate and blood pressure, and ensures that the patient feels well at each stage of the test. A GXT can be completed in less than 30 minutes and can provide several important pieces of information about heart health. Exercise testing is not foolproof, however—it is better at detecting more advanced blockages of the coronary arteries, such as those that are occluding more than 75 percent of an artery, but is less sensitive at detecting smaller blockages. That's why we recommend checking your coronary artery calcium score with the ultra-fast CT scan because it's better at detecting smaller blockages. The GXT can also provide a good indication of a person's aerobic conditioning and exercise tolerance. A baseline GXT is recommended for individuals over 40 years of age who have not been exercising previously in order to ensure that it is safe for them to begin exercising.

A standard exercise test is recommended for healthy individuals without heart-related symptoms. For patients with known cardiovascular disease or for individuals who have been experiencing undiagnosed chest pain, a thallium treadmill test is preferred. In this test, as soon as the maximum level of exercise has been achieved, the patient receives an intravenous injection containing radioactive thallium. Using an imaging scanner, doctors can compare the amount of blood flowing to the heart muscle during maximum exercise to the amount flowing at rest. Patients with blockages in the arteries often have adequate bloodflow at rest, but inadequate bloodflow during maximal exertion.

Some risks are associated with this test. These risks are rare and include low blood pressure, chest pain, arrhythmia (abnormal heart rate or rhythm), heart attack, and stroke. Having a trained physician with appropriate resuscitation equipment immediately at hand increases the level of safety.

THE END OF HEART DISEASE

You already have the knowledge to dramatically reduce your risk of heart disease. If you adopt all of the methods we have described in this chapter, you can reduce your risk of having a heart attack almost to zero, regardless of your genes. Once the new therapies discussed by our future selves are fully developed over the next 20 years, we will have easily available means to reverse the damage already done by atherosclerosis, and even by previous heart attacks. We really do have the means to overcome our genetic legacy!

3

DIGESTION

"As a result of your program, my blood pressure fell in 6 months from 135/82 to 110/62, my asthma of 30 years totally disappeared, my sinuses cleared up for the first time in my life, my edema went away, and my digestion improved from poor to perfect. I think everyone MUST follow your program if they want a long, pain-free life."

KIRIL (60), IDAHO

Choosing the right foods can have a profound impact on your health and longevity. But no matter how good that food is, it will do you little good if it isn't digested properly—and proper digestion is hardly the norm. A study published in 1993 in the journal *Digestive Diseases and Sciences* estimated that 70 percent of American adults suffered from some form of gastrointestinal distress resulting from poor digestion. More recently, the Consumer Healthcare Products Association estimated that total U.S. sales of over-the-counter gastrointestinal medications in 2006 were well over $2 billion—and that doesn't include the billions spent on prescription medications taken by those who found no relief from over-the-counter products.

Some of those digestive ailments, such as lactose intolerance (the inability to digest milk sugar), result from genetic defects. Many problems, however, simply result from excess stress placed on the digestive system by a poor diet. In some cases, specific supplements or medications may help to correct a malfunctioning system, but in all cases, eating the right foods is essential to restoring and maintaining gastrointestinal health.

HOW IT WORKS

The moment food passes your lips, digestion begins. As your teeth grind food into smaller pieces, the food mixes with saliva from your salivary glands. Saliva contains amylase, an enzyme that breaks starchy carbohydrates into simple sugars (glucose and maltose) for easier absorption later in the digestive process. Saliva also buffers (modifies the acid–alkaline balance) and lubricates the food, making it easier to swallow and preparing it for further digestion. The act of chewing doesn't appear to be very difficult, but inadequate chewing is not at all uncommon. The hurried pace of modern life leaves a lot of people literally eating food on the run, which can cause malabsorption, robbing your body of vital nutrients. Inadequate chewing also puts an extra burden on your digestive tract, causing it to overproduce digestive juices in an attempt to break down the large food particles. This can cause gas, bloating, and, over time, more serious damage to your digestive system. So relax, slow down (and sit down!), savor your food, and chew thoroughly—you'll enjoy your food more and you'll be doing your digestive tract a favor.

When you swallow, automatic rhythmic contractions of the muscles in your esophagus called *peristalsis* move the food down to your stomach. There, hydrochloric acid and other fluids secreted by the stomach lining dissolve it into a thick liquid called chyme. Stomach juice also contains digestive enzymes (pepsins) that begin to break proteins into amino acids for absorption further along the digestive tract. A protein called *intrinsic factor* is also secreted by the stomach to enable absorption of vitamin B_{12} in the small intestine.

After being processed by the stomach for 1 to 3 hours, chyme is slowly released into the small intestine, where the greatest amount of nutrient absorption takes place. In the small intestine, additional enzymes produced in the pancreas are released, further breaking down starches and proteins and beginning to break down fats in order to make absorption possible.

After passing through the small intestine, chyme enters the large intestine, also called the colon or bowel. At this point, most usable nutrients have already been absorbed by the small intestine. The primary function of the large intestine is to prepare what is left for elimination, a process that is aided by "friendly" bacteria that break down indigestible fiber such as cellulose.

Specialized cells in the colon also extract much of the water from the chyme, turning it into solid feces.

Interestingly, the bacteria in your colon are the most numerous type of cell in your body, and each of us harbors somewhere in the neighborhood of 2 or 3 pounds of these healthy bacteria. Our bodies actually have 10 times more cells of these bacteria than human cells with our own DNA. Beneficial colonic bacteria species, such as *Lactobacillus*, *Bifidobacterium*, and *Escherichia coli*, have several important functions. They help remove toxins from the gut, help produce nutrients, and, by their sheer numbers, crowd out pathological or toxic organisms.

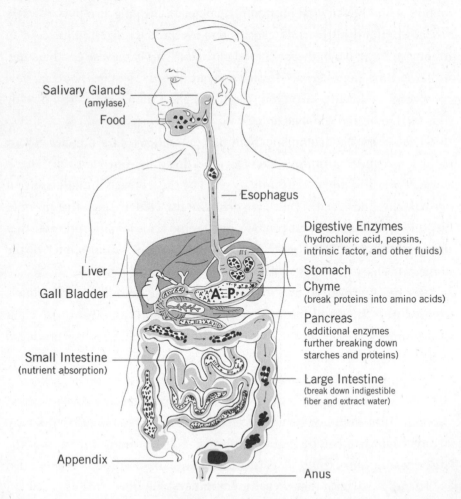

Salivary Glands
(amylase)

Food

Esophagus

Digestive Enzymes
(hydrochloric acid, pepsins,
intrinsic factor, and other fluids)

Liver

Stomach

Gall Bladder

Chyme
(break proteins into amino acids)

Pancreas
(additional enzymes
further breaking down
starches and proteins)

Small Intestine
(nutrient absorption)

Large Intestine
(break down indigestible
fiber and extract water)

Appendix

Anus

HOW DIGESTION GOES AWRY

As food makes its way through your digestive tract, there are many stops along the way where the process gets into trouble. If you suspect any of the conditions we describe here, our *TRANSCEND* program recommends that you always *Talk with your doctor* to determine the true source of the problem and your best treatment options. Following our *Nutrition* and *Supplementation* recommendations is also vital because it is an unhealthy diet that causes many digestive problems in the first place.

As we described above, when food reaches the stomach, intrinsic factor is secreted to facilitate absorption of vitamin B_{12}. If an insufficient amount of intrinsic factor is released, vitamin B_{12} deficiency can result. If not addressed, this can lead to a range of symptoms, including anemia, fatigue, tingling or numbness in your fingers and toes, poor sense of balance, depression, and dementia. However, insufficient intrinsic factor can be managed by supplementing with intrinsic factor and vitamin B_{12}. Keep in mind that vitamin B_{12} does not survive in the digestive tract, so you need to supplement it either through injections or sublingual tablets.

Helicobacter pylori (or *H. pylori*) infection is another potential problem at this stage of the digestive process. These bacteria can disrupt the fragile balance between the highly acidic gastric juices and the mucus that coats the stomach lining to protect it from being digested by those juices. Serious damage, such as peptic ulcers, is often the result of *H. pylori* infections. Note that at least one-third of Americans are colonized with *H. pylori*, many of whom are asymptomatic. Symptoms of peptic ulcers may include a burning pain or dull ache in the stomach, often 2 to 3 hours after eating or in the middle of the night, which typically goes away after eating. Should you exhibit such symptoms, your physician can administer a blood test for antibodies, an indicator of an *H. pylori* infection.

A more common stomach problem, which is often missed, is *hypochlorhydria*, a condition in which inadequate amounts of hydrochloric acid are produced. This can leave food insufficiently dissolved, resulting in the malabsorption of nutrients. Hypochlorhydria is a relatively common disorder, particularly among older people, with studies indicating that 20 to 30 percent of those over age 60 may be affected. Symptoms of hypochlorhydria are often mistaken for just the opposite—excess acid indigestion or acid reflux disease—and often result in people taking over-the-counter and prescription antacids, which only exacerbate the condition. The hair mineral analysis test described in the following section on Early Detection can suggest whether you have hypochlorhydria.

In the small intestine, a common genetic defect can result in insufficient production of the digestive enzyme lactase, which is needed to break down milk sugar or lactose. This results in a condition known as *lactose intolerance*, which can cause nausea, cramps, gas, and diarrhea when dairy products are consumed. Lactase production falls off after childhood in many people; more than 50 million Americans are affected, including about 50 percent of Hispanics, 80 percent of African-Americans, and 90 percent of Asians. Avoiding milk products altogether is one way to prevent symptoms. Many people also benefit by supplementing with lactase (available over the counter as Lactaid or other brands) when they consume milk products.

The small intestine is also where *leaky gut syndrome* can develop. Caused primarily by poor diet, it can also result from long-term use of NSAIDs (nonsteroidal anti-inflammatory drugs), such as aspirin or ibuprofen. Leaky gut is a progressive condition whose incidence increases with age, leaving most people over 50 affected to some degree. Over time, constant irritation of the lining of the small intestine results in chronic inflammation. The inflammation opens microscopic gaps between the cells, which allow minute food particles, gastric juices, microbes, and toxins to seep into the bloodstream. This can cause malabsorption of vital nutrients, strain the liver's ability to detoxify the blood, and, most important, lead to autoimmune reactions to normally healthy foods. When fragments of partially digested foodstuffs enter the bloodstream before they're fully digested, the immune system mistakes them for foreign invaders and forms antibodies against them. These antibodies, in turn, can cross-react with normal tissues, resulting in increased risk of asthma, arthritis, and other autoimmune diseases.

Treatment of leaky gut syndrome primarily involves ending the poor dietary practices that are at the root of the problem, thereby reducing inflammation so the intestinal lining can repair itself. Follow our *TRANSCEND Nutrition* recommendations for a low-glycemic-load, high-fiber diet (see Chapter 11) and avoid alcohol, caffeine, NSAIDs, and other foods and medicines that can irritate the intestine. In addition, testing for food allergies can determine whether specific foods, such as wheat, dairy, or citrus, are irritating your digestive system.

Often linked to leaky gut syndrome is an overgrowth of unhealthy bacteria or yeast in the intestinal tract. This can be addressed by using herbal medications and prescription drugs, as well as by *Supplementation* with probiotics (supplements that help replace unhealthy intestinal bacteria with healthy strains, such as acidophilus). In such cases, we also recommend taking 2 to 5 grams of FOS (fructooligosaccharides) a day; this fiber supplement provides nutritional support to the healthy intestinal bacteria. Enzyme supplements may also help break down foods for easier digestion. Supplemental garlic, bioflavonoids, and aloe vera may also be useful in promoting healing.

One of the most common disorders of the lower gastrointestinal tract is *irritable bowel syndrome* (IBS). There are no specific diagnostic tests for IBS. It is considered a "diagnosis of exclusion" after tests fail to confirm evidence of food allergy or intestinal damage and when other specific diseases have been ruled out (for example, Crohn's disease, ulcerative colitis, colon cancer, gastroesophageal reflux disease, or other inflammatory conditions). Cramps, pain, bloating, diarrhea, constipation, feelings of incomplete bowel movements, and excessive passing of gas can all be signs of IBS, and it is often associated with simultaneous upper gastrointestinal symptoms, such as heartburn, nausea, and excessive belching.

A specific cause for IBS, which is thought to affect 10 to 20 percent of the population, has not been fully established, but several factors have been found to contribute. A bout of food poisoning, a bacterial infection, such as with *Salmonella* or *Shigella*, or a stomach virus that leaves the lower gastrointestinal tract irritated and inflamed can sometimes lead to longer-term IBS symptoms. More frequently, the cause of that irritation and inflammation is long-term consumption of an inflammatory diet high in unhealthy fats, refined sugars, and starch. IBS also appears to be stress-related, with symptoms exacerbated by increased levels of anxiety, somewhat like normal "butterflies" run amok.

Following our *TRANSCEND Nutrition* recommendations can help to reduce IBS symptoms (see Chapter 11). You may also find relief through

Supplementation with enteric-coated peppermint oil and Seacure (a supplement that contains amino acids derived from fish). Managing your stress through *Relaxation* (see Chapter 9) may also reduce the incidence and severity of symptoms.

EARLY DETECTION

Many tests can be done to assess your digestive system function. Most can be performed at home with kits supplied by a healthcare provider because they don't require blood sampling. If you are troubled by digestive symptoms, the tests below can provide a wealth of information about how well your body is performing its functions of digestion, absorption, and utilization of the food you eat. Most of these tests are "functional" tests—they evaluate how well your digestive system is functioning—but they fall into the realm of "alternative" or "complementary" medicine as almost no conventional physicians utilize them. Conventional testing emphasizes direct observation of the intestinal tract, such as through an endoscope (upper gastrointestinal tract) or colonoscope (lower gastrointestinal tract), or by way of imaging studies such as ultrasounds, computed tomography, or barium x-rays.

Test kits for the tests below can be obtained through complementary healthcare practitioners. These tests are not taught in medical school, so interpretation is best done with the assistance of a nutritionally oriented healthcare practitioner such as a naturopath, chiropractor, or alternative and complementary medicine physician.

A **breath test for methane** can determine whether you are lactose intolerant. If so, you can avoid dairy products or take lactase supplements, which will break down lactose and will reduce symptoms when taken whenever dairy products are consumed. This test is available from a complementary doctor, and all you have to do to perform this test is exhale into a balloon. The gas in the balloon is then examined in the laboratory for the presence of methane, a gas that is not normally found in the colon but is formed by the bacteria in your bowel by fermentation of undigested lactose.

The **intestinal permeability test** can help you determine whether you have leaky gut syndrome. In this test, patients are given small amounts of two sugars, lactulose and mannitol, to drink, and a urine sample is collected some hours later. Normally, very little lactulose is absorbed by a healthy small intestine, so very little should appear in the urine. Mannitol, on the other hand, is very readily absorbed. A relatively high level of lactulose compared to mannitol indicates that the lactulose entered the bloodstream because the patient has leaky gut syndrome.

Hair mineral analysis provides two types of information. It can tell you about the levels of various toxic metals in your body and also indicate levels of essential minerals needed for optimal health. Hair mineral analysis can provide you with semi-quantitative data about essential minerals, such as calcium, magnesium, strontium, chromium, vanadium, manganese, selenium, zinc, and others. To perform this test, a few snips of hair, totaling about 1 gram, are collected from the back of the head and sent to the laboratory for analysis.

If a person has low levels of many essential minerals, the cause is often hypochlorhydria, or low stomach acid. To absorb beneficial minerals such as calcium, magnesium, chromium, and zinc from the food you eat, it is necessary to have adequate amounts of stomach acid. The hydrochloric acid in your stomach helps to ionize minerals to facilitate their absorption. Many people, particularly as they age, develop hypochlorhydria. Paradoxically, a common symptom of decreased stomach acid is heartburn—just the opposite of what you'd expect—and individuals typically treat their heartburn symptoms with antacids, which decreases stomach acid even further. This can lead to mineral malabsorption and is reflected in low levels of minerals in the hair analysis and can lead to symptoms such as chronic fatigue or frequent infections. If you perform a hair mineral test and find that many of your minerals are low, supplementation with betaine hydrochloride (an inexpensive form of hydrochloric acid) can help reverse the underlying hypochlorhydria and facilitate mineral absorption. (Caution: Betaine hydrochloride should never be taken if you have a history of peptic ulcer disease or are taking aspirin or NSAIDS [nonsteroidal anti-inflammatory drugs].)

We'll be the first to admit that collecting and preparing specimens for the **comprehensive stool analysis** (CSA) isn't particularly pleasant, and it isn't helped by the fact that the laboratory wants you to collect your stool specimen in a soft cardboard container that looks like it was intended to hold a small order of French fries. But, it's amazing how much information you can get by analyzing your stool. A typical CSA report provides information about all aspects of the digestive tract from the mouth to the colon. The report discusses four specific aspects of digestive tract function: digestion, absorption, metabolic markers, and colonic bacteria.

The *digestion* part of the CSA test looks at what happens from the time the food enters your mouth until the start of your small intestine. Maldigestion may be the result of improper chewing, inadequate amounts of amylase in your saliva, or hypochlorhydria. It includes a measurement of chymotrypsin as well as the number of undigested meat and vegetable fibers in the stool. Chymotrypsin is an enzyme excreted by the pancreas that helps to digest protein. An abnormally low amount of chymotrypsin in the stool suggests inadequate pancreatic function and the possible need for pancreatic digestive supplementation. A high level of this enzyme is associated with rapid transit time of food throughout the digestive tract, perhaps as a result of food allergies. Undigested vegetable and meat fibers in the stool suggest improper breakdown of food due to inadequate chewing or lack of stomach acid.

The *absorption* part of the CSA test measures how well your digested food is absorbed through the walls of the small intestine. Total fecal fat, cholesterol, long-chain fatty acids, and triglycerides are among the tests done. These tests provide information about the adequacy of the liver and gallbladder at making bile acids to help emulsify fat, the pancreas at producing lipase for fat breakdown, and the integrity of the walls of the small intestine for absorption. Liver, gallbladder, pancreatic, and small intestine dysfunction can all be diagnosed through this analysis.

Metabolic markers included in the CSA include several tests of large-bowel or colon health. The tests include beta-glucuronidase, n-butyrate, beneficial short-chain fatty acid distribution, and more. Beta-glucuronidase is an enzyme produced by some of the bacteria living in the colon, and elevated levels are associated with increased risk of colon cancer. N-butyrate is the main nutrient needed by the cells of the colon and help keep these cells healthy. Beneficial short-chain fatty acids are produced by the fermentation of nonabsorbed dietary fibers. They help to keep pathogens out of the bowel and also help produce as much as 30 percent of the energy used by the body.

Patient data from Terry's clinic suggest that the overwhelming majority of people do not have adequate amounts of normal *colonic bacteria*. As a result, they often develop intestinal dysbiosis, the overgrowth in the colon of abnormal organisms such as alpha hemolytic streptococcus, *Klebsiella*, or yeast. Treatment of dysbiosis includes restoration of normal bacteria through the use of a *probiotic supplement*, as well as an herbal product designed to kill the abnormal organisms. Performing a CSA can provide vital and lifesaving information—as Hippocrates observed centuries ago, "Death begins in the colon."

Reader: In the future, I hope you guys have come up with a better way of analyzing digestive function than collecting poop.

Terry2023: Even in your day, there were new technologies that allowed doctors to view the gastrointestinal tract with amazing clarity. As early as 2001, some doctors in Israel had developed a "pill camera," a camera the size of a large vitamin C capsule that patients would swallow. The camera would travel through the gastrointestinal tract and took photos as it traveled throughout the gut for up to 8 hours. This allowed doctors to get a clear view of the esophagus, stomach, and small bowel.

Reader: It sounds like it's been around for almost 10 years already. How come I haven't heard of it?

Ray: As we've said before, most physicians are very slow to adopt new technologies. Endoscopy, passing a flexible tube either through the mouth into the stomach or from the bottom up through the colon, has been how doctors looked at the gastrointestinal tract since the 1980s.

Terry2023: The swallowed camera was much more convenient, less uncomfortable, and less dangerous for the patient. Even so, it still took until the middle of the 20-teens to become the standard of care.

Ray2034: With the advent of the nanobiotic cameras not much bigger than blood cells, there was another paradigm shift again in the 2020s. These microscopic cameras move around on their own and explore the colon, taking pictures and performing chemical and biological tests. The latest version will perform immediate repairs when it detects a problem.

4

HORMONE OPTIMIZATION

No one would ever accuse Bill of not being masculine in every way. Now in his fifties, he was still one of the strongest men in the gym. He had briefly played professional football when he was younger and was still able to bench-press over 350 pounds. Despite exercising regularly, eating a prudent diet, and following almost all of the aspects of our program, he still had started to gain weight and was dismayed by the spare tire he was now carrying around his midsection. Bill ran a very successful business, had many friends, traveled often, and enjoyed many hobbies and interests. Despite all of his successes, a deep sense of failure and despair had begun to affect him, and he complained of difficulty concentrating. One of the happiest and most outgoing people anyone had ever met, Bill now found himself becoming withdrawn and depressed. His wife, Mary, complained that she couldn't remember the last time they had made love, and she was frustrated.

In the course of his comprehensive health evaluation, he was shocked to find that his testosterone level was very low. He started on testosterone replacement therapy, and almost immediately, he began to feel better. The excess weight he had gained literally melted away, and he was back to being the "life of the party" everywhere he went. He was now bench-pressing over 400 pounds, and some of the other men in the gym were complaining that he was using too many of the weight plates—over 1,200 pounds—for his leg presses. Within 3 months, Mary wanted to know what she could do to increase her sex drive so that she could keep up with her husband!

Hormones get blamed for a lot these days. The "raging hormones" of youth, "testosterone poisoning" in men, and "that time of the month" in women all allude in not-so-flattering terms to the role of hormones in making us who we are. And in the world of sports, wide-ranging "doping" scandals expose professional athletes trying to gain unfair advantage by ingesting or injecting, among other things, anabolic steroids and human growth hormone.

Don't be put off by all the bad press, though. Hormones are essential to life. These chemical messengers travel throughout the body to trigger and regulate an array of critical processes, including cell growth, digestion, sleep cycles, cholesterol levels, reproduction, and sex drive.

As you get older, production of some hormones declines while production of others increases, and the delicate balance between hormones needed for health can be lost. In fact, declining hormone levels are a key way that our outdated genetic code works to destroy our vitality and causes us to age prematurely. Hormone levels can also be affected by stress, disease, and diet. These changes in hormone levels have long been known to contribute to a number of conditions we associate with aging, including loss of muscle mass, increased body fat, cardiovascular disease, osteoporosis, cognitive defects, and cancer. Gaining a basic understanding of how hormones work in your body is crucial to living long and well.

There are actually hundreds of hormones, such as aldosterone (which controls blood pressure), ADH (which regulates body water), and digestive hormones (e.g., pancreozymin and cholecystokinin). In this chapter we will discuss just a few: thyroid hormone, our master hormone regulator, as well as those we have named the hormones of aging (cortisol and insulin), the hormones of youth (dehydroepiandrosterone [DHEA], growth hormone, and melatonin), and the sex hormones (estrogen, progesterone, and testosterone). Each of these hormones is critical to optimal health, and following our *TRANSCEND* recommendations will safely optimize and balance your hormone levels to stay youthful and live longer.

THYROID HORMONE— MASTER REGULATOR

Hormones are chemical substances created in one part of the body (typically in glands such as the ovaries or adrenals) that control or regulate bodily functions locally or at a distance. While all of the hormones of the body must work together in harmony for optimal health, thyroid hormone occupies a role of master regulator. When your thyroid function is out of kilter—which appears to be the case for one in 10 of us—health status is adversely affected on multiple levels.

The two main thyroid hormones produced by the thyroid gland are T_3 (triiodothyronine) and T_4 (thyroxine). T_4 is 20 times more prevalent in the bloodstream than T_3, and T_3 is three to four times more powerful, but both are essential. Both T_3 and T_4 are iodine-containing hormones, and in earlier times, before iodination of salt, people living in regions of the world located far from the ocean (and, thus, commonly deficient in iodine) were occasionally afflicted with goiters, or enlarged thyroid glands.

Thyroid hormones are concerned with heat regulation, so people who don't have adequate thyroid function (hypothyroidism) often complain of being cold all the time. These hormones also help control basal metabolic rate (the rate at which calories are burned), so fatigue and difficulty controlling weight are common hypothyroid symptoms as well. Other hypothyroid problems include constipation, memory problems, and sluggishness. People with hypothyroidism often have very high cholesterol levels (350 or higher) as well, predisposing them to coronary atherosclerosis. Excess thyroid activity (hyperthyroidism) is characterized by the opposite types of problems: trouble maintaining weight, feeling excessively warm, frequent bowel movements, nervousness, and palpitations. Other functions controlled by thyroid hormones include metabolism of fat, protein, and carbohydrates, as well as protein synthesis.

Many people are unaware that their thyroid function is abnormal. The Colorado Thyroid Disease Prevention Study looked at the thyroid blood test results of nearly 26,000 participants in a statewide health fair in 1995. Ten percent of the people were found to have abnormal levels of thyroid hormone *without their knowing it*—8.9 percent were hypothyroid and 1.1 percent were hyperthyroid. This suggests that nearly 9 million people in the United States may have elevated cholesterol levels as a result of occult or hidden hypothyroidism. In addition, even among people receiving thyroid medication for previously diagnosed hypothyroidism, this study revealed that 40 percent were taking inadequate amounts of medication.

Thyroid function is also interrelated to control of other hormone levels. Abnormal levels of thyroid hormone can place stress on your adrenals—and vice versa—as well as affect sex hormone levels. Later in this chapter we discuss three tests of thyroid function—the thyroid-stimulating hormone (TSH), free T_3, and free T_4—that you want to have checked regularly as part of your program of early detection.

CORTISOL AND INSULIN— TOO MUCH OF A GOOD THING?

Some hormones stimulate tissues in your body to break down (*catabolize* in medicalese) body tissues. As you gain in years, your levels of these catabolic hormones of aging can increase and the balance between hormones in your body can shift, leading to a range of age-associated conditions. Cortisol and, under some conditions, insulin and estrogen (in men only) can have this effect. The consequences of these hormonal changes can be severe, but there are ways to slow or even reverse the process.

CORTISOL

Without cortisol, you wouldn't be here. Over the eons, part of the reason your ancestors were able to survive and pass their genes on to the next generation was cortisol. When faced with potentially deadly perils, a sufficient jolt of cortisol from their adrenal glands set a series of bodily changes in motion that almost instantaneously prepared them to defend themselves or to escape. Your body still reacts to acute emergencies with this same fight-or-flight mechanism. Cortisol increases your breathing and heart rate to boost your strength and endurance, dilates your pupils to help you see better, and increases blood sugar levels for quickness of thought. At the same time, your reproductive, digestive, and immune systems are suppressed, which frees up internal resources to address the immediate threat.

In modern times, the occurrence of life-or-death situations is thankfully rare for most of us, but a low level of nearly continuous, chronic stress has taken its place. The economy, kids, politics, work, health, war—the list goes on—can induce a constant flow of cortisol when it really isn't needed, and the results can be toxic. Your digestive and immune systems must constantly work against the suppressive effects brought about by chronically elevated cortisol levels, resulting in digestive disorders and leaving you more susceptible to disease. Cortisol raises sodium and sugar levels in the blood, putting strain on your cardiovascular system. Bone and muscle tissue is broken down, while mental abilities are impaired. Cortisol, DHEA, testosterone, and estrogen are all made from the same cholesterol building blocks; therefore, higher cortisol production can deplete the raw material needed for producing these other hormones, lowering their levels and affecting the overall balance of hormones in your body. All of these work together to result in accelerated aging.

Many of the *TRANSCEND* principles outlined in this book will help to bring your cortisol production under control and maintain a more favorable hormone balance. For example:

- Manage your stress (see Chapter 9).

- Get a good night's sleep every night (see Chapter 1).

- Maintain a regular and vigorous exercise routine (see Chapter 14).

- Eat a low-sugar, low-glycemic-load diet (see Chapter 11).

- Supplement with DHEA.

- Take herbs such as natural licorice (used in Chinese herbal medicine) or ashwaganda (used in Ayurveda, traditional medicine native to India).

After sustained periods of chronic stress, your adrenal glands can become exhausted and no longer able to help you when you actually are subjected to periods of high levels of stress. There is a simple test you can do to assess the health of your adrenals. You can determine your cortisol levels throughout the day and assess your progress in reducing stress and maintaining a healthier hormone balance by performing the *adrenal stress index* test discussed later in the chapter.

INSULIN

Insulin is the hormone that moves glucose or sugar from the bloodstream into the cells. At optimal levels, insulin is your friend, functioning as an anabolic hormone to stimulate tissue growth. Unfortunately, in excess, it becomes your enemy. Excessive insulin can lead to a "spare tire" of unsightly and unhealthy body fat around the waistline. Too much insulin raises the risk of cardiovascular disease and hinders the work of your hormones of youth. Extra insulin also stimulates production of cortisol, which stimulates production of more insulin, thereby creating a vicious cycle. Excessive insulin, which is a consequence of high-sugar and high-glycemic-food diets so common today, does more to accelerate the aging process than just about any other contributor.

Specialized cells in the pancreas secrete insulin to help regulate blood sugar levels. The more sugary or high-glycemic carbohydrates you consume, the higher the sugar level in your bloodstream. If your body doesn't need all that sugar to fuel various cell functions, your pancreas releases insulin to bring sugar levels back under control. This surge in insulin converts the unused sugary carbs into triglycerides or fat, and stores it in fat cells throughout your body; it is also a major contributor to the deposition of plaque in the walls of your arteries.

To make matters worse, as you get older, your cells gradually become less sensitive to insulin's effects, so insulin levels must rise even more to produce the same results. This leads to a state of *insulin resistance*, which leads to even more insulin production. Over time, these increasing insulin levels, especially when combined with a poor diet, promote the weight gain and increase in body fat so typical of the aging process.

It's not a pretty picture, but this deterioration is far from inevitable. Commonsense lifestyle changes can help you keep your insulin levels under control: Following the major themes of our *TRANSCEND* program will help you keep your insulin at low, youthful levels. Maintaining a regular vigorous exercise routine burns blood sugar and drives it into your muscle cells, decreasing your body's need for insulin. Following our recommendations for a low-sugar, low-glycemic-load diet will lower insulin levels further. Finally, controlling your stress will lower your cortisol level and avoid the vicious cycle of cortisol-raising insulin.

HORMONES OF YOUTH

From conception into young adulthood, our bodies are flush with anabolic hormones, chemical messengers in the bloodstream that build you up. Among other things, these hormones play a vital role in building muscle and bone, providing energy and increasing sex drive, urging us to procreate. The main anabolic hormones of youth are DHEA (dehydroepiandrosterone), human growth hormone, and testosterone. We'll discuss testosterone along with the sex hormones below, and, because of its important effect on promoting deep, youthful, restful sleep, we'll also discuss the sleep hormone, melatonin, in this section. Although levels of these hormones naturally decline as you age, you can take steps to replenish your supplies and maintain a more youthful balance.

DHEA

Once thought to be just a component used in the production of other hormones, DHEA is now known to play a crucial role in a range of youth-promoting processes. However, as your years go up, your supply of DHEA goes down. By your late twenties, natural DHEA levels have already peaked, and it's all downhill from there—levels are about 50 percent of peak in your forties and just 5 percent of peak levels in your eighties. But don't despair! Researchers have come up with some very interesting discoveries.

For example, studies published in 2002 in the *Journal of Gerontology: Biological Sciences* and in 2003 in *Experimental Biology and Medicine* found that DHEA supplementation can slow some aging processes in lab animals. Higher DHEA levels have been found to correlate with lower risk of cardiovascular disease in men. As a precursor to other hormones, such as testosterone, DHEA supplements have been shown to boost libido, particularly in women. DHEA helps your body convert food into energy and burn excess fat. A 2002 study published in the *European Journal of Endocrinology* demonstrated the role of DHEA in lowering levels of IL-6 (interleukin-6) and TNF-α (tumor necrosis factor alpha), potent chemicals that can trigger inflammation in the body. When stress boosts your cortisol levels, which then depresses your immune system, DHEA can balance the effects of cortisol and improve immune response. Some research has even pointed to possible uses for DHEA in cancer prevention or treatment because of its ability to inhibit uncontrolled cell growth. DHEA is available without a prescription as an inexpensive nutritional supplement. With all these benefits, DHEA supplementation seems like a no-brainer.

But not so fast! Some of the glitter came off of the DHEA rose after a Mayo Clinic study published in 2006 showed that DHEA supplementation had no effect on body composition, muscle strength, physical endurance, bone density, insulin sensitivity, or quality of life. In addition, taking

DHEA can't be approached in the same off-hand manner as popping a low-dose daily multivitamin. This is a potent hormone with potential side effects, and more isn't necessarily better. If you elect for supplementation, you should have your current DHEA-S (DHEA-sulfate) level tested before taking DHEA. Suggested DHEA-S levels are 200 to 250 for men and 150 to 200 for women. To reach those levels, men can start with 15 to 25 milligrams of DHEA per day, and women with 5 to 10 milligrams per day. After supplementing for 6 to 8 weeks, have your DHEA-S level checked again. You can decrease or increase your dosage as needed to achieve the ranges above.

Caution: Men must be especially careful when taking DHEA. As a precursor to testosterone, DHEA can increase your PSA (prostate-specific antigen), a key marker for prostate cancer. Before you start taking DHEA, have your PSA level checked. Then check it regularly (every 6 to 12 months) while you are taking DHEA. If your PSA level goes up, stop taking DHEA and talk with your doctor.

There are two simple ways to raise DHEA levels naturally as part of your *TRANSCEND* program. Both regular exercise and caloric reduction help increase DHEA to more youthful levels naturally.

Human Growth Hormone

These days talk of human growth hormone (GH) conjures images of bulked-up bodybuilders and baseball stars who return from the off-season doubled in size. This is powerful stuff! Secreted by your pituitary gland, naturally occurring GH regulates your transition from infant to child to adolescent to adult.

Children and adults with GH deficiency syndrome have long been treated successfully with GH injections. In addition, numerous studies, many involving patients with GH deficiency, have confirmed the anti-aging power of GH. Research into GH has been intense, and a recent online search for the key words "growth hormone" found over *56,000 studies*. The consensus of these studies is that GH injections can decrease body fat, increase muscle mass, strengthen bones, and lower blood pressure. GH supplementation can also improve cholesterol levels, lipid profiles, cardiovascular function, and the age-related decline in insulin sensitivity.

So what's not to like? Simply put, costs, inconvenience, and side effects. GH replacement therapy can set you back $3,000 to $10,000 per year, and most health insurance won't cover GH except for specific conditions where other more cost-effective treatments are not available. GH therapy also requires daily injections, not something you'd want to add frivolously to your routine. Of most concern, however, are some adverse effects that can result from GH injections, including diabetes, glucose intolerance, carpal tunnel syndrome, aching joints, and edema. The studies suggest that GH doesn't cause cancer, but, as a powerful growth factor, it seems to stimulate the growth of existing tumors.

That leaves your GH levels dropping with each passing year and replacement therapy of questionable value when weighed against the costs and risks. Is there any way to reap the anti-aging benefits of GH without the downsides? Yes! By changing your lifestyle and following our *TRANSCEND* recommendations, you can boost your GH levels naturally:

- **Deep sleep** increases GH, so try to get a good night's sleep every night.

- **Exercise,** particularly strength training, raises GH, so be sure to maintain a regular exercise routine that includes weight lifting.

- **Protein consumption** increases GH production, while high-glycemic carbohydrates and sugars decrease it, so follow our recommendations for a low-sugar, low-glycemic-load diet.

- **The amino acids** (arginine, glutamine, glycine, and ornithine) stimulate the pituitary to secrete GH held in reserve, so consider supplementing with these if your levels remain low despite following the measures above.

- **Supplemental DHEA** can also increase GH.

Remember, however, that GH is a powerful hormone even when produced naturally in your own body. You want to achieve an ideal level, not just increase it indiscriminately. Before trying to raise your GH level, you should test for your blood level of IGF-1 (insulin-like growth factor-1), a better indicator of GH in the body than GH itself because it provides an average value. On the basis of your age and sex, your physician can help you determine your optimal IGF-1 level and follow your progress with subsequent testing as you strive for a healthy balance of GH in your system.

MELATONIN—NOT JUST FOR SLEEP

The increasing number of TV commercials for sleep medications is a good indicator of the increasing number of people suffering from sleep disorders. Numerous studies in recent years have described an "epidemic" of sleep deprivation. For example, a 2002 National Sleep Foundation survey found that 47 million American adults were not getting adequate sleep, with grave consequences to individual health and significant costs to the economy from accidents and missed work. Prolonged sleep deprivation stresses the body and suppresses the immune system. But counting on over-the-counter or prescription sleep aids to get you through the night can bring its own set of problems, including dependence, sleep walking, and daytime drowsiness.

Melatonin, a hormone produced in the pineal gland found in the recesses of your brain, is crucial to your ability to enjoy deep restorative sleep. Triggered by the daily cycle of sunlight and darkness, melatonin levels start to rise in the evening, crest around midnight, and decrease toward morning so, ideally, you wake refreshed.

Just like the other hormones of youth, however, melatonin production drops with age. From peak capacity around age 7, levels decline steeply in your teens and continue to fall from there. By the time you reach 60, melatonin levels are half what they were in your twenties, so it's no wonder that around half of the U.S. population 65 and older complains of sleep disorders.

But there's more to melatonin than just regulation of sleep patterns. Studies indicate that, as a powerful antioxidant, it may be of use in cancer treatment, particularly for breast cancer. And melatonin also plays a key role in the overall aging process and may help retard brain aging. As levels decline over time, systems throughout your body respond to that reduction by also slowing down in a cycle that accelerates aging. Your immune defenses weaken and you become more susceptible to autoimmune disorders, infections, and cancer. Testosterone production declines in men. And in women, estrogen production falls off and the ovaries shut down with the onset of menopause. As your systems and organs decline, including your ability to secrete melatonin, the aging process gathers speed.

Sounds pretty grim, doesn't it? But all is not lost. The evidence suggests that you can slow the progression with melatonin supplements. As with all hormones, melatonin can be potent and a little goes a long way. When used exclusively for its anti-aging properties, we suggest taking 0.1 to 0.5 milligram in the sublingual form about a half-hour before going to bed at night. Dosage can be increased to 1 to 3 milligrams if well tolerated. This should be all that is needed if you are in good health with no sleep concerns.

If you are having problems getting the rest you need, though, and want to avoid the use of prescription sleep medications, melatonin supplements may help you there as well. A study of patients over the age of 55 published in the *American Journal of Medicine* in 2004 found that melatonin replacement therapy significantly improved quality of sleep.

Melatonin taken sublingually (under the tongue) is particularly effective as a sleep aid because it is absorbed rapidly into the body. Between 1 and 3 milligrams taken just before bedtime should be sufficient in most cases, although some people need up to 9 milligrams. Melatonin has also been found to relieve symptoms of jet lag in people traveling across time zones. Taking 1 to 3 milligrams at bedtime on the first few nights at your destination can help reset your body's internal clock to the new location.

SEX HORMONES

There's probably no other word in the English language as loaded as "sex." Points of reference range from the interpersonal, moral, and medical to the cultural, religious, and commercial (and a whole lot more in between), but the sex hormones—estrogen, progesterone, and testosterone—are the driving force behind it all. The primary evolutionary purpose of these hormones is to ensure the survival of the species, to guarantee each succeeding generation by creating strong urges in men and women to reproduce.

Beyond that absolutely critical task, however, sex hormones are active players in regulating many physiologic functions. As with the previously discussed hormones of youth, maintaining optimal levels and ratios of sex hormones can play a vital role in maintaining optimal health and longevity. This is no easy task, however, because medical opinion on the subject changes frequently and, at times, reverses itself completely. Recommendations also vary for men and women and by age. We'll try to sort out the latest thinking, realizing that this information is subject to change as new research becomes available.

As most of us recall with a mixture of both relief and regret, our sex hormones start at very high levels during adolescence, reach a lower but steady-state level throughout our twenties, and then begin a slow decline for the rest of our lives. And in the midst of it all, at around the age of 50, women experience menopause, literally the cessation of menstruation, at which time their levels of estrogen and progesterone drop precipitately and fluctuate while testosterone continues its slow downward spiral. It is the sudden drop and fluctuation in estrogen and progesterone that causes many of the disconcerting symptoms, such as hot flashes and mood swings, experienced by many women around the time of menopause.

In addition to stimulating interest in sexual thoughts and activities, the sex hormones also exert profound physiologic effects on the body. No substances found in the human body are more closely associated with youthfulness than these three hormones. Having lower levels of estrogen, progesterone, and testosterone plays a key role in the evolutionary blueprint that we've inherited from our Stone Age ancestors. Declining hormone levels serve two main purposes: First, they make us less interested in mating, reducing the chance of overpopulation; but, second, these lower hormone levels destroy our youthful vitality, which in Stone Age times helped lead to our demise sometime around 30 years of age or shortly thereafter. Remembering that we are genetically almost identical to cavemen and cavewomen and that our ancient (and outdated) genetic makeup was designed for survival of the human species during an earlier time when day-to-day existence was very much hand-to-mouth, lower hormone levels that come on with age helped ensure that very few members of the tribe lived much past 30 years of age.

Higher levels of sex hormones not only make us, if you'll pardon the phrase, horny, they also keep us healthy. They help keep our bones and muscles strong, and they improve our eyesight and hearing. Stone Age people needed all of their human faculties to operate at a high level all of the time just to survive from one day to the next. Many of the changes associated with the aging process, such as loss of muscle mass (and its exchange into fat), osteoporosis, weaker eyesight, and diminished hearing are directly related to decreasing hormone levels. This helps explain why most cave people weren't able to survive very far into their thirties; they weren't able to compete against younger members of the clan who had higher levels of these hormones.

The idea that we can regain some of our youthful vigor by increasing our hormone levels through supplementation originated millennia ago. We think it began in ancient Egypt, when the pharaohs would have all the teenage boys in the community stand in a circle and pool their urine, which they would leave to sit in the hot desert sun. After the water evaporated, they would roll the remaining precipitate into little balls, which were rich in male hormones, and take these as an early form of testosterone replacement therapy. It may sound unappetizing, but our modern forms of hormone therapy (HT) haven't really changed all that much in the 4,000 years since the times of the pharaohs. Even today, Premarin, a commonly prescribed form of estrogen for hormone therapy in menopausal women, is derived from and named after its source—literally PREgnant MARe's urINe. It is important to note that as the name suggests, Premarin is not natural human estrogen, it's horse estrogen, and there is reason to believe that many of the side effects discussed below from estrogen substitution therapy result from the fact that the conventional treatment for women fails to use bioidentical hormones.

ESTROGEN AND PROGESTERONE— VERSION 2.0(09)

There is probably no other field of medicine in which the dominant sentiment of the medical community changes so often as it does with sex hormone therapy. Estrogen and progesterone supplementation for the treatment of menopausal symptoms has been around for more than 50 years, but medical opinion regarding its value has fluctuated on a regular basis. The current change began in 2002, when the results of the Women's Health Initiative (WHI) were published. This study began in 1997 at a time when 38 percent of the menopausal women in the United States were taking some type of hormone therapy, with most of them receiving Premarin. This study was conducted under the auspices of the National Institutes of Health (NIH) and involved almost 17,000 menopausal women. WHI was designed to explore the risks versus benefits of Prempro, which was the most commonly prescribed combined hormone preparation in the United States. Prempro is a combination of Premarin and Provera (a synthetic progesterone-like drug also known by its generic name, medroxyprogesterone). In the fall of 2002, the NIH abruptly terminated the study because they felt that the risks of taking Prempro seemed too great. They had found that Prempro users had 29 percent more heart attacks, 41 percent more strokes, a 26 percent increase in breast cancer, and almost twice the chance of developing a blood clot. On the positive side, the study also confirmed some benefits of HT that had been seen in previous studies: A 33 percent decrease in the risk of hip fracture, a 37 percent decrease in the risk of colorectal cancer, and effective relief of menopausal symptoms, such as hot flashes and vaginal atrophy. Yet, because of the negative effects found, within 1 year, prescriptions for the combo product Prempro fell by 66 percent and those for Premarin alone by 33 percent. The tide of opinion had reversed direction once again.

In the past few years, the data from the WHI trial have been examined more closely. It now appears that the severe adverse effects reported were found much more often in older women, that is, those more than 15 years after menopause. In fact, upon further analysis, an article in the *Journal of the American Medical Association* (*JAMA*) in 2007 reported that women who were less than 10 years from menopause had a 24 percent *decreased risk* of cardiovascular disease. It's usually only younger women who suffer symptoms such as hot flashes, mood swings, and irregular bleeding around the time of menopause, and HT appears relatively safer for them. As these findings have started to filter into the medical community, the tide is changing back once again, and more doctors have started to prescribe HT for their patients to help control symptoms around the time of menopause. In the latest analysis, it may even be that HT may decrease the risk of cardiovascular disease in women under 60 years of age rather than increase it after all.

Yet, many other symptoms and effects that result from lower levels of sex hormones do not go away with age, and they continue to wreak havoc on a woman's body. These include loss of bone mass, increasing the possibility of osteoporosis; atrophic vaginitis, creating itching and painful intercourse; menopausal weight gain; thinning of the skin; and decreased sex drive. These are classic symptoms of the aging process, and menopausal women of all ages have found that HT can help them with many of these symptoms and age-related changes as well. Yet, the large WHI study seemed to show that taking HT was more dangerous for older women.

Current medical thinking is coming around towards acceptance of the use of HT for up to 10 years after menopause if needed to control classic menopausal symptoms. Since the average age of menopause is around 51 years, the risk–benefit ratio seems to weigh in favor of HT use up to, say, 60 years of age. After age 60, heart attacks, strokes, blood clots, and breast cancer—the problems associated with Prempro use in the WHI study—become more common for women in general, and as the WHI study showed, more common still for women receiving Prempro.

However, one problem with the WHI study may have been the investigators' choice of Prempro rather than another form of hormone therapy. This formulation contains forms of estrogen foreign to a woman's body—remember Premarin is conjugated equine estrogen and is derived from horses, and Prempro also contains Provera, which is not progesterone but rather a progesterone-like compound known as a *progestogen* (also foreign to a woman's body). Medroxyprogesterone has subsequently been shown to counteract some of the cardioprotective benefits of estrogen and is also known to increase the risk of blood clots.

As a result, a movement has been sparked within the medical community away from artificial estrogens and progestogens such as conjugated equine estrogen and medroxyprogesterone and towards usage of estrogen and progesterone in their more natural forms. Historically, doctors were hesitant to use natural estrogen and progesterone, which are commonly called their *bioidentical* forms. Bioidentical estrogen and progesterone have been available for several decades as topical gels and creams, but these have never been as popular as oral formulations. Yet, oral bioidentical hormones couldn't be used because they are easily destroyed by the acid in the stomach when they're taken by mouth. Thanks to recent advances in pharmaceutical technology, however, bioidentical hormones are now widely available as *micronized* formulations through compounding pharmacies (see next paragraph for further discussion of these pharmacies). A micronized preparation is able to bypass the stomach acid and be absorbed into the bloodstream intact in the intestinal tract. Bioidentical estrogen is also now widely available at most conventional pharmacies as topical gels, patches, or vaginal preparations under brand names such as Climara, Estraderm, Vivelle, and Estrace. Bioidentical progesterone is available orally as Prometrium, as a topical gel as Pro-Gest, and vaginally as Crinone and Ultrogestan.

You can also obtain micronized oral forms of bioidentical estrogen and progesterone as generic preparations through compounding pharmacies. Unlike most conventional pharmacies, which just use prepackaged formulations, compounding pharmacies can create any strength or combination of these hormones, enabling them to tailor a personalized formulation for each individual.

Still, the question lingers—is bioidentical HT safer? Is it possible for a woman to get the same benefits without experiencing the same risks found with artificial, chemically altered hormones by taking bioidentical estrogen and progesterone? Unfortunately, there haven't been any large, double-blind, placebo-controlled trials such as the WHI to prove this beyond a shadow of a doubt, but several studies suggest that this is the case.

In a study published in 2005 in *International Journal of Cancer*, researchers followed 54,548 postmenopausal women and compared the risk of their developing breast cancer in relation to different types of bioidentical and conventional hormone therapies. Women were followed for 5.8 years, and the average duration of hormone therapy was 2.8 years. The type of progesterone that the woman took seemed to play a significant role in their risk of breast cancer. When women took estrogen along with a synthetic progestin, such as medroxyprogesterone, their breast cancer risk increased 40 percent; however, when micronized bioidentical progesterone was used, risk decreased 10 percent. This was a very large study, but since it wasn't a double-blind, placebo-controlled trial, its conclusions aren't as compelling as those of the WHI. Still, they do strongly suggest that bioidentical HT is safer.

A study done at the University of Connecticut in 2000 suggested that bio-identical estrogen could provide the same degree of osteoporosis protection in women over 65 as compared with conventional HT. Several small studies have indicated that bioidentical HT does not increase cardiac risk and has beneficial effects on blood lipids. A recent heart catheterization study showed that bioidentical estrogen didn't lead to progression of coronary artery disease in postmenopausal women already known to have heart disease.

The American College of Obstetricians and Gynecologists doesn't disavow the use of HT in menopause. Their recommendation is that HT should be administered with the lowest dose possible for the shortest possible period. We believe that this recommendation is prudent but also reflects the symptom control bias of conventional medicine, to which we don't subscribe. We feel medicine can do more than merely treat symptoms; rather, we can (and should) use it to aggressively address our underlying biochemistry as well as the aging processes. Many menopausal women want HT to do more than simply control uncomfortable symptoms. They want to use it to help them reverse the downward spiral in their health and well-being that is a result of their outdated Stone Age genes that force them to spend the last several decades of their lives with suboptimal hormone levels.

Further review of the 2007 *JAMA* analysis of the WHI study showed that cardiovascular risk increased 10 percent in women who were 10 to 19 years postmenopause and increased 26 percent in women 20 to 29 years postmenopause. This means that women 60 to 69 years of age had four more heart attacks and strokes per 10,000 women-years, and women between 70 to 79 years of age had 17 more events per 10,000 women-years. Women need to factor in this increased risk if they elect to continue HT past 60 years of age.

An increased risk of blood clots was also found in the WHI study. All types of estrogen substitution therapy—both conjugated equine estrogens and bio-identical estradiol—as well as artificial progestogen therapy appear to be linked to a higher risk of blood clots.

In an abstract presented at the 2004 annual meeting of the American Society of Hematology, however, Kenna Stephenson and colleagues found that the use of bioidentical progesterone cream did not increase blood markers associated with increased clotting. We suggest that to reduce the risk of blood clots to a minimum, bioidentical progesterone should be used in almost all cases. In addition, to decrease the risk even further, be sure to include ample amounts—such as 2 or more grams—of EPA/DHA (fish oil) per day, as well as the natural blood thinner nattokinase. Take one or two capsules (1,440 fibrin units) twice daily to safely thin your blood and decrease your risk of blood clots. As a final step, consider taking two 81-milligram aspirin tablets a day as well.

In addition, we feel it is important to mimic the natural ebb and flow of hormone levels when taking HT. A young woman does not make the same amount of estrogen and progesterone every day of the month. When a woman is still menstruating, she will go through a cycle of hormonal changes throughout a given month. The first half of the month, her ovaries will make estrogen. Then, midway through her cycle, she will ovulate and experience higher levels of progesterone as well as estrogen. Then, unless pregnancy occurs, her levels of both estrogen and progesterone fall and she has her menstrual period. Many doctors who prescribe bioidentical HT try to mimic this natural cycle. Women will take estrogen for the first 2 weeks of the month, then add progesterone as well for the next 2 weeks, and then, finally, won't take either hormone for a few days at the end of the month, corresponding to the lower hormone levels of the menstrual period. Estrogen and progesterone are designed to balance one another, and taking HT in this fashion allows all the woman's tissues to experience the benefits of youthful levels of hormones in a balanced fashion and reduces the total amount of hormones taken.

To increase the margin of safety and enable you to use HT later in life, we recommend that the hormone doses used decrease with age. For example, if a woman took bioidentical estrogen at a dose of, say, 1 milligram of estradiol per day during her fifties, a more appropriate dose might be 0.5 milligram per day during her sixties.

It is worth pointing out that testosterone substitution therapy in men also started out with forms of testosterone that were not bioidentical. As we discuss below, negative side effects of testosterone supplementation went away when the change was made from artificial to bioidentical preparations. This is another suggestion that bioidentical hormones are superior in general.

Terry2023: I hate to butt in, guys, but things are about to change quite a bit in how you'll be prescribing HT starting in the next couple of years. In 2010 genomics testing to determine a woman's risk of blood clots, breast cancer, heart attack, and stroke became commercially available.

Reader: So I can find out exactly if it's safe for me to take HT after menopause?

Terry2023: It's not exact, as nothing in medicine is 100 percent, but it's a lot better than what you've got now. Unfortunately, it still took several more years for most doctors in the mainstream medical community to start using genomics testing before they would prescribe HT. But, luckily, there were some forward-thinking physicians who began to use this therapy right out of the box in 2010. How old are you, by the way?

Reader: I just turned 48, but I'm still having my period every month, and I haven't had any hot flashes or other symptoms.

Terry2023: Sounds like you're still a few years off from when you might need HT, but it'll be nice for you to be able to know if it's a safe option for you when you might need it in the future.

Ray2023: The twenty-teens have also been the decade of the artificial organs. The artificial pancreas was first. So, very few diabetic patients need to bother with insulin injections or checking their blood sugars any longer.

Terry2023: In addition, we now have artificial ovaries and testes, which are closed-loop systems, able to monitor the levels of hormones in the bloodstream and adjust their levels by using bioidentical replacements on a minute-to-minute, day-by-day basis.

Women can also use some nonhormonal alternatives to treat menopausal symptoms. Isoflavones, such as genistein and daidzein, are plantlike estrogens that reduce hot flashes and help protect against osteoporosis, heart disease, and cancer. Women need to consume about 50 milligrams of isoflavones per day to achieve these benefits. This amount can be obtained by drinking 2 cups of soy milk, or eating 6 ounces of tofu or 4 ounces of a fermented soy product such as tempeh. Studies have also shown that the herb black cohosh can help reduce hot flashes and other menopausal symptoms.

TESTOSTERONE—NOT FOR MEN ONLY

A woman's levels of estrogen, progesterone, and testosterone peak at 20 years of age, and by 40 she is making only half of the amounts of these hormones. After menopause, many women elect for HT with estrogen and progesterone; however, because it is not FDA-approved for use in women, relatively few receive testosterone replacement. A woman's level of testosterone is only 5 percent of a man's level, but this small amount is critical to vitality and well-being. Testosterone is your body's main natural anabolic steroid and can help you maintain muscle mass, keep your skin from thinning, and keep your bones strong. While a man will typically receive doses of testosterone supplementation in the range of 50 to 100 milligrams per day or more, a woman is often helped dramatically by only 1 or 2 milligrams.

After vaginal dryness, the main sexual complaint of menopausal women is low sex drive and difficulty achieving orgasm. These can be helped significantly by supplemental testosterone. While testosterone is widely available and approved for use in women outside of the United States, it has still not received approval for use in women by the FDA. Just because testosterone has not yet received FDA approval for women does not mean that your doctor cannot write a prescription for it. Before prescribing a drug for an "off-label" use such as this, your physician needs to do a few things. First, she needs to explain that the medication is not FDA approved for this indication—in this case, that it is FDA approved only for use in men. Next, she needs to go over the benefits of testosterone use, as well as the risks, such as breast tenderness, acne, increased aggression, and facial hair, which can usually be controlled by lowering the dose. Finally, she will have you sign an informed consent form indicating that you understand the side effects and risks of using an unapproved drug and realize that the use of testosterone in women is an off-label use and not approved by the FDA. Unfortunately, the odds are good that your doctor won't do this for you. Most doctors are unwilling to prescribe anything that is not FDA approved because of fear of reprisals from their local medical boards. Forward-thinking physicians in your area who are willing to prescribe testosterone for women can be found on the American College for Advancement in Medicine's Web site, www.acam.org, as well as through the American Academy of Anti-Aging Medicine at www.worldhealth.net.

For men, testosterone is the quintessential hormone. The decline in a man's testosterone level that comes along with aging is part of the genetic program that has been inherited from our caveman ancestors. Like other parts of this outdated software, this decline is designed to destroy health and hasten death, typically beginning by age 30. But, unlike in women, for whom the drop in hormone levels is sudden and abrupt, the decrease in a man's testosterone levels—referred to as *andropause*—is much more gradual but lasts throughout a man's lifetime.

Men with declining levels of testosterone experience a decrease in their muscle mass, an increased tendency to gain fat, particularly around their belt line, decreased sexual desire, erectile dysfunction, fatigue and depression, thinning of the skin and bones, and a decrease in their energy and intellectual function. Exercise, particularly weight-lifting exercise or strength training, is one of the best ways for a man to increase his level of testosterone naturally. Some men, however, even though they exercise regularly, still have levels that are too low for optimal health. For them, the benefits of testosterone replacement therapy can be dramatic, as in the example given at the beginning of this chapter.

Testosterone replacement therapy for men is most commonly prescribed as a topical gel or patch applied to the skin every day or as a weekly injection. Conventional pharmacies carry topical testosterone gel as Testim, or as a patch applied to the skin under the brand name of AndroGel. Topical testosterone gel is also available in its generic form from compounding pharmacies. These topical formulations work well for many men, but some don't seem to achieve adequate levels of hormone in their bloodstreams unless they use injections. These injections are done at home, typically about once a week, and are almost painless. All of these formulations are bioidentical.

A few precautions must be observed when men take testosterone. Regular prostate cancer screening must be done with both digital rectal examinations and blood tests for PSA. Current thinking is that testosterone replacement does not increase the risk of prostate cancer; however, it seems able to speed the growth of an existing cancer, so regular prostate cancer screening is essential. Testosterone therapy can also thicken the blood and raise levels of estrogen. As a result, it is important to regularly check your hematocrit (a measurement of blood thickness) as well as your estradiol level.

Men often notice that they are stronger after starting to use testosterone, but this may be a beneficial side effect of the therapy, not the desired goal. Ethical physicians do not prescribe testosterone for their patients just to increase muscle size or to improve athletic performance. When used properly to treat documented testosterone deficiency, as measured by blood tests and andropausal symptoms, testosterone can be an important adjunct in maintaining the health and well-being of men as they grow older.

As we noted above, the story of testosterone substitution therapy for men is similar in one regard to the story of estrogen substitution in women. Testosterone substitution started with an oral artificial form of testosterone that was not identical to human testosterone. Although benefits were noted in terms of increased muscle mass, side effects, including abnormal liver function tests, were also noted. Bioidentical testosterone became the standard therapy a few years ago, and studies have shown benefits in terms of both sexual and heart health, without the side effects noted for the earlier artificial formulation. This is another suggestion that we will see a similar benefit in switching from artificial to bioidentical forms of estrogen and progesterone for women.

MAKE LOVE, LIVE LONGER

There is a wide range of what constitutes "normal" when it comes to sexual activity. We know of couples who make love twice a day. For the majority, it's more like two or three times a week. For others, it's twice a month, and, for some, twice a year . . . or never. All of these are "normal" for these individuals or couples, but experiencing orgasm on a regular basis can play an important role in your wellness program and help you live longer. In other words, orgasms are good for you.

A study published in the prestigious *British Medical Journal* "examine[d] the relation between frequency of orgasm and mortality" in men. The sexual habits of 918 men 45 to 59 years of age from the town of Caerphilly, South Wales, and five adjacent villages were followed for 4 years. The study found that men with more orgasms per year experienced half the rate of death from heart attack and other causes compared with men who were less sexually active. The study authors concluded that "sexual activity seems to have a protective effect on men's health." While no similar studies on the relationship between sexual activity and mortality have been performed on women to date, many positive health benefits have been reported in women as well as men. Among the documented advantages associated with regular sexual activity for both sexes are the following:

- **Stress reduction**—A study from Scotland published in the journal *Biological Psychology* found that both men and women who engaged in regular sexual activity experienced lower blood pressure in response to stressful situations such as public speaking compared with people who were abstinent.

- **Increased intimacy**—Orgasm triggers the release of the neurotransmitter oxytocin, which promotes feelings of intimacy and nurturing.

- **Boosts in immune function**—Researchers in Pennsylvania found that having sexual activity once or twice a week increases levels of immunoglobulin A, an important component of the immune system.

- **Prostate cancer prevention**—British researchers have found that frequent ejaculation reduces a man's risk of prostate cancer.

- **Pain reduction**—The surge of oxytocin that precedes orgasm increases levels of prostaglandins, short-acting hormones associated with intense pleasure and diminution of pain, such as that associated with premenstrual syndrome.

Yet the frequency of sexual intercourse among married couples has fallen in recent years. When Alfred Kinsey performed his studies on married couples in the late 1940s and early 1950s, the average frequency of intercourse was twice a week. Social scientist Morton Hunt found that it had increased to approximately 3.25 times weekly by the mid-1970s, which he felt was due to societal changes related to the availability of the birth control pill and the "sexual revolution." More recent studies, however, have found that the frequency has fallen and now averages less than twice weekly. Regular sexual activity, either with a partner or masturbation, offers health benefits and can play an important role in maintaining health.

Reader: As long as we're on the topic of sex hormones, I've got to ask—what's sex like in the future?

Terry2023: Most of us still have sex the same way as we did in your day, but robotic sex has become quite popular for many people.

Reader: Sounds weird.

Ray2023: You'd have a hard time recognizing your partner as a robot today, at least physically. The skin texture of sexual surrogates today is quite realistic, and they are programmed to respond to sexual activity in manners that are very similar to human men and women.

Ray2034: Now that we're able to have experiences in virtual reality in 2034, they're just as realistic and compelling as in real reality, and, as you might imagine, sex is a whole new ball game.

Reader: So I'll be able to have virtual sex with my favorite celebrity?

Reader's spouse: Hey, wait a minute!

Ray2034: You'll be able to have any type of experience with anyone, ranging from a business meeting with colleagues from around the globe to sensual and sexual encounters in virtual environments. And it's just as easy as using your cell phone.

Reader: So exactly how does this work?

Ray2034: Nanobots in our bloodstream enter our brains through the capillaries. If I want to enter a virtual reality environment, the nanobots in my brain shut down the signals coming from my real senses and replace them with the signals that my brain would be receiving if I were actually in the virtual environment. We have millions of environments to choose from, and design of virtual reality environments is a thriving new art form. So to my brain it feels like I am actually in the virtual environment, and it incorporates all five of my senses. If I go to move my arm, it is my virtual arm on my virtual body that moves. So I don't have to have the same body that I have in real reality. An interesting experience for many couples is to take on the body of the other in the virtual environment and explore the relationship from their partner's point of view.

Reader's spouse: You'll finally get to see what I've been talking about all these years.

Ray2034: Yes, see and feel too.

EARLY DETECTION—HORMONES

Having optimal hormone levels can play a key role in maintaining health and youthfulness throughout life, yet very few people have had their levels of any of these hormones measured. Getting a basic hormone profile, which includes measurements of the above hormones (other than melatonin, which we don't typically check), can provide you with critical information as to how well your body is handling the changes the aging process is trying to throw at you.

THYROID TESTING

As we mentioned earlier in this chapter, three blood tests will give you a comprehensive picture of your thyroid function: TSH, free T_3, and free T_4. TSH is actually a hormone produced in the pituitary gland in the brain, which stimulates your thyroid to secrete the thyroid hormones T_3 and T_4 into the bloodstream. There is an inverse relationship between TSH blood levels and thyroid function. When the levels of T_3 and T_4 are low, the pituitary releases more TSH, which tells the thyroid gland to release more thyroid hormone into the bloodstream. Historically, this has been the most common way of detecting abnormalities of thyroid function. High TSH corresponds to hypothyroidism, necessitating more supplemental thyroid, while low TSH suggests a hyperthyroid condition, which usually needs to be treated with surgery or medication.

More recently, doctors have begun to follow the free or unbound levels of the thyroid hormones themselves—free T_3 and free T_4. More than 99.5 percent of the thyroid hormone that circulates in your bloodstream is bound to proteins, which renders it inactive. The small fraction of a percent that is unbound or "free" is what actually does the work of thyroid hormone, and this is measured by the free T_3 and T_4 levels. Until recently, these tests cost several hundred dollars each and thus were done infrequently. Thanks to newer technologies, free T_3 and T_4 levels can now be measured accurately and inexpensively, enabling you to know the precise levels of these hormones in your bloodstream. They can also help your doctor prescribe the precise dose of thyroid medication to help optimize your thyroid function. Unfortunately, many doctors still rely exclusively on the TSH and the total T_4 level, so be sure to ask for the free T_3 and T_4 as well.

Most conventional physicians prescribe L-thyroxine, which is synthetic T_4. Many complementary doctors, on the other hand, prefer to use a natural form of thyroid, such as Armour Thyroid or Westhroid, which are desiccated porcine thyroid products (dried pig thyroid glands). Conventional doctors prefer the synthetic T_4 because the amount of medication in each pill is exactly the same, while the natural thyroid product may have some slight variation from pill to pill. Complementary physicians use natural thyroid because it contains both T_3 as well as T_4 in a 4:1 ratio, as is normally found in the body, while L-thyroxine contains only T_4.

DHEA Testing

DHEA is the hormone found in the greatest concentration in the human body, and it serves as a precursor to most of our sex hormones. DHEA is made by the adrenal glands; levels peak around 29 years and then begin a slow decline throughout life. There is currently debate about the value of DHEA supplementation in improving health. Several recent studies have failed to show that DHEA improves aspects of the aging process. There is still value in knowing your DHEA level whether you plan to supplement or not, because it can serve as a marker to let you know how well your adrenal glands are helping you with the aging process. You also can monitor your level to see how it is affected by your exercise program.

The test you want is the DHEA sulfate (DHEA-S); it is inexpensive and an accurate indicator of DHEA levels in the bloodstream. We like DHEA-S levels of 200 to 250 for men and 150 to 200 for women.

GROWTH HORMONE TESTING

Unlike many anti-aging gurus, we are not advocates of GH replacement as a mainline age management strategy. There is a major difference in opinion between longevity researchers and practitioners of anti-aging medicine in regard to growth hormone. Animal experiments performed in the laboratories of longevity researchers have shown that *lower* levels of IGF-1, the hormone in the body that does the work of growth hormone, are associated with greater longevity, while higher levels decrease life expectancy. In human studies, higher levels of IGF-1 also seem to be associated with increased risk of certain cancers. We recall that caloric restriction is the only proven method of increasing longevity in animal experiments. It appears to do so, in part, by decreasing IGF-1 activity. As a result, the feeling among some longevity researchers is that GH supplementation is something to be avoided and that lower IGF-1 levels may even be preferable.

Some practitioners of anti-aging medicine, on the other hand, look to do just the opposite and want to increase GH and IGF-1 levels of older adults to their levels when they were much younger. Their feeling is that young people have higher levels of GH and that increasing the levels in older people to what they had when they were younger confers greater youthfulness. These physicians recommend GH injections or other means of raising levels. Their approach is based on a landmark study completed in 1990 by Dr. Daniel Rudman, who worked at the University of Wisconsin. It showed that middle-aged men who were given GH injections appeared to become more youthful. Other studies performed since then have shown that some changes associated with the aging process—most notably increased body fat and decreased muscle mass—can be reversed by GH administration.

In this book we are concerned not only with maintaining youthfulness in our later years but, more important, with youthful longevity—living long enough to live forever. Our concern is that higher GH levels brought about by

GH injections have not been adequately shown to promote longevity, while some animal experiments even suggest they may decrease it. We feel that a middle-of-the-road approach is most prudent as regards GH and IGF-1. While we do acknowledge the value that higher levels of GH seem to play with regard to some aspects of the aging process, we suggest that you maintain your GH level naturally by getting adequate sleep, eating fewer high-glycemic-index carbohydrates, and engaging in regular exercise. We do *not* advocate GH injections.

Even though you don't plan to use GH injections, there is value in knowing your GH level. The test you want is the IGF-1 rather than the GH level itself. Growth hormone is naturally secreted by the pituitary gland in a pulsatile fashion, so levels rise and fall every few minutes. After it leaves the pituitary, GH travels via the bloodstream to the liver, where it is converted into IGF-1, and levels of it are more stable. Like GH, IGF-1 levels peak during the teenage years, then slowly decline throughout life. There is a syndrome known as adult growth hormone deficiency (AGHD), which is usually diagnosed through specialized tests done by endocrinologists. AGHD can also be diagnosed if your IGF-1 level is very low. The current consensus is that IGF-1 levels less than 77 are diagnostic of AGHD; if you find that your IGF-1 is in this very low range, this is one case where you should consult with your healthcare provider or an endocrinologist to see if you are a candidate for GH therapy.

CORTISOL TESTING

Cortisol is the main stress hormone. When you are under stress, specialized cells in the adrenal glands squirt a burst of cortisol into your bloodstream. The chronic stress of modern life results in nearly continuous release of cortisol from the adrenals. When you experience a high level of stress for long periods, it is possible for the adrenals to become fatigued and become unable to keep up with the constant demand for cortisol. Chronic fatigue, predisposition to infections, and a host of other problems are then possible. The two tests that can help tell how well your adrenal glands are doing at their job of producing cortisol are the blood cortisol level and the adrenal stress index (ASI).

Blood cortisol levels are usually highest in the morning, and most tests are done around 8:00 a.m. A low cortisol level suggests that your adrenals are exhausted or that you may have some other condition. Levels above the normal range can be due to excessive stress or, again, may be symptoms of another problem that needs further evaluation.

The ASI test is completed at home. A physician or health practitioner who practices complementary medicine can provide you with a test kit that you will use at home to analyze saliva samples at four intervals during the day—at breakfast, lunch, supper, and bedtime.

In general, you would want to see higher levels of cortisol shortly after you awaken in the morning, with levels gradually declining through the day with each successive sample. When a person first begins to experience stress on a chronic, long-term basis, cortisol levels will typically remain high throughout the day and night. After a prolonged period of continuous stress, however, it's not unusual for the adrenal glands to become fatigued. In this case, even though the individual is still under stress, the test will show low levels of cortisol throughout the day and night. The test also measures the ratio of DHEA to cortisol. The expectation is that the ratio would be higher when you are young and fall off as you age, but following our *TRANSCEND* recommendations can dramatically slow that decline.

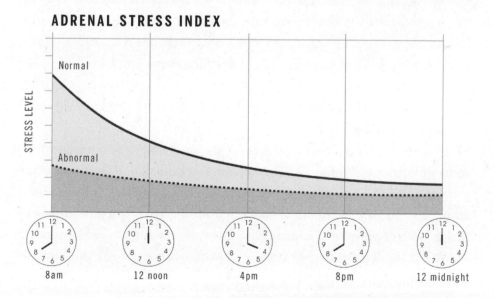

ADRENAL STRESS INDEX

A healthcare practitioner oriented towards natural treatments, such as a naturopath, chiropractor, or physician affiliated with the American College for Advancement in Medicine (see www.acam.org), can help you restore abnormal adrenal function towards normal.

INSULIN TESTING

Whenever you eat a sugary food or any type of carbohydrate, specialized islets of cells in your pancreas secrete insulin, a hormone that helps move the sugar from your bloodstream to the interiors of your cells, where it can be burned as energy (*a good thing!*) or changed into fat for storage (*not so good!*). Sugary foods and high-glycemic-index carbs need more insulin than the lower-glycemic-index carbs we recommend. Having high levels of insulin floating in your bloodstream can lead to a host of problems, including high blood pressure, weight gain, and increased risk of heart disease. The more of these high-glycemic-index foods you eat, the more insulin you produce; since high levels of insulin are so toxic, however, your body tries to protect you by creating a condition of *insulin resistance*. But if you develop insulin resistance and continue to eat these high-glycemic-index foods, you just keep making more and more insulin, compounding the problem.

Although insulin is rarely checked by conventional physicians, you can learn a lot by checking the level of insulin in your blood. This test should be done in the morning after you have been fasting for at least 12 hours. Optimal fasting insulin is less than 3, and acceptable values are less than 5. A fasting insulin of 12 doubles your risk of heart attack, and anything over 15 triples it.

THE SEX HORMONES—TESTING
Estrogen Testing

Many types of estrogen are found in the body, but the most powerful, and, in some ways, the most important, is estradiol. Estradiol, as the powerful estrogen it is, acts like a double-edged sword. On the one hand, it helps build strong bones and prevent osteoporosis, but in women, it can lead to abnormal growth of breast and endometrial tissue, increasing the risk of cancer. Men need small amounts of estradiol to help balance testosterone and for optimal brain function, but higher levels have been linked to excessive growth of the prostate gland, often resulting in urinary problems, male pattern baldness, and increased risk for prostate cancer. Therefore, both men and women should have their estradiol levels checked. Men and postmenopausal women can check their levels at any time, but women still having periods should have their testing done between days 17 and 21 of their menstrual cycle. Typical levels of estradiol during this time of the month are 50 to 180 ng/mL. Typical estradiol levels for menopausal women are < 32 and levels between 50 and 100 are a reasonable target for women on bioidentical hormonal therapy. An optimal level of estradiol for men is less than 40.

Progesterone Testing

Estrogen causes growth of tissues such as breast and endometrium (the inner lining of the uterus), and progesterone stops it. Therefore, estrogen and progesterone work hand in hand to keep hormone-sensitive tissues such as the breasts and uterus healthy and balanced. By providing this balance, progesterone has a protective effect against cancers of the breast and uterus. It is important to have progesterone any time there are elevated estrogen levels in the body. Estrogen is released throughout the entire menstrual cycle, while progesterone is released after ovulation in midcycle; levels remain high until menstruation and then drop. Progesterone contributes to the function of building new bone tissue just as estrogen does.

In women who are menstruating, progesterone levels should be checked on days 17 to 21 of a 28-day cycle or at least 1 week before their period. Postmenopausal women can check their levels at any time unless they are receiving HT, in which case their levels should be checked between days 14 and 21. Men do not need levels checked. Optimal levels are between 15 and 25 ng/dL.

Testosterone Testing

In men, testosterone is responsible for muscle and bone mass, sexual drive, sperm production, male hair distribution, and a general sense of well-being. Testosterone is produced mainly by the testes, with a small amount produced by the adrenal glands. Testosterone also reduces abdominal fat, enhances red blood cell production, and is involved in cholesterol and blood sugar levels. In women, this hormone is produced by the ovaries and adrenal glands but in a much smaller quantity than in men. It helps the female body maintain muscle and bone mass and sex drive. Healthy levels of testosterone are vital for our well-being. Testosterone, similar to our other hormones, declines with age. Both men and women should have their testosterone levels checked. Optimal levels are 500 to 800 ng/dL for men and 45 to 85 for women.

EARLY DETECTION—OSTEOPOROSIS

It's the natural order of things—part of our Stone Age genetics—to lose bone mass with age. Women lose bone earlier in life than men do, but both sexes will experience thinning of their skeletal bone mass eventually. Using the two simple tests discussed below—the NTX and hair minerals analysis—can help you determine whether you are already losing bone so that you can begin to take active measures to slow down and reverse this process.

Osteoporosis is a common disease of the skeleton in which the bone becomes thinner and more susceptible to fracture. While almost everyone experiences some degree of bone loss with aging, osteoporosis, a more severe form of bone loss, affects 10 million Americans; 18 million more have a pre-osteoporosis condition known as osteopenia. Fifty-five percent of cases occur after age 50, and 80 percent occur in women. According to a 2007 article in the *Journal of the American Academy of Orthopaedic Surgeons*, more women die of fractures from osteoporosis than from breast cancer.

Postmenopausal women are most susceptible to osteoporosis, and it is estimated that the average woman loses between one-third and one-half of her total bone mass over a lifetime. This is largely the result of decreasing levels of estrogen and progesterone after menopause. Men are also susceptible to osteoporosis, which is the result of declining testosterone levels, and they typically develop bone loss later in life than women do. Osteoporosis is more common in women of white, Hispanic, or Asian origin, less so among African-American women. The bone loss of osteoporosis is worst in the bones of the spine, hips, and wrists, and these are the regions most commonly susceptible to fracture.

Major risk factors include personal or family history of fracture, cigarette smoking, and low body weight (less than 127 pounds). Other risk factors include physical inactivity, heavy alcohol consumption (more than four drinks daily), and low calcium intake. Thyroid or kidney disease, diseases of malabsorption such as celiac disease (gluten sensitivity), vitamin D deficiency, anorexia nervosa, and other conditions that result in poor nutrition or low caloric intake also increase risk. Numerous medications, including steroids, diuretics, and seizure medications, contribute to bone loss as well.

The standard test for osteoporosis screening is the dual-energy x-ray absorptiometry (DXA). The DXA scan uses low-dose x-rays (1/30th the dose of a chest x-ray) to take pictures of the hip and spine to evaluate bone density. The test is covered by Medicare and most insurance plans for women only.

A consensus of the American Academy of Family Physicians, the American Academy of Orthopaedic Surgeons, and the World Health Organization is that women without any risk factors should begin screening for osteoporosis at 65 years of age. They suggest that women with any of the risk factors listed above be screened around the time of menopause, while women at high risk, those with multiple risk factors, can be screened at any age. For example, a 25-year-old woman who weighs 112 pounds and has a thyroid condition should get DXA screening in her twenties. No specific screening guidelines are given for men. In addition, even though DXA is regarded as the gold standard for the diagnosis of bone loss, it has recently come under attack. An article published in the journal *Bone* in 2007 contends that DXA produces results that are "unreliable, misdirected, and misrepresented."

Another problem with these recommendations is that too often they diagnose osteoporosis and osteopenia after the fact. We recommend two indirect tests—the NTX and hair minerals analysis—to help determine whether you are *at risk* of osteoporosis or osteopenia and should get DXA screening no matter what your age.

The NTX is a blood or urine test that measures a specific type of bone collagen—N-telopeptide cross-linked collagen type 1. Higher levels of this collagen are associated with increased risk of developing bone loss in the future. As part of our *TRANSCEND* philosophy of *early* detection, we recommend the NTX be used to screen people at a younger age for bone loss. We recommend that NTX screening begin at age 35 for women and 45 for men. (It seems men are typically at least 10 years behind women in lots of things, and bone loss is just one of them.) If you have an elevated NTX, then you should follow up with a DXA scan.

A hair minerals analysis involves taking a small sample of your hair from the nape of your neck close to the scalp, which is then sent to a lab to be evaluated for mineral elements. Any time we see high levels of the major bone minerals—calcium, magnesium, and strontium—on the hair analysis, our suspicion for increased bone loss goes way up. The only way large amounts of these three bone elements can get into the hair is through bone breakdown, which implies that the individual is at risk of osteoporosis. If your hair analysis shows high levels of two or three of the three bone minerals, we recommend that you follow up with a DXA scan. Many physicians offer NTX testing, but not for osteoporosis screening. Only a small fraction of physicians perform hair minerals analysis or understand how to interpret the results. Since NTX testing can be done on urine and the hair minerals test simply requires a snip of hair, these are tests that you can do at home (in which case the lab will provide an interpretation of the results) or through a healthcare provider as part of your personal program of early detection.

5

METABOLIC
PROCESSES

"Although I have always been in excellent health, your program helped me
achieve even higher levels. As a commander of over 400 military personnel,
my physical stamina and performance are at such a level that I consistently
out-perform troops that are 30 plus years my junior. Additionally, my blood
pressure and body chemistry remain that of a 25-year-old. Your program is
fantastic!"

JIM (62), NASHVILLE

We have often turned to the common theme that our genetic code, our
biological software, is not well adapted for life in the modern world. It was
written thousands of years ago and is long overdue for an update. While we
patiently wait for the biotech and nanotech engineers to provide us with human
body versions 2.0 and 3.0, we need to use our contemporary knowledge to
reprogram our biochemistry. *Glycation, inflammation,* and *methylation* are
three important metabolic processes that were designed to perform specific
functions that were much better adapted to life in a primitive world than the
world of today. Let's look at each of these processes in turn and see what we
need to do to optimize their function for modern-day life and life extension.

There are also three simple laboratory tests that can provide you with the
needed information as to how well your body is performing these critical
metabolic functions. *Hemoglobin A_{1c}* provides a direct measure of how well
your body is processing sugar in the blood, *C-reactive protein* measures
inflammation, and *homocysteine* levels indicate the status of processes called
methylation.

GLYCATION: NO SUGAR COATING

Glycation and glycosylation refer to the process of attaching a glucose or sugar molecule to another molecule such as protein or fat. Glycosylation occurs under the direction of specific enzymes, while glycation occurs spontaneously. The more sugar there is circulating in the bloodstream, the more sugar molecules there are to stick to proteins and fat, in other words, to glycate them. Many of the chronic complications of diabetes, such as blindness, neuropathy, and renal failure are due to glycation of proteins or lipids (fats). Some proteins need to be either glycosylated or glycated in order to work properly. For instance, it is theorized that one of these processes is needed as a trigger for puberty in adolescent girls. Some scientists feel that the high-sugar diets of many children today may be related to why many young girls now start puberty at younger and younger ages.

You control your rate of glycation by the amount of sugar and high-glycemic-index foods that you eat. Chronic consumption of sugary and high-glycemic-index foods leads to persistent elevations in blood sugar levels, which means that more glucose is available to glycate proteins. This is also manifested in the body as increased formation of aptly named AGEs (advanced glycation end products), which are sticky conglomerations of sugar and protein. AGEs are toxic debris that gum up your enzymes, add to the burden of toxic waste that must be handled by your detoxification system, and accelerate the aging process.

You also can consume preformed AGEs in the foods you eat. How food is cooked has a direct effect on glycation and the food's AGE content. One glycation reaction is known as the Mallard or "browning" reaction, so named because this is what makes cooked foods turn brown. When foods are cooked at high temperatures, their outer surfaces become brown and hard. Any time foods turn brown during the cooking process, you know that the Mallard or browning reaction—and AGE formation—has occurred. The crust that forms when bread is baked in an oven is the result of AGE formation. Baking, barbecuing, broiling, frying, and roasting increase the AGE content of food. Frying, which occurs at temperatures of 375°F, and broiling, typically done around 500°F, are associated with the creation of large amounts of toxic

AGEs. Boiling or steaming, which never lets the food get above the boiling point of water or 212°F, produces fewer AGEs.

The goal is moderation—to achieve just the right amount of glycation (and glycosylation) so that the proteins that need to be glycated do so, but not so much that excess AGE formation occurs. There are a few simple tests you can do on yourself to determine your rate of glycation and AGE formation.

- Do you have high blood pressure? Some scientists feel that high blood pressure may be a result of abnormal glycation. Glycated collagen molecules in blood vessel walls are less flexible, so blood pressure rises as blood travels through these stiffened vessels.

- Do you have cataracts? Cataracts in the lens of the eye are another example where the crystal-clear proteins in the lens of the eye become damaged—by ultraviolet rays of the sun—and form dark conglomerates of AGEs that interfere with vision.

- Do you have a non-zero coronary calcium score? You'll recall that high levels of glucose in the blood increase the glycation of low-density lipoprotein (LDL) cholesterol, a key step in the formation of the pathological foam cells that is the first step on the road to the formation of vulnerable plaque, the precursor to heart attacks.

EARLY DETECTION OF GLYCATION— HEMOGLOBIN A$_{1C}$

Eating a low-glycemic-index diet and avoiding most sugary foods and high-glycemic-index carbohydrates is a cornerstone of our dietary recommendations. In addition to all of the other health problems associated with eating these foods, they also cause the tissues of the body to age more quickly by coating them with sugar. Hemoglobin A$_{1c}$ (HbA$_{1c}$) is a laboratory test that tells you how much glycation has been going on in your cells for the past 4 months. Hemoglobin is the molecule found in your red blood cells that carries oxygen to the tissues. Hemoglobin molecules have a life span of about 120 days, and sugar molecules get stuck to them throughout their entire

4-month life span. In general, since diets containing more sugary foods and high-glycemic-index carbohydrates expose the hemoglobin molecules to higher amounts of sugar in the bloodstream, this translates to higher HbA_{1c} levels.

The HbA_{1c} results are expressed as percentages, so a HbA_{1c} level of 5.0 means 5 percent of your hemoglobin molecules are glycated or coated with sugar. Each 1 percent rise in HbA_{1c} is associated with an average blood sugar level increase of 30 points over the past 4 months.

TABLE 5-1: RELATIONSHIP BETWEEN HbA₁c AND BLOOD SUGAR LEVELS

HbA_{1c} LEVEL (PERCENTAGE)	AVERAGE BLOOD SUGAR (MG/DL)
4	60
5	90
6	120
7	150
8	180
9	210

We feel that an optimal HbA_{1c} level is less than 5.8, which corresponds to average blood sugar values of 114. In people without diabetes, HbA_{1c} levels above 6.3 are worrisome and suggest both the need for additional testing for diabetes as well as urgent measures to lower blood sugar levels. Most endocrinologists try to have their patients with diabetes keep their HbA_{1c} levels less than 7, which we feel is a realistic goal. The HbA_{1c} is an inexpensive blood test that is widely available. Recent research has suggested that it may be counterproductive for patients with type 2 diabetes to lower their HbA_{1c} levels to the "normal" levels of people without diabetes (6 or less). This observation may result from the harsh medications that are often used to lower blood glucose levels. We suggest that patients with type 2 diabetes keep their HbA_{1c} levels between 6 and 7 and emphasize natural approaches, especially reducing high-glycemic-index foods. We discuss this further in the chapter on nutrition.

Ray2034: The good news is, you won't need to be so careful with your diet for much longer. We now have nanobiotic macrophages (robotic white blood cells), which regularly scour our blood vessels, cleaning up AGEs as soon as they form. Thanks to these devices and fat insulin receptor blockers, we're able to eat pretty much whatever we want whenever we want these days.

Reader: Hot fudge sundaes, rib eye steaks, pepperoni pizza?

Ray2034: Yeah, pretty much anything you want. Only most of these foods are now made in desktop molecular nanotech factories. Steaks don't come from cows anymore.

Reader: Sounds worth waiting for.

Terry2034: And more motivation to stay healthy the old-fashioned way back in your day.

Ray2034: The cows were pretty happy about it, too.

Ray2023: Actually, the cows are happy in the early 2020s with new food production technology. Although we are not yet producing food in desktop nano factories, meat in the 2020s is produced in factories using in vitro cloning.

Reader: Okay, you're going to have to explain that.

Ray2023: Already back in your day, we were able to clone organs and body tissues on a small experimental basis. This included the cloning of animal muscle tissue. Today, in 2023, we can produce enormous quantities of meat very inexpensively in robotic factories using this cloning technique, no animals involved. People for the Ethical Treatment of Animals (PETA), the animal rights organization, has supported this from the start. We can produce every type of meat, including varieties that have been genetically modified to provide an improved nutritional profile, such as in vitro beef with omega-3 fats instead of saturated fat.

Terry2023: On another front, abnormal glycation has always been a big problem for people with diabetes, but now that it's been a few years since the artificial pancreas was perfected, people with type 1 diabetes no longer need to bother with checking their blood sugar or giving themselves insulin shots. A "closed-loop" system is created that continuously monitors blood sugar levels while an insulin pump continuously administers tiny amounts of insulin to keep the blood sugar level perfect at all times, just like a natural pancreas. Blood sugar control is much better and glycation remains normal.

INFLAMMATION: FRIEND AND FOE

The literal definition of inflammation is "to set on fire." There are two types of inflammation found in the body, but only one of them is fiery. *Acute inflammation* is what happens when you scrape your knee. The area becomes hot, red, swollen, and painful, as if burning. Acute inflammation, although temporarily uncomfortable, is one of the body's primary mechanisms for dealing with trauma or infection. It was critically important to our survival eons ago when our ancestors lived in a world without antiseptics or antibiotics. This helps explain why many doctors have come around to the position that treating the fever or pain of an acute inflammatory process such as an injury or infection with medications such as aspirin or ibuprofen is actually counterproductive to healing. The bottom line is that acute inflammation is mostly beneficial.

But good inflammation also has its own badly behaved cousin—*chronic inflammation*. Chronic inflammation often produces no symptoms for many years, is low grade, and is not associated with the classic signs of acute inflammation, such as heat, redness, pain, or swelling. In fact, it's mostly silent for many years yet may erupt suddenly and catastrophically. You might recall that most people have no symptoms or signs of a cardiac problem until the very moment that they suffer a heart attack, at which time soft or vulnerable plaque—the result of many years of silent inflammation—erupts with sudden fury. Although it is silent and symptomless, and most people don't even know they have it, chronic inflammation has been found to play a significant role in almost all chronic diseases and our leading causes of death. Heart disease, cancer, diabetes, and Alzheimer's disease have all been associated with chronic, silent inflammation. Assessing whether you have any significant degree of silent inflammation within your body—through the use of the

C-reactive protein test discussed below—and taking aggressive action to control and reverse the process provides you with powerful tools to combat these major killers.

Early Detection of Inflammation

C-Reactive Protein

C-reactive protein (CRP) is a protein made in the liver that indicates the amount of inflammation in the body. Like elevated homocysteine, an elevated CRP level increases a person's vulnerability to cardiovascular disease, Alzheimer's disease, and cancer. Elderly people with high CRP readings are more likely to have serious disease or disability than their elderly peers who have low CRP levels.

CRP can be elevated because of infections, inflammation, trauma, diabetes, or certain medications. In the absence of underlying factors such as these, an elevated CRP is predictive of future heart attack risk. Both the American Heart Association and the Centers for Disease Control and Prevention acknowledge that high levels of CRP are associated with increased risk for heart attacks, strokes, peripheral artery disease, and death from vascular disease. The most accurate test is the high-sensitivity CRP (hs-CRP) assay, and it's the test you want. The following table provides suggested values for hs-CRP.

TABLE 5-2: SUGGESTED HS-CRP VALUES

HS-CRP	RISK OF CARDIOVASCULAR DISEASE
< 1.0	Low
1.1–2.9	Average
> 3.0	High

The average CRP for middle-aged Americans is 1.5, and approximately one in four American adults has an elevated CRP (greater than 3.0). Your goal should be to keep your CRP under 1.

Essential Fatty Acids Testing

Essential fatty acid (EFA) testing offers you another way of assessing your inflammation status; EFA testing can tell you whether you've got the proper ratio of all the various EFAs. Since we now realize that inflammation plays a critical role in the development of so many disease processes, ranging from heart disease to cancer to Alzheimer's disease, and that inflammation is largely controlled by the relative amounts of EFAs in our bodies, knowing your EFA levels can provide critical information.

The EFA test measures total amounts of the main families of fats—saturated and unsaturated—and then breaks them down into subsets: omega-3, omega-6, and omega-9 fats, so that you can see the ratios between them. It also measures the levels of individual fatty acids, such as the proinflammatory omega-6 fatty acids, linoleic acid, gamma-linolenic acid, and arachidonic acid; the anti-inflammatory omega-3 fatty acids; alpha-linolenic acid; eicosapentaenoic acid; docosahexaenoic acid; and the omega-9 fatty acid oleic acid (from olive oil). Imbalances can then be corrected with appropriate dietary modifications and supplementation to keep inflammation at a minimum. Most conventional physicians are unacquainted with the EFA test and don't offer it at their offices. You will need to see a nutritionally oriented physician affiliated with the American College for Advancement in Medicine (www.acam.org) to arrange for EFA testing.

STEPS TO CONTROL SILENT INFLAMMATION

The chief methods that you can use to reduce silent inflammation include diet, lifestyle modifications, supplements, and medication.

Diet

Inflammation in the body is largely controlled by localized, very short-acting hormones known as prostaglandins. Prostaglandins are products of fatty acids, and both prostaglandins and fatty acids are classified as proinflammatory or anti-inflammatory. The two main families of fatty acids associated with inflammation are omega-3 and omega-6. There are two fatty acids that are classified as essential nutrients—alpha-linolenic acid, an omega-3 fat,

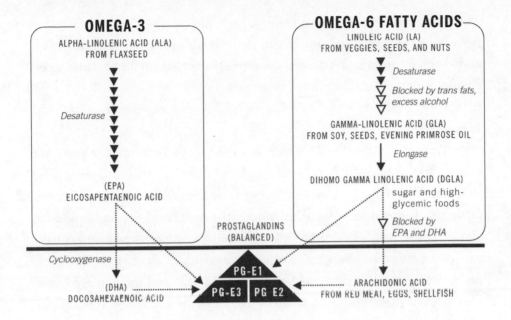

and linoleic acid, of the omega-6 family. Omega-3 fats result in production of the anti-inflammatory prostaglandin, prostaglandin-E3, while omega-6 fats lead to formation of prostaglandin-E1 and prostaglandin-E2, prostaglandins that tend to increase inflammation. Until the last hundred years, consumption of omega-3 and omega-6 fatty acids in the human diet was about equal. As a result of "modern" agricultural techniques and diets, consumption of inflammation-producing omega-6 fatty acids now exceeds that of anti-inflammatory omega-3 fatty acids by a ratio as high as 25:1. Is it any wonder that so many of us suffer the consequences of silent inflammation?

Fish and flaxseed are rich sources of anti-inflammatory omega-3 fatty acids, while vegetable oils, such as soy, sunflower, canola, or corn oil, are very high in omega-6 fats. To help shift the balance in favor of anti-inflammatory omega-3 fats and favorable prostaglandin production, you want to increase consumption of fish, fish oil, nuts, and flaxseed oil, while simultaneously decreasing consumption of vegetable oils. You'll also notice in the diagram above that when you consume sugar or high-glycemic-index carbohydrates, you increase production of the proinflammatory intermediary, arachidonic acid, which is a precursor to the proinflammatory prostaglandin-E2.

Therefore, your second step is eating a high-fiber, low-glycemic-index diet (that is, low in sweets and simple starches). In addition, red meat, egg yolks, and shellfish contain large amounts of preformed arachidonic acid, so you should limit the consumption of these as well.

Lifestyle

Obesity, a sedentary lifestyle, cigarette smoking, and even poor sleep have been found to raise CRP levels, the key blood test for inflammation. The first step in treating an elevated CRP level is to correct these types of risk factors. Other sources of inflammation in the body, such as gingivitis (gum disease) or arthritis, should be treated as well. Deep sleep and exercise have each been found to lower levels of inflammation in the body. Even a few nights of reduced or disturbed sleep have been found to increase blood measures of inflammation. Obstructive sleep apnea, which reduces the amount of time spent in deep sleep, has been associated with high blood pressure and cardiovascular complications. Recent research suggests that sleep apnea increases inflammation, which, in turn, leads to increased cardiovascular risk.

Exercise has been found to lower inflammation. A 2006 study in the *European Heart Journal* found that moderate to intense physical exercise was able to significantly decrease CRP levels in sedentary individuals.

Supplements and Medications

One of the most important nutritional supplements you can take to reduce silent inflammation is fish oil. Fish oil contains high amounts of eicosapentaenoic acid (EPA), which is a precursor to prostaglandin-E3, an anti-inflammatory prostaglandin, as well as docosahexaenoic acid (DHA), a key brain nutrient. In addition, when you consume EPA and DHA in your diet in the form of fish or fish oil, formation of the proinflammatory, undesirable prostaglandin-E2 is inhibited. The RDA for fish oil is 1,100 milligrams a day for women and 1,600 milligrams a day for men, and our optimal nutritional allowance is 1,000 milligrams of EPA and 600 milligrams of the DHA daily. Eating fish at least three times a week, along with a daily fish oil supplement, will easily meet these requirements. Vegetarians can take flaxseed oil as their best alternative.

Indian cuisine is also anti-inflammatory. Curcumin is the active ingredient in turmeric, a spice commonly used in Indian cuisine and curries. Eating curry and taking curcumin as a supplement are other natural—and enjoyable—ways of reducing inflammation.

Conventional practitioners lower CRP levels and the underlying inflammation in the body by recommending aspirin, anti-inflammatory medications, and statin drugs. The statin drugs, used primarily to lower elevated cholesterol levels, also reduce inflammation, and research suggests that this effect of statin drugs is even more important than its cholesterol-lowering effects. These drugs can be used as a second line of therapy if the natural treatments don't work well enough. If you do take statin drugs, remember to supplement with coenzyme Q_{10} (see Chapter 12) because statin drugs deplete the body of this vital nutrient.

METHYLATION: SWITCH ON, SWITCH OFF

The lightest organic, or carbon-containing, compound is methane, a flammable, colorless gas with the chemical formula CH_4. Take away one hydrogen atom and you get CH_3, a methyl group. Methylation, quite simply, is the attachment of a methyl group to other organic compounds. All living things use methylation as a type of biochemical switch to turn various processes on or off. For example, methylating a gene in your DNA might make it active (that is, turn it on), while without this methyl group, the gene remains silent. Because of some common genetic variants, between 10 and 50 percent of the population—depending on ethnicity—demonstrate abnormal methylation. Abnormal methylation has been found to increase risk of coronary heart disease, stroke, Alzheimer's disease, several types of cancer, and other conditions. Luckily, it is very easy to test how well your body is performing its job of methylation with a simple blood test—the homocysteine level—as discussed below.

OPTIMIZING METHYLATION

Nutritional supplementation and diet have been shown to lower elevated homocysteine levels. Vitamins B_6 and B_{12}, folic acid, and trimethylglycine are the usual supplements that have been used. Yet, because of the genetic variations mentioned above, there is often as much as a *hundredfold difference* in the amounts needed by different people. Normal homocysteine levels may be easily achieved in one person with only RDA amounts of nutrients, such as 2 milligrams of vitamin B_6, 0.6 micrograms of vitamin B_{12}, and 400 micrograms of folic acid, yet someone else with a genetic mutation may need to take 200 milligrams of vitamin B_6 (10,000 percent of RDA), 1,000 micrograms of vitamin B_{12} (160,000 percent of RDA), 2,000 micrograms of folic acid (500 percent of RDA), and several grams of trimethylglycine to achieve the same degree of control.

There is controversy among researchers about the value of lowering homocysteine with vitamin supplements. Several large randomized controlled trials have shown that even though it is relatively easy to lower homocysteine levels with supplements, this has not translated into a reduction in cardiovascular risk—either heart attack or stroke.

There is also an association among methylation status, elevated homocysteine levels, and cancer risk. Higher homocysteine levels are associated with higher cancer risk. Some positive results of supplementation are that higher blood levels of the nutrients used to lower homocysteine such as vitamin B_{12} and folic acid have been linked to a decreased risk of colorectal, breast, and cervical cancer. These same nutrients have, however, also been associated with an increased risk of prostate cancer. Therefore, using nutritional supplements to lower homocysteine is currently controversial. Once again, talk with your doctor.

Dietary methods of reducing homocysteine, on the other hand, are safe. Homocysteine is a naturally occurring breakdown product of the amino acid methionine. Red meat, turkey, and chicken contain relatively large amounts of methionine, so we suggest that individuals who have elevated homocysteine levels limit their consumption of these animal products. Fish, fruits, and vegetables are low in methionine and should be emphasized instead.

These changes are beneficial for other reasons; for example, they also reduce chronic inflammation, as we discussed above. A vegan diet can lower homocysteine levels 13 percent without any supplementation. Moderate alcohol consumption of one or two alcoholic beverages a day decreases homocysteine, but drinking larger amounts will increase it.

EARLY DETECTION OF METHYLATION— HOMOCYSTEINE

Homocysteine is created in the body from the essential dietary amino acid methionine. Elevated homocysteine levels indicate that *methylation* processes are not operating properly and has been linked to higher risk of heart attack, stroke, peripheral arterial disease, bone fractures, and Alzheimer's disease. Blood testing for homocysteine is readily available, although Medicare still doesn't include it as a covered service. If your homocysteine level is above 8.0, we recommend the dietary and lifestyle strategies mentioned above: sharp reduction or elimination of red meat, turkey, and chicken; increased fish, fruits, and vegetables; and moderate alcohol consumption.

6

CANCER

Despite recent significant advances in treatment and survival, cancer remains an all-too-common and deadly diagnosis. Rates for many of the most frequently diagnosed cancers have not declined substantially in more than 60 years, with incidence of lung cancer actually rising dramatically during that period. The American Cancer Society found that nearly 1.5 million people were diagnosed with cancer in the United States in 2008, with 560,000 deaths. This makes cancer the number two overall killer, exceeded only by heart disease. In addition to the terrible physical and emotional toll on millions of cancer patients and their loved ones each year, the economic costs are enormous. According to the National Institutes of Health, the total dollar loss to cancer in 2007 exceeded $209 billion in the United States ($89 billion for direct medical costs and $120 billion for indirect costs due to lost productivity).

More than 100 different conditions are defined as cancers and named according to the organ or system of the body that is initially affected (for example, lung cancer, lymphoma, colon cancer). All cancers, however, are characterized by the uncontrolled replication of cells. Normal cells reproduce only as needed, such as to build new tissues in a growing child and to replace cells that have gotten sickly or have died. Cancerous cells, on the other hand, are linked to DNA (genetic) mutations that cause them to divide continuously, creating a continually growing number of new cancerous cells. Unstopped, these cells can damage or starve the affected organ or system. Cancerous cells also sometimes spread (metastasize) from the original site in the body to cause cancerous cells to form in other distant organs or systems.

It is this process that accounts for the majority of the deaths from cancer.

Your risk of contracting cancer is determined to some extent by your genes, as evidenced by certain cancers, such as breast cancer, that seem to run in families. However, the principal cause of the DNA mutations that trigger such uncontrolled cell growth is exposure to free radicals. These are unstable molecules that steal electrons from other molecules in your body, including your DNA molecules. This can seriously damage the attacked molecule, sometimes causing mutations that lead to cancer.

Thousands of free radicals are created every day in your body when nutrients are oxidized by your cells to produce energy. You are also exposed to free radicals from a variety of external sources, such as tobacco smoke, radiation (sunlight, x-rays, etc.), deep-fried foods, and toxins in the air, food, and water (lead, mercury, pesticides, ozone, plastics, etc.). One toxin, tobacco, is estimated to cause almost a third of all cancer deaths.

By the time cancer is diagnosed, the progression of the disease has typically been under way for years. In most cases, the earlier in its development a cancer is discovered, the better the odds are of survival, and we'll discuss some early detection guidelines later in this chapter. Better still, however, is to prevent cancer from developing in the first place.

AN OUNCE OF PREVENTION

You can significantly reduce your risk of cancer by reducing your exposure to the free radicals that cause the DNA mutations that cause many cancers. You want to support and strengthen your body's own mechanisms for neutralizing toxins, free radicals, and cancerous cells. Following our *TRANSCEND* recommendations should be a major part of your program to accomplish this, and in this chapter we underscore the most important changes to make.

FOODS THAT HURT, FOODS THAT HELP

In 2004, the journal *Public Health Nutrition* published a report analyzing a large number of studies that had examined the relationship between diet and the risk of cancer. The evidence was compelling: What you eat has a profound impact on your risk of cancer. For example, consumption of red meat and pre-served meats (cold cuts, hot dogs, etc.) is associated with higher risk of stomach and colorectal cancer, while a diet high in fruits and vegetables is associated with lower risk of cancers of the mouth, esophagus, stomach, and colon.

Other research supports similar findings. A study reported in 2003 in the *New England Journal of Medicine* found that the traditional Mediterranean diet—low in red meat and high in fresh fruits, vegetables, fish, and whole grains—reduced incidence of cancer by 24 percent. Studies have found that consuming olive oil lowers the risk of breast, colon, and skin cancers. Toma-toes, especially as tomato sauce, have been found to protect against prostate cancer. And soy-based foods and green tea have both been found to contain powerful anticancer substances, perhaps explaining the lower incidence of cancer in Japan compared with the United States.

While eating healthy foods lowers your cancer risk, eating unhealthy junk foods has the opposite effect. Of particular concern is sugar. A 2002 study pub-lished in the *Journal of the National Cancer Institute* found that participants who ate a diet high in sugars and refined starch had three times the average risk of pancreatic cancer. Another study published in the same journal in 2004 found that a diet high in sugars and starch significantly raised the risk of colorectal cancer. Cancer cells are known to function differently than normal cells, relying almost exclusively on sugar to fuel their growth and proliferation. By drastically limiting sugar and refined starches in your diet, you can limit the amount of energy available to cancerous cells and inhibit their growth.

When it comes to cancer prevention, what you eat clearly does matter. You must avoid consumption of toxic foods that can increase your cancer risk. And you must eat plenty of the kinds of foods that provide your body with the vitamins, minerals, and healthy proteins, fats, and phytochemicals it needs to neutralize free radicals and cancerous cells. That can be obtained only by eating a healthy diet rich in fresh vegetables, fruits, fish, nuts, and olive oil. We recommend that you follow our nutritional guidelines discussed in detail in Chapter 11.

LIFESTYLE MATTERS

In addition to fortifying your body with healthy foods, you can make a number of lifestyle choices to reduce your risk of cancer. These include:

• **Lose excess body weight.** A study reported in 2003 in the *New England Journal of Medicine* found that being overweight or obese accounted for 14 percent of cancer deaths in men and 20 percent in women. A study that was part of the Nurses' Health Study reported in 2008 that abdominal obesity, the type also associated with cardiovascular risk, increased cancer mortality in women 18 to 63 percent depending on the amount of excess abdominal fat. (See Chapter 13 for our weight loss recommendations.)

• **Maintain a regular exercise routine.** Studies have linked a sedentary lifestyle with higher risk of cancer. Exercise also helps you attain and maintain your optimal weight, another factor that reduces your risk. (See Chapter 14 for our exercise recommendations.)

• **Keep stress under control.** Higher levels of stress are associated with a higher risk of cancer. (See Chapter 9 for our stress management recommendations.)

• **Avoid *overexposure* to sunlight.** There is a fine line when it comes to being out in the sun. Moderate exposure has been found to have benefits, including a reduction in cancer risk. A study reported in the journal *Cancer* in 2002 found that exposure to UVB (ultraviolet B) radiation is associated with a lower risk of developing cancer of the breast, colon, ovary, prostate, and lymphoma and with reduced mortality rates for cancers of the bladder, esophagus, kidney, lung, pancreas, rectum, and stomach. Overexposure to bright midday sun, however, is linked to skin cancer. Limited exposure without the use of sunscreen is beneficial before 10:00 a.m. and after 4:00 p.m. At all other times, when out in the sun for extended periods of time, cover up and use sunscreen on exposed skin.

• **Don't smoke and don't expose yourself to secondhand smoke.** Not only is smoking by far the primary cause of lung cancer, tobacco use is also linked to cancers of the mouth, throat, larynx, kidney, bladder, cervix, and pancreas. Of course, there are numerous other reasons not to smoke, including (to name just a few) heart disease, stroke, emphysema, and COPD (chronic obstructive pulmonary disease). If you currently smoke, quit! We know this is easier said than done because tobacco is extremely addictive, but a number of medications and therapies are available to help you kick the habit.

• **Avoid exposure to agricultural chemicals.** The pesticides, herbicides, and other chemicals used in industrial agriculture and in many commonly available yard and garden products have been linked to dramatically increased risk of cancer. For example, research published in the *International Journal of Occupational Medicine and Environmental Health* in 2001 found that agricultural workers had a 40 percent higher risk of contracting cancers of the prostate, stomach, and larynx. To reduce your exposure, use nontoxic pesticides and herbicides around the home, eat organic produce (grown without toxic chemicals), and limit your con-

---

HELP WITH QUITTING SMOKING

Nicotine is an addictive drug, and most people who quit smoking experience unpleasant withdrawal effects, but there are both prescription and over-the-counter aids to help reduce these significantly. Use of medications has been shown to double the probability that people will be able to quit. The two prescription medications approved by the FDA are sustained-release bupropion (available as a generic) and Chantix. Bupropion is an antidepressant that affects levels of neurotransmitters in the brain to reduce craving, while Chantix blocks nicotine receptors in the brain. You can ask your doctor for a prescription for either (or both) of these. Over-the-counter aids include various forms of nicotine delivery— inhaler, nasal spray, gum, patch, and lozenge. These help former tobacco users gradually wean themselves from the addictive effects of nicotine.

sumption of red meat (which concentrates these chemicals in the fat). If your work exposes you to toxic agricultural chemicals, always use appropriate safety gear and precautions.

• **Avoid exposure to other environmental toxins.** Many products used at home and on the job contain dangerous chemicals. These can include solvents, cleaners, glues, and paints. Use nontoxic alternatives whenever possible, and use appropriate safety gear and precautions when there is potential for exposure.

Supplemental Insurance

A number of synthetic and naturally occurring substances have been found to be chemopreventive, meaning that they can reduce your risk of cancer. These include the specific nutrients that give the foods we discussed earlier in this chapter their cancer-preventive characteristics. Taking supplements that contain higher concentrations of these substances is generally not as effective at reducing cancer risk as is consuming these substances as they naturally occur in food sources. However, the following supplements can provide an added boost to a healthy diet:

• *Vitamin C* is a potent free-radical scavenger (antioxidant) that has been shown to be of value in preventing and treating cancer. For example, a study reported in 2002 in the journal *Pharmazie* found that a combination of vitamin C and selenium can stimulate precancerous cells to remain benign. For cancer prevention, we recommend a dosage of 500 milligrams a day.

• *Selenium* is a crucial mineral that is required by your body to activate an important antioxidant enzyme. In 2002, the journal *Cancer Epidemiology Biomarkers and Prevention* published a study that found that taking selenium lowered incidence of cancer, especially prostate cancer. For cancer prevention, we recommend a dosage of 200 to 400 micrograms a day.

• *Coenzyme Q$_{10}$* supports production of crucial antioxidant enzymes that are needed to control free-radical damage. Because cancerous cells are especially adept at creating free radicals, the need for coenzyme Q_{10} supplementation rises when cancer is present. For example, breast tumors exhibit extremely low levels of coenzyme Q_{10}, and supplementation is often recommended for patients with breast cancer. For cancer prevention, we recommend a dosage of 60 to 200 milligrams a day of coenzyme Q_{10} or 50 to 100 milligrams a day of reduced coenzyme Q_{10} (ubiquinol) for healthy adults.

• *Curcumin* is found in the spice turmeric, which has been used for centuries in traditional Chinese and Indian food and medicine. Curcumin possesses as many as a dozen different chemopreventive properties. For example, several studies published in 2008 showed that curcumin was able to inhibit cancer growth. One study found that the combination of curcumin and a green tea flavonoid (epigallocatechin gallate) suppressed breast cancer cell growth in test tubes and in mice. Another found that a curcumin derivative (diphenyl difluoroketone) suppressed colon cancer cells grafted to mice. Yet another showed that curcumin suppressed bladder cancer cells in test tubes and the development of urinary tract tumors in rat bladders. For cancer prevention, we recommend a dosage of 900 milligrams a day along with liberal use of the spice turmeric in your food.

• *Melatonin* is a naturally occurring hormone that regulates sleep patterns, among other functions. Numerous studies have found that melatonin also has powerful anticancer properties, especially in the case of breast cancer. For cancer prevention, we recommend a dosage of 0.1 to 3 milligrams a day at bedtime.

• *EPA/DHA* (eicosapentaenoic acid and docosahexaenoic acid) are essential fatty acids found primarily in fish, walnuts, and flaxseed. These healthy fats are powerful natural anti-inflammatory agents that inhibit production of the inflammatory chemicals found in high concentrations in certain cancers, such as colon and breast. For cancer prevention, we recommend a daily minimum dose of 1,000 to 3,000 milligrams of EPA and 700 to 2,000 milligrams of DHA.

• *Vitamin D* has been shown to be chemoprotective against many types of cancer. A 2005 study showed that supplementation with 1,000 IU of vitamin D a day (substantially higher than the Recommended Daily Allowance of 200 to 400 IU) cut the risk of colon, breast, and ovarian cancer in half. A major study published in the *American Journal of Clinical Nutrition* in 2007 showed a 60 percent lower cancer risk over 4 years in women who took 1,000 IU of vitamin D supplementation. We think most healthy adults should check their 25(OH) vitamin D level and supplement to achieve a level greater than 50 ng/mL.

• *Folic acid* is also known as vitamin B_9. In 2001, the journal *Swiss Medical Weekly* published a review of 34 studies that found that lower than normal levels of folic acid were associated with breast and colon cancers. Some newer research, however, links increased folic acid fortification of food with higher rates of colon cancer. Therefore, we'll have to wait for further studies and clarification on the role of folic acid (if any) in cancer prevention.

• *Beta-carotene* (not for everyone) is a powerful antioxidant that reduces free-radical tissue damage from ultraviolet (UV) light and protects skin from cancer. However, several studies have shown that supplementing with beta-carotene actually *increases* the incidence of lung cancer among participants who smoked cigarettes. Of course, if you smoke and are concerned about cancer, you need to quit smoking. If you do smoke, however, you should *not* supplement with beta-carotene.

A report published in the *Journal of the National Cancer Institute* in 2007 reported a correlation between multivitamin use and increased risk of advanced prostate cancer. Yet, this same study, which involved nearly 300,000 men, did not find any association with early-stage prostate cancer. Many people took the results of the study to mean that there were significant dangers associated with multivitamin use. Our interpretation of the data, on the other hand, was that men with advanced prostate cancer were more likely to have already developed symptoms and therefore had begun a program of nutritional supplementation. In addition, men who take supplements tend to be individuals who seek a more aggressive approach to their own health and, thus, would be more likely to be diagnosed.

EARLY DETECTION: CANCER SCREENING AND PREVENTION GUIDELINES

In addition to the cancer prevention strategies discussed above, early detection remains a critical component of a comprehensive program to reduce cancer deaths. The main reason that cancer is such a lethal threat is its tendency to spread silently and insidiously throughout the body through the process known as metastasis. When cancer is detected before it has metastasized or spread, the probability that it can be cured is dramatically higher. Survival statistics for metastatic cancer are much lower, so the goal is detection as early as possible.

In 2007, 213,000 people were diagnosed with lung cancer, and 160,000 people died of this disease, making it the leader in cancer deaths. That same year, there were 149,000 cases of colorectal cancer but only 50,000 deaths. This means that about 75 percent of the number of people diagnosed each year with lung cancer die each year, while the figures for colorectal cancer are only one in three. This is because, in part, it is much easier to detect colorectal cancer at an early treatable stage than lung cancer. As part of your program of early detection, there are a number of tests that you should have performed regularly to increase the odds of finding any type of cancer at the earliest possible stage, dramatically increasing your odds of cure.

The following guidelines are a summary of prudent age- and sex-specific cancer screening recommendations. We begin with the recommendations of the American Cancer Society, the National Cancer Institute, the American College of Obstetricians and Gynecologists, and the U.S. Preventive Services Task Force. We also include some additional, more comprehensive *TRANSCEND* recommendations of our own, which we feel will help detect more cancers at an early stage, when cure is still possible.

CANCER SCREENING GUIDELINES FOR BOTH WOMEN AND MEN

Basic Screening Guidelines

Everyone should have routine, periodic physicals. These should include a physical examination, detailed health history, health and preventive medical counseling, nutrition instruction, and laboratory work that the primary care provider deems appropriate to the person's condition at the time of exam.

The physical exam should include evaluation of the skin and oral cavity for abnormalities, assessment of the lymph nodes, a genital exam (for men and women), and an evaluation of the thyroid. Self-testing as discussed in Chapter 10 is important. Women should learn breast self-examinations and men should learn testicular exams from their primary healthcare providers; everyone should learn how to examine the skin for suspicious lesions. Preventing cancer by utilizing good nutrition, avoiding sugar and excessive alcohol use, eliminating smoking and/or drug use, and maintaining a good physical exercise program is a great place to start.

Lung Cancer Screening Guidelines

Currently, there are no standard screening guidelines for lung cancer. The most important factor is to stop smoking cigarettes. Some individuals, via their occupations, are exposed to environmental pollutants and chemicals that increase the risk of lung cancer. For those at higher risk because of either smoking or pollutant/chemical exposure, routine chest x-rays should be obtained as prescribed by your primary care provider. Spiral computed tomography (CT) is being used to detect early lung cancer in former and current smokers, although there is no consensus about the value of routine CT screening of smokers for lung cancer. Some new studies suggest the use of blood markers to detect early lung cancer. Carcinoembryonic antigen (CEA), retinol-binding protein (RBP), alpha-1 antitrypsin (AAT), and squamous cell carcinoma antigen (SCCA) are four markers being studied, but they are not being routinely used at this time.

Colorectal Cancer Screening Guidelines

About 50,000 Americans will die of colorectal cancer this year, but this number could be cut in half if people followed the American Cancer Society testing recommendations. African American men have higher rates of colorectal cancer, with 67.6 of 100,000 men developing the disease. This is more than twice the rate of American Indians and Alaska natives at 32.6 per 100,000. The U.S. Preventive Services Task Force issued the following recommendations for colorectal cancer screening in late 2008. Each of the following three protocols was found to be equally effective:

- **High-sensitivity fecal occult blood testing** every year

- **Sigmoidoscopy** every 5 years with high-sensitivity blood testing every 3 years

- **Colonoscopy** every 10 years

Screening should begin at age 50 (African Americans should begin screening at age 45). Screening can be stopped after age 76 in people who have had consistently negative screening results since age 50. No specific recommendations were made with respect to newer technologies, such as fecal DNA testing and CT colonography (virtual examinations). Regular screening is important because early colorectal cancer often has no symptoms. When colorectal cancer is detected at an early stage, patients have a 90 percent chance of cure.

Screening should begin earlier for those at high risk of colon cancer because of race, a personal history of polyps or cancer, a family history of colorectal cancer, or hereditary colorectal cancer syndrome in a first-degree relative who was 60 years of age or younger when diagnosed or in two first-degree relatives who developed colorectal cancer at any age. A personal history of chronic inflammatory bowel disease (such as Crohn's disease or ulcerative colitis) also increases a person's risk of colorectal cancers. Discuss with your primary care provider when screening should begin if you meet any of these high-risk criteria.

Any positive results on fecal occult blood testing, fecal DNA testing, flexible sigmoidoscopy, CT colonography, or barium enema need to be followed up with a colonoscopy. Many people defer colon cancer screening because of the expense and discomfort associated with full optical colonoscopy, even though this procedure is not that uncomfortable. As an alternative, radiographic examination of the colon with CT scanning called virtual or CT colonography is now available in many locations. Virtual examinations are more comfortable and better tolerated by patients, but they do involve exposure to increased amounts of radiation. In addition, the most recent studies suggest that standard optical colonoscopy is slightly better at detecting earlier and smaller polyps and cancers—the desired goal of early detection—while both procedures are equivalent at detecting more-advanced cancers. For this reason, standard optical colonoscopy remains the gold standard for early detection of colorectal cancer. However, a benefit of virtual colonoscopy is that it will also scan the abdominal organs for cancer and other diseases at the same time. For the small subset of people who are simply unwilling to undergo the discomfort of colonoscopy or the radiation associated with colonography, another newly developed option is fecal DNA testing. In this test, a stool sample is submitted to the laboratory and screened for the presence of abnormal DNA associated with colorectal cancers. This test is not as accurate as colonoscopy or virtual colonography, but it is better than doing nothing or simply doing fecal occult blood testing alone.

Skin Cancer Screening Guidelines

The best way to begin screening for skin cancer is to inspect your own skin on a monthly basis. You want to watch for any changes that occur, in particular, unusual moles that vary in shape, size, height, or color or have ragged edges. An acronym to help you remember what to look for is ABCDE.

- **A** = asymmetry: shape is not uniform

- **B** = borders: uneven, ragged, or irregular

- **C** = color: different colors are present and may include black, brown, tan, white, gray, red, pink, or blue

- **D** = diameter: greater than a pencil eraser

- **E** = enlargement

Notify your primary care provider if any skin lesion or mole itches, oozes, or bleeds and if moles become lumpy or hard. Screening with your healthcare provider should start at age 35 unless you have a family history of skin cancers, in which case you should discuss with your doctor what screening plan to adopt.

Head/Neck/Oral Cancer Screening Guidelines

Many types of cancer arise in the head or neck region, including the nasal or sinus cavities, mouth, lips, salivary glands, throat, and larynx (voice box). People who both smoke and drink alcohol on a daily basis are at increased risk for these types of cancers. Individuals who use smokeless tobacco have an increased risk for mouth cancers and should have an exam with a dentist at least every 2 years to evaluate the inside of the mouth for any abnormalities.

Symptoms of head and neck cancers can include a thickening or lump in the cheek; patches of red or white tissue on the tongue or inside of the cheek, gums, tonsils, or mouth lining; a sore throat or feeling like you have a lump in your throat; and difficulties swallowing or moving your jaw or tongue.

CANCER SCREENING GUIDELINES FOR WOMEN

Breast Cancer Screening Guidelines

Breast cancer screening for women younger than 20 years of age should include breast self-exams after each monthly menstrual cycle. Although breast cancer is extremely rare in this age group, beginning at this age will help women get into a lifelong habit of performing routine self breast checks.

Women in the 20- to 40-year-old age group without a family history of breast cancer should include a clinical breast exam by their primary care provider, preferably at the same time as their Pap and gynecologic exam, no less often than every 3 years, along with continuing to perform monthly breast self-exams. If a woman is at high risk for breast cancer because she has a first-degree relative with early-onset breast cancer (age younger than 50 years at onset), she may be a candidate for genetic testing. This testing would include the breast cancer genetic markers *BRCA1* and *BRCA2* and should be discussed with your healthcare provider. The conventional recommendations are that women age 40 and over should have a mammogram as well as a clinical breast exam by their healthcare provider each year until age 70. There are no clear-cut guidelines for mammography after 70, and the U.S. Preventive Services Task Force recommends that women "decide for themselves." (Our advice: Talk with your doctor.)

Women who are deemed high risk because of a family history of breast cancer should get an MRI (magnetic resonance imaging) scan in conjunction with their annual mammogram. The conventional recommendations are for high-risk women to start having mammograms at age 35 rather than 40.

However, we disagree with these conventional recommendations, which do not include any specific imaging tests in women under 40 years of age. Breast cancer is not a rarity among women in their thirties and is even found in women in their twenties. Breast cancer found in younger women also tends to be more aggressive, so we feel that waiting until a lump is big enough to be felt in a younger woman allows a potentially curable cancer to grow for several years undetected. We recommend *thermography*, which does not expose patients to any radiation, in this age group. Thermography measures the temperature of various regions of the breast, and is based on the tendency of breast cancers and precancerous tumors to have a higher temperature than normal breast tissue. It is completely noninvasive, and the patient simply sits in front of the measuring device with no contact to the body—and no radiation. We recommend that women get a baseline thermography screening at 25 years of age with follow-up screening intervals determined by the results.

We actually prefer to include thermography in addition to mammography in all age groups. Some women prefer this test because it doesn't expose the breast tissue to radiation and the breasts don't need to be pressed between plates as in conventional mammography. Studies comparing thermography with mammography have found the two procedures to be very close in their ability to detect breast cancer. Most conventional physician organizations, such as the American Medical Association and the American Cancer Society, still do not recommend thermography because they say that there aren't enough studies proving its effectiveness for breast cancer screening, even though thermography has existed for over 45 years and numerous published studies have reported a very high degree of diagnostic accuracy.

Mammography is a multibillion-dollar business, and any time there is a move toward a paradigm change, political and financial issues come into play as well. Because it hasn't received official approval from the American Medical Association, another problem is that most insurers regard breast thermography as investigational or experimental and do not cover it as a benefit.

We suggest using thermography as part of an integrated program for early detection of breast cancer. We recommend the following schedule as a cost-effective way to maximize early detection while simultaneously reducing a woman's exposure to radiation.

- 25 years: baseline thermogram

- 26–40 years: follow-up thermograms at intervals recommended by the thermographer

- 40 years and up: alternate mammograms with thermograms on an annual basis (a mammogram one year and a thermogram the next). Following this cycle reduces radiation exposure by half but still maintains mammography as a key element of the screening process. There are other advantages to alternating types of testing since thermography and mammography look at different markers for breast cancer, and there are cases where thermography picks up a breast cancer that was missed on mammography and vice versa.

Reader: Can I get some free medical advice?

RayandTerry: Absolutely. That's what we're here for.

Reader: I actually had a thermogram a few months ago. They told me I was class 3 in one breast, which they called medium risk. My previous thermograms had been 1 and 2, or low risk. This means I am at increased risk for developing breast cancer in the future, so I want to know what I can do now.

Terry: Following all the aspects of the *TRANSCEND* program can help a lot.

A diet high in fresh fruits and vegetables is protective, and moderate consumption of soy can help.

Exercise has been shown to decrease breast cancer risk.

Not taking antioxidant vitamins is a risk factor for breast cancer. A study in the journal *Cancer Research* showed that premenopausal women who did not take supplemental vitamin C had 4.8 times the risk of breast cancer, while women who didn't take vitamin E had 3.8 times higher risk.

Women have been using supplemental iodine to treat fibrocystic breast disease for many years, but new findings suggest it can help prevent breast cancer as well. Taking 1 to 5 drops of Lugol's solution (a mixture of calcium, potassium, and iodine) in a glass of juice or water is all that is needed. (Be careful not to take antiseptic iodine orally as it contains wood alcohol, which is poisonous.)

Reader: Sounds easy enough.

Terry: Follow this program carefully and repeat your thermogram in 6 to 8 months, and let's hope you'll be back to class 1 or 2.

Breast cancer seems to be related to a woman's own estrogen. A blood test known as the 2:16 estrone ratio is used by some complementary physicians to gauge the types of estrogen in a woman's body. There are many types of estrogen, and one particular type, known as 2-OH estrone, appears to be protective against breast cancer, while another (16-α-OH estrone) increases risk. Measuring the relative amounts of these two types of estrogen in the urine—the 2:16 estrone ratio—can provide prognostic information as to whether a woman is at increased cancer risk. The test is done by collecting a first-morning urine specimen and is available from complementary physicians. Adverse ratios can be corrected by increasing consumption of soy products, cruciferous vegetables, and supplements such as indole-3-carbinol (I3C) or diindolemethane (DIM), and decreasing exposure to pesticides and plastics.

Cervical Cancer Screening Guidelines

The American College of Obstetricians and Gynecologists recommends that cervical cancer screening begin 3 years after a woman initiates sexual intercourse but no later than the age of 21. Gynecologic and liquid-based Pap exams should continue every 1 or 2 years. The screening should include DNA testing for human papilloma virus (HPV), a virus associated with cervical cancer.

After 30 years of age, a woman who has had three consecutive normal Pap screens can be screened every 2 to 3 years. Women at high risk should continue to have annual exams. The high-risk group includes women exposed to diethylstilbestrol (DES) before birth, those with a compromised immune system due to presence of HPV, those with a history of organ transplantation, those who are receiving or have received chemotherapy, and those with a chronic illness or who use steroids regularly.

At the age of 70, a woman who has had three consecutive negative Pap screens in the previous 10 years may choose to discontinue any further cervical cancer screening exams. High-risk women should continue to have annual exams.

A woman who has had a hysterectomy, in which both the cervix and uterus were surgically removed, may choose to stop having cervical cancer screening. Women who have had a hysterectomy need to be aware that on rare occasions, the vaginal cuff (suture site inside of the vaginal canal after a hysterectomy) can still develop cancer because some cervical tissue is sometimes left. Such women should consider this when making the decision to discontinue exams.

Endometrial Cancer Screening Guidelines

Endometrial cancer is a cancer of the inside lining of the uterus. It is most common in the postmenopausal years and must be considered any time a menopausal woman has any abnormal vaginal bleeding. Risk factors include obesity, high blood pressure, polycystic ovary disease, nulliparity (no pregnancies), early onset of menstruation, and late menopause. The American

Cancer Society recommends that perimenopausal and postmenopausal patients report any abnormal vaginal bleeding or discharge to their doctor immediately because this can be a sign of endometrial cancer. The easiest way to diagnose endometrial cancer early in its course is with ultrasound examination of the uterus using a vaginal probe. Women receiving menopausal hormone therapy should have this test done regularly as per the recommendations of their prescribing physicians—usually once each year. Endometrial cancer screening is not recommended for postmenopausal women not receiving hormone therapy in the absence of abnormal vaginal bleeding.

Ovarian Cancer Screening Guidelines

Ovarian cancer is the fifth leading cause of cancer deaths among women in the United States. It becomes more common with age, with one-fourth of cases occurring between 35 and 54 years of age and half between ages 55 and 74. There are no screening tests specifically recommended for ovarian cancer at this time, although a gynecologic exam can sometimes detect this type of cancer. The provider performing the exam can sometimes feel a mass if it is large enough, and a transvaginal ultrasound can visualize any abnormal growths in the pelvic area. Unfortunately, ovarian cancer has few warning signs until it is fairly advanced.

Risks for this cancer include family history, obesity, advanced age, menopausal hormone therapy for more than 5 years, and use of fertility drugs. Birth control pill use, on the other hand, has been shown to decrease ovarian cancer risk. The Centers for Disease Control and Prevention's Cancer and Steroid Hormone (CASH) Study has shown that use of birth control pills can cut a woman's risk of ovarian cancer in half and that decreased risk is found even in women who have taken an oral contraceptive for as little as 3 to 6 months.

The CA-125 test is often used as a screening blood test in women in high-risk groups; however, many false-negative results are associated with this so-called tumor marker, and many women with ovarian cancer have normal CA-125 levels. A 2008 article in the journal *Clinical Cancer Research* suggests that a new generation of blood tests in phase III clinical trials will be able to detect ovarian cancer at an early stage with 99 percent accuracy.

CANCER SCREENING GUIDELINES FOR MEN

Prostate Cancer Screening Guidelines

Prostate cancer is the third most common cause of cancer death in men overall, with 230,000 men in the United States diagnosed each year. It is the most common cause of cancer deaths in men over 75 years of age, and young men often have prostate cancer without knowing it. According to the Detroit Autopsy Study, 29 percent of 30-year-old men were found to have cancerous changes in their prostates, and prostate cancer was even found in 2 percent of men in their twenties.

Men are considered high risk if they have a family history of prostate cancer in a first-degree relative (e.g., father or brother) diagnosed before age 65 or are of African American descent. Occupational exposure to agricultural and industrial chemicals increases risk; and painters, tire factory workers, and farmers appear to be at higher risk as well.

The two screening tests used for prostate cancer include a blood test called the prostate-specific antigen (PSA) and the digital rectal exam (DRE), performed by a physician. Conventional recommendations are for low-risk men to begin having annual DREs beginning at age 40 and annual PSA blood tests beginning at age 50. Conventional recommendations for high-risk men are the same for DREs, but they should get a baseline PSA at age 40 and, if the result is negative, annual PSA screening beginning at age 45.

Since prostate cancer affects a significant number of men in their thirties, we feel these conventional recommendations are inadequate for early detection. Our *TRANSCEND* recommendations are that men should obtain a baseline PSA at age 30 and then every 2 years until age 50 if normal. After 50, PSA should be checked at least every year.

Conventional opinion is that a normal PSA value is less than 4. We feel that this is much too high and that PSA should be less than 1. Any time it is higher than this, further evaluation is needed. This represents a recent change in our thinking but reflects new data showing that men who have a (conventionally) "normal" PSA of 2 have a 500 percent increased risk of being diagnosed with prostate cancer within 10 years. Men with the highest "normal" PSA of 4 have a ninefold increase in risk.

Reader: That's all you have to say about prostate cancer screening?

Ray: We have to move on. We have a lot more to cover.

Reader: I'm very concerned since my last PSA was 1.4. My doctor said this was fine for my age.

Terry: We don't believe in "age-specific PSA," the idea that it's okay to have a higher PSA if you're older. We feel that men need to keep their PSAs under 1 no matter what their age.

Reader: 1.4 isn't that bad, is it?

Terry: We don't have the exact statistics, but your risk of prostate cancer is probably double what it would be if your PSA was lower, say, 0.7.

Reader: That's upsetting. What should I do?

Terry: There are several things we suggest. First, it's critical that you follow all the aspects of our *TRANSCEND* program. Elevated PSA is most commonly the result of prostate inflammation, so you want to eat a diet that reduces inflammation—less red meat, dairy, and egg yolks. Get regular exercise, control stress, and take a natural anti-inflammatory such as fish oil. Next, take a supplement specifically designed to support prostate health. I also suggest you get a prescription for finasteride, either Propecia, usually prescribed for male pattern baldness, containing 1 milligram of finasteride, or Proscar, used for symptoms of an enlarged prostate, which has 5 milligrams. These drugs have been found to lower PSA. And, finally, recheck your PSA in 3 to 4 months.

Reader: What prostate supplement should I take?

Ray: We provide specific product recommendations on our book Web site (www. rayandterry.com/transcend), or see Resources and Contacts on page 424. The odds are good that doing all these things will lower your PSA below 1.

Reader: Thanks a lot!

Testicular Cancer Screening Guidelines

This is a cancer found most commonly in men ages 20 to 39. Young men in their teens should begin monthly testicular exams in which they roll their testicles between their fingers feeling for any lumps and report any suspicious findings to their healthcare provider immediately. Symptoms may include a lump, enlargement, swelling, pain, or discomfort in the testicle or scrotal sac, but sometimes the only symptom is lower abdominal or back pain. Three blood tests that can serve as tumor markers of testicular cancer are alpha-fetoprotein (AFP), beta-human chorionic gonadotropin (BHCG), and lactate dehydrogenase (LDH), and they can be used in conjunction with a physical exam. An ultrasound of the scrotal region can help determine whether abnormalities are due to infections, a collection of fluid, or a tumor.

Men who are at high risk are those with congenital abnormalities of the testicles or penis, as well as those who have inguinal hernias. A man who has a history of testicular cancer or a family history is at increased risk as well.

Reader: In the early 21st century, cancer edged out heart disease and became the leading cause of death in the United States. So what's the story with cancer in 2023?

Ray2023: We've made a great deal of progress. But let's first step back to your time. Back in 2009 there was a lot of excitement about stem cells.

Reader: And some controversy too.

Ray2023: Well, the controversy had to do with embryonic stem cells. Cancer is caused by a different type of stem cell.

Reader: As an aside, are embryonic stem cells still controversial?

Terry2023: There was a breakthrough back in your day on that. In 2008 we discovered how to create the equivalent of embryonic stem cells by injecting special genes into ordinary cells such as skin cells. This technique panned out beautifully and has given us an unlimited supply of stem cells. These modified skin cells work just like embryonic stem cells in that they can convert themselves into any type of tissue. They have the patient's own DNA and do not require the use of any embryos.

Reader: So what are they used for?

Terry2023: These newly created stem cells have been used to rejuvenate our tissues so heart patients can, for example, grow new heart cells and diabetic patients can grow new pancreatic islet cells.

Reader: Yes, I read about these experiments. I'm glad that worked out. So, what about cancer?

Ray2023: These "rejuvenating" stem cells have an evil cousin: the cancer stem cell. A single one of these cells can start a cancer tumor. It's like the queen bee of a bee colony, with the routine cancer cells being like worker bees. If you kill all the worker bees, the queen can simply repopulate the hive.

Reader: So that's why cancer kept coming back.

Ray2023: Exactly. Chemotherapy and radiation did work, but often they failed to kill the cancer stem cell, which would then generate new tumors.

Reader: I have read that the cancer stem cell was identified in 2008.

Ray2023: Well, yes and no. Many of the cancer stem cell projects were embarking on the wrong foot. They were identifying the cancer stem cell based on chemical markers on the surface of the cell. A popular one in 2008 was called CD135. But it turned out that these markers were not correct, and identifying the cancer stem cell was more complex.

Reader: So what happened?

Ray2023: We finally identified the cancer stem cell for most types of cancer. It turned out to be a genetically mutated stem cell that originated when the patient was a fetus or small child. In a fetus, this special form of stem cell could start a fetal organ like the colon, but in an adult it could start a tumor.

Reader: Do you mean that a cancer tumor is actually a fetal organ?

Ray2023: Yes, that's a very good insight and a good way to look at it. The cancer tumor starts and grows just like a fetal organ. A colon tumor, for example, is basically a fetal colon growing inappropriately in a mature individual. But with the fetal regulatory system not in place, it doesn't stop at the size of a fetal organ and just keeps growing. I was involved in a key project at MIT to identify, characterize, and target these cancer stem cells back in 2008 and 2009 and had the opportunity to actually see these cells in action.

Reader: Okay, tell me more.

Ray2023: A related insight comes from Dr. Judah Folkman, the pioneer of antiangiogenesis.

Reader: Antiangiogenesis?

Ray2023: "Angio" means blood vessel. "Genesis" means creation. So "angiogenesis" means creating new blood vessels, which is something a cancer tumor needs to do in order to grow beyond a small size. *Anti*angiogenesis drugs are designed to stop the creation of these blood vessels. These drugs were working in the sense that they succeeded in arresting the growth of cancer tumors. But Dr. Folkman made a disturbing discovery before his tragic death in 2008. These drugs also appeared to increase metastasis, which is the process where new tumors start growing.

Reader: That sounds rather counterproductive.

Ray2023: Indeed, and we now know why this happened. By cutting off the blood supply, the antiangiogenesis drugs were creating what are called anaerobic, "without oxygen," conditions. Well, it turns out the cancer stem cells thrive in exactly these conditions. They are anaerobic. Many of the other chemotherapy drugs have the same problem, killing cancer cells but encouraging the reproduction of cancer stem cells.

Reader: So, how did you deal with that?

Ray2023: We realized that we had to specifically target the cancer stem cell. Once the cancer stem cell was identified, unique properties of its reproductive machinery were also identified, which could be targeted with special drugs.

Reader: So was that sufficient, to kill off the cancer stem cells and then let the tumor take care of itself?

Ray2023: Not quite. If the tumor was at a mature size where it was not going to grow, then indeed it was sufficient to just kill the cancer stem cells. The tumor would stabilize and within a year or so would disintegrate. But in the majority of cases, even with metastasis stopped, the growth of the primary tumor could still endanger the patient. So we had to do both: kill the "worker bees" (that is, the cancer cells) as well as the "queen bee" (that is, the cancer stem cell).

Reader: So we're still using chemotherapy in 2023?

Ray2023: Sometimes, yes. A variety of means are used to kill ordinary cancer cells once we've killed the cancer stem cells. Dr. Folkman's original concept of antiangiogenesis has become more popular since it is more benign than most chemotherapy drugs.

Reader: I thought antiangiogenesis drugs would only stop the growth of a tumor and not kill the cancer cells.

Ray2023: In most cases that is sufficient. The killing of the cancer stem cell prevents new tumors from forming. And the antiangiogenesis drugs stop the growth of the existing tumors. With the cancer stem cells unable to start new tumors, the existing tumors just disintegrate within about a year. In other cases, we still use some combination of surgery, radiation, and new forms of chemotherapy to deal with the cancer cells. These new targeted forms of chemotherapy are much milder than they were back in 2008 and don't make you sick or cause your hair to fall out. But the ability to destroy the cancer stem cells has made all the difference in the world.

Reader: So cancer is cured?

Terry2023: Not quite. For many patients, killing both the cancer cells and the cancer stem cells does mean the end of the cancer. But it turns out that there are many types of cancer, each of which has a different genetic profile. So, in many cases, the killing of the cancer cells and the cancer stem cells is incomplete. It's enough to prevent the cancer from killing the patient but not sufficient to call it a cure.

Reader: Sounds like the situation with HIV and AIDS here in 2008.

Terry2023: That's a good way of putting it. Cancer is now a treatable condition. Most patients still need to take long-term drug therapies, but in most cases we can prevent it from being a death sentence.

Reader: That sounds like a lot of progress. I guess these are Bridge Two biotech therapies. What about Bridge Three nanotech breakthroughs?

Terry2034: Good question. Here in 2034, we've had another decade of progress, and we can finally declare victory in the war on cancer.

Reader: Gee, that's only 62 years after President Nixon declared the war on cancer.

Terry2034: Yes, sometimes progress needs to wait until all the enabling factors are in place.

Reader: So what turned the tide?

Terry2034: Today we have nanobots that act as artificial white blood cells, except they're a lot smarter than our natural white blood cells. These robotic "microbivores" can detect and destroy virtually any dangerous pathogen, including viruses, bacteria, cancer cells, as well as the cancer stem cells. So, tumors are stopped when they are just a handful of cells, long before they become a tumor. And we can download new software into these microbivores when we discover new pathogens. That's one of the responsibilities of doctors here in 2034.

Reader: So, I won't have to get any of those uncomfortable colonoscopies and prostate exams in 2034?

Terry2034: Indeed, the nanobots do that for us as well.

Reader: Okay, I guess that's another reason for me to take care of myself the old-fashioned way relying on the *TRANSCEND* Bridge One strategies a little bit longer.

7

GENOMICS

A new genomics company, Complete Genomics, says it will offer sequencing of a whole human genome (~3 billion base pairs) by mid-2009 for $5,000, which would be a major price breakthrough. It will be able to do so by optimizing the computation of genetic sequencing, building one centralized, giant super sequencer, and plans to sequence 1 million human genomes within the next 5 years.

KURZWEILAI.NET, FEBRUARY 6, 2009

Genomics is the study of your genetic material—the DNA in your genes and chromosomes. The majority of your genetic information is contained within the double-stranded DNA molecules found in the nuclei of your cells. Because of the unique double-stranded structure of the DNA molecule, it is able to unzip itself in various places along its length and reproduce reverse-image complementary copies of itself as so-called RNA messenger molecules. These RNA molecules then travel from the nucleus into the remainder of the cell, where they serve as blueprints to carry out the genetic instructions specified by the master DNA molecules.

DNA contains just four molecular structural variants: adenine (A), guanine (G), thymine (T), and cytosine (C), which are cross-linked one to another like the rungs of a ladder. Adenine always links to guanine, and cytosine only binds to thymine. RNA is similar, only it uses uracil (U) instead of thymine. Using just these four letters, our 23 chromosomes spell out the genetic book of life. Genomics is the new medical field that seeks to read and interpret this book.

DNA AND RNA

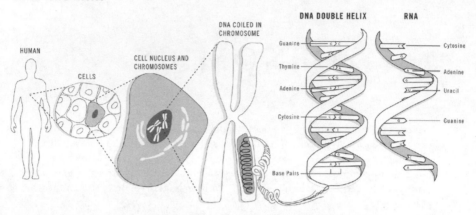

YOUR GENES, YOUR MISSION

Since it is such a new field of medicine, genomics is not yet used by most healthcare practitioners. One reason doctors have been hesitant to even dip their toes into the waters of this new ocean of knowledge is because the sheer volume of information and data that are potentially available appears overwhelming. Most physicians also feel the field is too new and uncertain for clinical use in their practices. Yet, genomics has been around for a number of years, and there are a few simple inexpensive genomics tests that you can do today to help you tailor your personalized wellness program.

For instance, a 2008 study showed that several common genetic variations were related to whether an individual would have relatively more or less good high-density lipoprotein (HDL) or bad low-density lipoprotein (LDL) cholesterol. Unfavorable gene types were shown to increase LDL cholesterol by an average of 19 points, a significant amount. Each unfavorable gene type also was shown to cause a 15 percent increase in an individual's risk of developing a heart attack. Yet, in an editorial comment at the end of a review of this article, the authors concluded, "More study is needed [of these genes] to determine their clinical usefulness. For now, genetic tests do not need to be part of cardiovascular risk evaluation."

This conservatism is typical of conventional medical thinking, so that, even though many genomics tests have been available for several years, this testing hasn't yet entered mainstream medical practice and is still not used *at all* by most practicing physicians. To use the power of genomics—at least for now—you'll either need to take matters into your own hands and order your testing directly from the Internet or establish a relationship with one of the minority of physicians who have started to use genomics testing in their practices. However, as we'll discuss below, it's now quite simple to access a wide range of genomics test panels on your own to help guide you with your personalized health program.

GIFTS FROM YOUR PARENTS

A comprehensive health evaluation begins with a careful health history taken by your doctor. Your family history, with a thorough listing of the diseases that have affected other family members, is one of the most important components of your health history. It can help direct you or your healthcare provider to drill down to look for specific health risks and challenges. You are, after all, genetically, half your mother and half your father. (Actually, you are slightly more your mother, since the genes in your mitochondria, the energy power centers of your cells, are inherited exclusively from your mother.)

The new medical specialty known as *predictive genomics* can provide you with more detailed information. Where your family history paints the broad strokes of your genetic inheritance, genomics testing, the analysis and interpretation of the genes you have, allows you to see the details. Genomics testing can help you look at the fine print of your genetic inheritance and determine how your genes interact with one another and with your environment to create preconditions for health or disease. Predictive genomics has already advanced to the point that many genomics tests are now commercially available to help predict your genetic predisposition to many serious—but preventable or modifiable—diseases, such as heart disease, Alzheimer's disease, and cancer.

One problem is that many people are frightened by the prospect of knowing the genes they carry. Much of this fear is unfounded since, in almost all cases, genes merely express *tendencies*. The latest research has shown that your lifestyle choices play a much larger role in determining what actually happens to you than do the genes you carry. You *can* overcome even the strongest genetic tendency to diseases such as heart disease and cancer by carefully following our *TRANSCEND* program.

Dr. Dean Ornish and his colleagues at UCSF and the Preventive Medicine Research Institute recently published two landmark studies showing that changing lifestyle changes our genes. In the first, a combination of improved nutrition, moderate exercise, stress management techniques, and increased social support caused the expression of over 500 genes to be changed in only 3 months—in effect, upregulating or "turning on" disease-preventing genes and downregulating or "turning off" genes that promote heart disease, cancer, inflammation, and oxidative stress. This study was published in the *Proceedings of the National Academy of Sciences*.

In the second study, they found that these lifestyle changes increase telomerase, an enzyme that repairs and lengthens damaged telomeres.

Telomeres are the ends of our chromosomes that control how long we live. This was the first study showing that any intervention can increase telomerase and, thus, telomere length. It was published in *The Lancet Oncology* in collaboration with Dr. Elizabeth Blackburn, who discovered telomerase. Even drugs have not yet been shown to accomplish this.

WHAT ARE THE ODDS?

Gregor Mendel, the father of modern genetics, felt that the genes you were born with would determine your fate in an absolute fashion. This thinking has given way to the newer concept that most genes express themselves as tendencies or probabilities. There are a few genes, such as those for cystic fibrosis or Huntington's disease or even blue eyes or blond hair, that do relegate an individual to an almost certain fate, but these are the exceptions. They represent only a tiny fraction of the millions of variants possible in the human genome. As a result, current thought is that the best way to think of genes is as probabilities. Whereas physics has its Heisenberg's uncertainty principle (which says that we cannot precisely determine the location and velocity of a subatomic particle), biology has its own genetic uncertainty principle. We can, however, use the lifestyle choices we make every day—the food we eat and the supplements we choose to take, for example—to affect and modulate these tendencies. Making proper lifestyle choices plays a fundamental role in your personal wellness program, and we refer to doing so as "reprogramming your biochemistry."

IMPROVING YOUR ODDS

In the past few years, the new field of *epigenetics* has evolved. Epigenetics looks at how environmental factors such as your lifestyle can change how your genes are expressed—in other words, whether these genes, or tendencies, become manifest as health or disease. In fact, it is now believed that your lifestyle choices are 80 percent of the reason that you'll get a given disease, while your genes contribute only 20 percent of the risk.

In the game of life, your genes are the cards you've been dealt. As any good player knows, however, how you play your cards is more important than the actual cards you've been dealt. That's why a good card player will invariably defeat someone less skilled. The same is true in the game of life. But, until the Human Genome Project was completed in 2002, you had to play the card game of life without being able to see the cards you'd been dealt. As any poker player knows, if you don't know what cards you hold, every bet is a bluff! Try deciding whether to take a hit or double down at a blackjack table without seeing either of your hole cards. You'll play a much better game if you can see the cards you are playing.

DRAMATIC COST REDUCTIONS

Like most new technologies, prices for genomic testing have been dropping at an exponential rate. Many individual genomics tests that look at common genetic variations are now available for less than $20 each. Less than 10 years ago, many of these tests cost hundreds or thousands of dollars. At the other end of the spectrum, if you have the money, you can decode your entire genetic code today—all 25,000 genes—for $350,000. Sure, it's a lot of money, but in 2002 it had cost $3 billion to sequence the full genome of James Watson.

A few years later the price fell tenfold, down to $300 million when Craig Venter of Celera Genomics published his genome sequence in Plos Biology in September 2007. Yet, as is characteristic of the double exponential growth rate of technological advances, shortly afterwards, the price fell by a factor of almost 1,000 to $350,000. Because of the acceleration in technological change, the price will continue to fall very rapidly in years ahead, and in a few years, it will be possible for you to sequence your entire genetic makeup for less than $1,000.

Terry2023: As we discussed back in 2009, exponential progress in genetic sequencing just kept on going. Soon after this book came out, a team won the Archon X Prize and collected $10 million for being the first to sequence 100 human genomes within 10 days at a recurring cost of no more than $10,000 per genome, and by 2013 the "thousand dollar genome" became a reality. Today, it costs less than $100 (in 2009 dollars), so almost everyone has full genome sequencing done.

Ray2023: But it wasn't until several years later that the computer folks were able to crunch the awesome amount of data contained in each genome and began to make sense of it all. It takes a lot of computing power to sort out the 3 billion base pairs of one person's genome. We still have quite a ways to go, but most people today know with far greater accuracy than ever before many of the disease processes to which they are most susceptible.

Reader: Isn't it scary knowing what's going to happen to you years ahead?

Terry2023: You need to bear in mind that, in most cases, genes don't predict the future. They're only tendencies. Even today, the *TRANSCEND* choices you make are much more important than your genes. It's true that there are a handful of genes that lead to diseases that we still can't prevent or cure, but these are few and far between.

Reader: But those are the ones I'd worry about finding I had.

Ray2023: You're not alone. Most people feel the same way and choose not to know if they have a really bad gene that they can't do anything about. In these cases, the data are simply hidden so that no one knows—not the doctor or the patient. This is one place where ignorance is still bliss.

Terry2034: Not anymore. With the more robust gene therapies available in the 2030s, there are virtually no genetic problems that can't be effectively erased.

Reader: I'd worry about my insurance company trying to cancel or refuse to renew my policy if they knew I was genetically predisposed to a disease that might cost them a lot of money down the line.

Terry2023: But we are now able to treat a few diseases that were incurable in your day by inserting new healthy genes to replace missing or defective ones. This is called gene therapy, and after a somewhat rocky start in your day, gene therapies are now commonplace.

Ray2023: It began with RNAi (RNA interference) in 2002 before we were able to change genes altogether. Doctors began by blocking the effects of harmful or dangerous genes by interfering with the RNA that the genes were supposed to make. This proved to be much easier than correcting the genes themselves. An entire industry of RNAi drugs came on the market during the late 2010s and revolutionized the way medicine was practiced.

Terry2023: Doctors don't write nearly as many prescriptions for "drugs" these days. Instead, they simply correct the underlying problem. After finding what potentially harmful genes a patient has, they first try to compensate by utilizing our modern-day versions of *TRANSCEND* therapies. With some of the more dangerous genes, in addition, they will often perform some type of gene therapy where they insert copies of healthy genes.

GETTING TESTED

There are three levels of entry into the world of predictive genomics. You can choose which is most appropriate for you based on both your budget and how much information you'd like to receive.

Individual Gene Tests

There are two basic types of genomics tests: DNA sequencing, in which every bit of the DNA of a gene is tested, and SNP testing, which is more of a screening test, typically done as part of genomics panels. Today, full gene DNA sequencing remains relatively expensive and can cost thousands of dollars per gene. This is the type of genetic testing needed, for instance, by women with strong family histories of breast cancer who need to know if they carry the *BRCA1* gene. SNP testing is far less costly and is the more common type of screening done today.

SNP Panels

SNPs (single nucleotide polymorphisms, pronounced *snips*) are genetic mutations at one point on a gene. SNPs are very common, and every individual may have hundreds of thousands. More affordable, although much less thorough information, is available by looking at common SNPs. Having these mutations can influence your propensity to many diseases and conditions. Several personal genomics start-up companies, such as 23andMe (cofounded by Anne Wojcicki, wife of Sergey Brin, cofounder of Google), Navigenics, and DecodeMe, have recently begun offering SNP panel testing online direct to consumers. The test panels they offer don't specifically decode your personal DNA; rather, for a few hundred dollars, they test to see whether you have any of several dozen common SNPs that affect your individual susceptibility to disease. The SNP panels offered by these companies test for genetic susceptibility to diseases such as age-related macular degeneration, asthma, Alzheimer's disease, atrial fibrillation, cancer (breast, colorectal, and prostate), celiac disease, Crohn's disease, diabetes, heart attack, multiple sclerosis, obesity, psoriasis, restless legs syndrome, and rheumatoid arthritis.

Among the most important SNPs identified to date and for which we have quite a bit of data is the *APO E* gene, which can play a role in your susceptibility to Alzheimer's disease. *APO E* comes as *APO E2, APO E3,* or *APO E4.* Sixty percent of the population is *E3/E3,* meaning they have two copies of the normal *E3* gene, one from each parent. Forty percent of us, however, have one or two copies of mutated *APO* genes, either *E2* or *E4.* The *E2* SNP, carried by about one person in six, dramatically reduces risk of Alzheimer's disease and is associated with increased longevity. It is an example of a highly beneficial SNP. About one-fourth of the population, however, has at least one copy of *APO E4,* which increases the tendency towards both Alzheimer's and cardiovascular diseases. Finding out that you carry the *E4* gene can serve as a call to action to take aggressive action to prevent the expression of this genetic tendency. We need to remember also that most potentially harmful genes, such *APO E4,* also have beneficial aspects. Otherwise, these genetic variants wouldn't have endured the rigors of natural selection and survived. If you find you carry the *APO E4* gene, the good news is that you have a decreased risk of developing macular degeneration, the leading cause of vision loss in older people.

To perform basic genomics testing on yourself, you can go to the Web sites of the companies listed above (www.23andme.com, www.navigenics.com, www.decodeme.com), and you can order a test kit that uses a saliva sample to test your DNA. No physician referral is needed—in fact, that's one of the main problems with genomics testing today—you're pretty much on your own when you get your results, since your doctor probably won't know what they mean.

Full DNA Sequencing

As mentioned above, anyone willing and able to pony up $350,000 can obtain their full genome—all 3 billion base pairs of nucleotides—through a service offered by Knome Genomics in Cambridge, Massachusetts. If you're interested, you'd better hurry; their initial offering is limited to the first 20 to sign up each year. This service will tell you the exact DNA sequence of every one of your 25,000 genes and is delivered to you on an 8-gigabyte thumb drive in a velvet-lined silver box. The price will come down rapidly and may be significantly reduced by the time this book is in your hands. In fact, on the final day of editing this book, we inserted the paragraph that begins this chapter, noting that Complete Genomics plans to offer full genome sequencing for $5000 by mid-2009, a 70-fold drop in price. Fully understanding your genome, however, is going to take longer. Reverse-engineering the human genome and the biological processes that result from the genome is now the goal of many thousands of researchers. Although a formidable task, progress will be exponential (that is, doubling about every year), as was the case with the genome itself.

PART II

THE PLAN

8

TALK WITH YOUR DOCTOR

THE TWO PILLARS: PREVENTION AND EARLY DETECTION

Two key pillars to our *TRANSCEND* program are *prevention* and *early detection* of disease. Preventing disease altogether is our goal, but, if you eventually do develop a disease process, you at least want to detect it early—when it's easier to treat and at a point where a complete cure may be more likely. Most people will implement much of the early detection strategies with the help of healthcare providers. Unfortunately, because of misguided financial considerations, most insurance companies don't pay for preventive healthcare (or pay very little) and most physicians have received little training in this specialty. An example of why we say they're misguided is that very few insurance companies will cover the $300 it costs for a coronary artery calcium test, which can detect coronary artery disease at its earliest stages, but they don't hesitate to pay tens of thousands of dollars for a hospitalization for a coronary artery angioplasty and stenting a few years later because the disease wasn't picked up earlier by more aggressive screening.

As a result, obtaining effective preventive healthcare isn't automatic. In all probability, you, not your doctor, will need to take charge. There are a few hospitals and clinics, such as Cooper Clinic in Dallas, the Mayo Clinics, and Terry's wellness center in Denver, that specialize in preventive healthcare and early detection of disease. But most people would prefer to receive their care at home and from their usual healthcare providers. Because of the fundamental conservatism of most conventional physicians and their lack of interest—and training—in preventive healthcare, you will need to take an active role in developing your own preventive health program. "Talk with your doctor" is one key to your *TRANSCEND* program, and you need to let her know what you want. Let's begin by discussing what a comprehensive preventive health evaluation with a physician should be like, so that you'll know what it is you want and what to ask for.

THE MEDICAL HISTORY

Most doctor–patient visits typically consist of three parts: the medical history, the physical exam, and laboratory testing. Your medical history is taken when your doctor first sits down with you and discusses all the aspects related to your past and current health. The history usually begins with a discussion of your *chief complaint*—what it is that brought you to the doctor—as well as the *history of the present illness*—the story of the events leading up to your current situation. For many people who have come in for a comprehensive preventive health evaluation, there really isn't much of a chief complaint or history of present illness because they may not have anything wrong with them. The reason they've come is they want to maintain their good health, and there's nothing wrong with that!

When you fill out the forms to see a new doctor, you're often asked to write down the main reason for coming. Terry has seen some interesting reasons from patients who've come to see him for comprehensive health evaluations, including "I want to live to 150," "I'm seeking immortality," or "I never want to die." Since our last book, *Fantastic Voyage: Live Long Enough to Live Forever,* was published in 2004, it would seem that many people now take the idea of living forever quite seriously! Regular comprehensive healthcare evaluations will help you get there.

For many brief and limited doctor visits, the chief complaint and history of present illness are the full extent of their medical history. For instance, if you see a doctor for a sore throat, that's your chief complaint. The doctor might ask how long you've had it, whether you have a fever, whether there is an associated cough, and a few other such questions. That's the history of present illness. Next the doctor proceeds to your physical exam. She looks at your throat, your ears, and listens to your heart and lungs—a limited physical exam. Perhaps a strep screen is done—the testing phase—and, if the results are positive, you're given a prescription for an antibiotic.

In a comprehensive evaluation—particularly one directed at the earliest possible detection of disease—there are several other parts to both the history and the physical that are needed. After discussing your chief complaint and history of present illness, the doctor will also want to ask about the medications and nutritional supplements you take. Information about prior illnesses and hospitalizations is needed, along with a detailed history of your lifestyle and habits: your vocation and hobbies, how you sleep, what kind of exercise you get, as well as how often and how long. A thorough diet history is taken, including your use and possible abuse of coffee, alcohol, and sugar. A detailed review of systems is completed ranging from head to toe—from your vision and hearing to your bowel movements. Finally, your family history is reviewed to drill down in detail as to diseases and risks that run in your family.

You can make the time you spend with your doctor much more efficient by organizing the information about your health history in advance of your appointment. Even if you're planning to do some of your evaluation and testing on your own, it's well worth your time to take a comprehensive history on yourself periodically. But even if you go the self-directed route, you should still see a physician at regular intervals to have physical examinations. First of all, you can't do all the aspects of a physical exam on yourself. Also, unless you've had specialized training, you wouldn't know what to look—and feel—for.

WHEN TO GET EXAMS

Because of the need for cervical cancer screening, women need to begin their examinations at 21 or within 2 years after having vaginal intercourse. Women will continue their gynecological exams, which should also include breast exams, blood pressure, and examination of heart and lungs, every 1 to 2 years until age 30, and, if these have all been normal, every 2 to 3 years until 70.

Men should get a baseline physical, including blood work, around 25 years of age. These should be repeated every 5 years until age 40. Digital rectal examination of the prostate should begin annually at age 40. Beginning at 50, both men and women need annual fecal occult blood tests for colorectal cancer screening.

In addition to the above, you should do periodic *comprehensive* health evaluations at regular intervals. These are designed to look at all aspects of your health in detail. A baseline comprehensive health evaluation should be done between 40 and 45 years of age for both men and women and repeated every 3 to 5 years until age 60. After 60, every 2 to 3 years is optimal.

There are several components of the physical exam—breast self-exams for women and testicular self-exams for men, blood pressure and heart rate measurement, body fat, waist-to-hip ratio, and fitness testing—that are done at home and that you should get in the habit of doing regularly. These self-assessment tests are discussed in Chapter 10.

PHYSICAL EXAM WITH A DOCTOR

A wealth of information can be obtained from a thorough head-to-toe physical examination. This is the hands-on part of your evaluation and takes only a few minutes. The following is the minimum of what you should expect in a thorough physical exam as part of your comprehensive wellness program. Most doctors begin at the top and work down.

- **GENERAL:** Your doctor will want to evaluate your appearance. Are you alert, oriented, or do you look distressed? Do you have good hygiene? Do you appear the same as your age or older or younger?

- **HEAD, EYES, EARS, NOSE, AND THROAT:** The doctor will look at the neck and thyroid gland and check to see that all of your facial features are symmetrical. This is where the doctor uses an otoscope to examine the ears, nose, and throat, and an ophthalmoscope to look into your eyes. Your neck is palpated or felt for abnormalities, including the thyroid, bones, and blood vessels. A stethoscope may also be used to listen to the arteries in your neck because a partial occlusion can be heard there. You will be asked to swallow as the clinician feels your thyroid, evaluating its symmetry, size, and whether there are any nodules. The head and neck will also be checked for any swollen lymph nodes or other abnormalities.

• **CARDIOVASCULAR:** The clinician will use a stethoscope to listen to your heart for its rate and rhythm. Any extra or unusual sounds, such as murmurs or other sounds, as well as abnormalities of rhythm such as disturbances in rate or regularity, will be evaluated.

• **LUNGS:** The doctor will look to see that your breathing pattern is regular, unlabored, and even. Breaths should not be too fast or shallow. The stethoscope is used again to listen to the lungs. They should be clear without any wheezes, congestion, or other unusual sounds.

• **BREASTS:** For both men and women, a visual inspection should be done to look for asymmetry, dimpling, or lumps. Palpation is then done to evaluate the breast tissue for lumps, masses, or nipple discharge. Both underarm areas should be checked for swelling, nodules, or lumps.

• **ABDOMEN:** The doctor will next listen to the bowel sounds in the abdomen using a stethoscope. Visual inspection is also done to evaluate for any obvious masses, rashes, discolored skin, or asymmetry. By palpating the abdomen, the clinician evaluates the size of the liver and spleen and checks for abnormal masses or lumps. Any pain or tenderness will also be noted, as will abdominal firmness. Sometimes tympany (using the fingers to tap on the belly) will detect abnormalities such as masses, increased air, or collections of fluid.

• **GENITOURINARY:** Men will have a visual inspection of the penis and scrotum, looking for lesions, discharge, redness, swelling, or asymmetry. The testes are palpated for lumps, masses, or swelling. This is when the clinician inserts an index finger alongside of one testis, then the other, into the inguinal canal as you are asked to turn your head and cough as he feels for a hernia. (The reason you're asked to turn your head, by the way, is simply so that you don't cough in the examiner's face!) If a hernia is present, the intestines may be felt pressing down as you cough. The groin of both sexes should be palpated to evaluate for any enlarged lymph nodes.

• **PELVIC EXAM** (WOMEN): A visual inspection of the exterior vagina and labia is done to evaluate for unusual discharge (clear or white discharge is normal for women), redness, swelling, irritation, skin discoloration, lumps, lesions, or asymmetry. A woman may be asked to bear down as if she is having a bowel movement to see how well her bladder is supported by her pelvic muscles in order to ensure there is no prolapse of the uterus. A metal or plastic instrument called a speculum will be inserted into the vagina, and cell samples and cultures are taken from the cervix (or vaginal cuff if the woman has had a hysterectomy) using a brush and cotton-tipped applicator. The samples will be sent to a laboratory for evaluation as part of the Pap test. The laboratory will look for abnormal cells, yeast, bacteria, and sexually transmitted diseases. A pregnant woman will have a more thorough laboratory panel. The speculum also enables the clinician to look inside the vagina for the same things that were looked for during the exterior inspection. The speculum is removed and then the index and middle finger are inserted to palpate the abdomen, uterus, and ovaries. This is done to evaluate the size of the uterus and ovaries, checking for fibroids, cysts, or masses in this area. This is sometimes a little uncomfortable but should not be painful.

• **RECTAL:** After the age of 40, men and women should have rectal examinations that involve the insertion of the clinician's gloved, lubricated index finger. This allows for palpation of the inside of the rectal vault for abnormalities such as lumps or masses, as well as sphincter tone. A small amount of stool may be obtained and tested for blood as a colon cancer screening test. In men, the physician also feels the surface of the prostate, checking for enlargement or abnormal masses.

• **NEUROLOGICAL:** The physician will take you through a series of verbal instructions to evaluate your range of motion of all of your movable body parts, as well as the innervation (nerves) that allow movement of the muscles and joints. Visual inspection will be done to evaluate symmetry as well. A reflex hammer will be used to test reflexes in the crook of your elbow, umbilicus, and lower extremities, especially the knees. You should be asked to perform some movements against resistance to test your muscle strength and tone. Various tests to evaluate your sensation of mildly painful and gentle stimuli will be done, along with tests of balance, such as standing on one foot, and rapid alternating movements (e.g., finger-to-nose testing).

• **MUSCULOSKELETAL:** Your joints will be evaluated through both observation and direct palpation to evaluate for pain, redness, swelling, tightness or laxity (looseness), pain, or crepitus (noise or creaking with joint movement). The clinician will look to be sure there is no asymmetry or abnormal loss of muscle mass.

• **SKIN:** Ideally, at some point during your exam, the skin on all the areas of your body will be examined. Specific attention will be directed at any suspicious or unusual lesions. The borders of moles, as well as the color, symmetry, or scaling, will be noted.

• **EXTREMITIES:** Visual inspection and palpation will be used to look for swelling, redness, uneven hair distribution, skin discoloration, and fluid retention, called edema. Pulses in the wrists, upper arms, groin, or feet will also be checked for their strength and regularity.

• **SPINE** (BACK AND NECK): Visual and tactile exam will evaluate the straightness of your spine, and you may be asked to walk to evaluate your gait. Palpation may be done to look for painful areas as well. You may be asked to bend over and move your neck and torso from side to side to evaluate your spine's range of motion.

Terry2023: Reading the above was very interesting. It reminded me how doctors examined patients years ago. Physicians in the 2020s almost never examine patients with their hands or even stethoscopes anymore. We have a wide variety of imaging devices that provide a much more accurate view of a patient's health today.

Reader: I thought that touching patients was an important part of medical care. Isn't medicine awfully sterile without any "laying on of the hands"?

Ray2023: Not at all. The interpersonal relationship patients have with their doctors is even stronger now that doctors have more accurate data about their patients' health.

Terry: Even today in 2009 we've started to move away from relying heavily on direct physical examination of patients and instead have begun utilizing various types of imaging studies.

Reader: You mean like CT scans and MRIs? Aren't they incredibly expensive and expose patients to serious amounts of radiation?

Terry: You're right that these types of diagnostic imaging devices are quite expensive, and CT scans do entail significant radiation exposure. As a result, doctors have begun to utilize ultrasound and thermography to a much greater extent.

Reader: I had an ultrasound when I was pregnant. That's how I found out that the baby was healthy and whether we were going to have a boy or girl.

Ray: Diagnostic ultrasound has been in use in medicine since the 1940s but actually fell out of favor beginning in the 1980s as CT scans and then MRIs became available and provided clearer images. But it's undergoing a resurgence today.

Terry: Exactly. For one thing, ultrasounds have the advantage of being dramatically less expensive. In addition, since they use sound waves, they're safe and don't expose patients to radiation. Thermograms measure the amount of heat being emitted from various regions of the body, which can help doctors detect a wide variety of pathological conditions from arthritis to cancer. We have begun to include total body ultrasounds and thermograms as part of our comprehensive evaluations. This allows doctors to get a very accurate picture of what's going on inside their patients' bodies.

There are also imaging techniques that look inside the brain. SPECT scans help doctors provide early detection of brain problems in patients.

Reader: SPECT?

Ray: Single photon emission computed tomography is a nuclear medicine imaging technique that uses gamma rays to obtain three-dimensional views inside the brain as well as other organs.

Terry: SPECT scans of the brain have traditionally been used to image physical brain problems such as the location of seizure centers, strokes, and traumatic brain injuries.

Ray: It's also very useful in the early detection of brain changes consistent with Alzheimer's disease, with some centers reporting a diagnostic accuracy of greater than 90 percent.

Terry: Some clinics have also begun to utilize SPECT to help diagnose and treat psychiatric illnesses such as anxiety disorders, ADD/ADHD, autism, bipolar disease, depression, and learning disorders.

Reader: This seems like the ultimate in early detection. Sounds like something I want to talk with my doctor about—for sure.

LABORATORY TESTING

After taking your medical history and completing your physical examination, your healthcare provider will often order some medical tests. Most common are blood tests, either performed in the provider's office or sent to an outside reference lab. Urinalysis is often done by conventional practitioners, while complementary clinics often also analyze samples of hair, saliva, and stool. Other types of testing include imaging studies such as x-rays and mammograms, computed tomography (CT), magnetic resonance imaging, and ultrasounds. Other imaging studies involve endoscopes—flexible fiberoptic instruments that allow physicians to look directly inside the body. Electrocardiograms provide information about the heart, while spirometry measures lung function.

COMPREHENSIVE HEALTH EVALUATIONS

To complete the *Assessment* portion of your personal *TRANSCEND* program, in addition to your routine health screening, you should consider undergoing periodic *comprehensive* health evaluations. Comprehensive evaluations are much more in-depth than the type of screening typically done by conventional physicians as part of the "annual physical." Not only is your height and weight measured, but your percentage body fat and muscle mass is determined as well. Rather than relying on your medical history to assess your exercise capacity, your doctor will have you get on the treadmill and will determine how fit you really are compared to other individuals of the same age and sex. A comprehensive health evaluation also includes many tests not included in routine checkups. These often include tests of digestive function, mineral analysis, vitamin and free radical levels, neurotransmitter levels, detailed cardiovascular tests such as coronary artery calcium screening or carotid intima-media thickness, and genomics testing.

Terry's clinic in Denver specializes in these evaluations, and they are also available at other centers, such as the Cooper Clinic in Dallas, the Mayo clinics in Rochester, Minnesota; Jacksonville, Florida; and Scottsdale, Arizona. A listing of physicians who specialize in antiaging medicine is available from the American Academy of Anti-Aging Medicine (www.worldhealth.net), and some of these physicians also offer comprehensive health screening.

HOW TO GET SPECIFIC TESTS DONE

Many of the tests discussed in this book (such as lipid profiles, homocysteine, and C-reactive protein) are routinely done by conventional physicians. Others (for example, hormone levels, blood cancer screening, vitamin levels) aren't done routinely but can be done if you ask (talk with your doctor). Some of the tests we recommend in this book (such as hair mineral analysis, stool microbiology, hepatic detox, environmental pollutants) are foreign to conventional doctors. You will need to get these tests through a holistic or complementary physician. To find these types of practitioners, a listing of holistic doctors is available through the American Holistic Medical Association (www.holisticmedicine.org), and complementary physicians can be found through the American College for Advancement in Medicine (www.acam.org).

You can arrange for certain tests on your own, such as the coronary artery calcium test, osteoporosis screening, Doppler examinations of the carotid arteries for stroke risk and of the legs for peripheral vascular disease. Mobile screening clinics periodically offer panels of these tests at very affordable rates.

KEEPING TRACK OF THE RESULTS

We recommend keeping a record of all of your test results yourself rather than relying exclusively on your healthcare providers. Having some of your records at one doctor's office, some with another, others at the hospital, and your own self-test records at home is not an optimal way to keep track of your health data. Despite this being an electronic age, the record-keeping methods used by most doctors are still in the Stone Age! It is now possible to assemble much of your health information by using free online programs. HealthVault is a free program available from Microsoft (www.healthvault. com) that allows you to keep track of many types of personal health data. You can enter multiple types of health information to create a personal health record that you can share with all your healthcare providers. You can choose what types of data you wish to enter into your health record, such as:

- Your health history
- Fitness activities, such as aerobic or strength training sessions
- Blood glucose and blood pressure
- Summaries from hospitalizations
- Lab test results
- Medications and allergies

You can also upload measurements directly from specific brands of weight and body fat scales, blood pressure devices, blood glucose meters, and pedometers to your personalized health program using a USB connection.

Google has a similar free service (www.google.com/health) to keep track of your health history. Either of these programs enables you to transfer any or all of your health records online to your healthcare providers.

Taking charge of your early detection program will require some effort on your part, but the rewards to you and your family will be measured in additional years of great health and extended youth . . . long enough to enjoy the next stage of health technology.

PITFALLS YOU MAY ENCOUNTER ALONG THE WAY

Heart attacks, the leading cause of death in the United States, are almost completely preventable if coronary atherosclerosis is diagnosed early in its course and then followed by aggressive corrective action. Cancer, the second leading cause of death, is far more treatable—and often curable—when it is detected early, before it has a chance to spread. Surgical removal of a malignant tumor before it has metastasized is often curative.

Medicine, as it is practiced today, focuses largely on diagnosing established disease. When damage has already been done, it is often irreversible. Only a small minority of medical students today, just like their older colleagues before them, have much interest in the specialty of preventive medicine, which incorporates prevention and early detection as its cornerstones. Who can really blame a budding medical student for being drawn toward more exciting and higher-paying careers like the surgical specialties? It is inherent in the nature of medical students to be more excited about learning how to do complex interventions when compared to making sure that their patients are eating enough servings of vegetables or getting their colonoscopies on time. For decades, preventive medicine has remained in the medical backwaters, marginalized and avoided.

The situation isn't helped by the fact that most conventional medical practitioners and organizations either don't advise or actively oppose routine use of many of the tests we recommend below for early detection. Why conventional medicine is very slow at adopting new ideas would be the subject of another book. There are indeed different viewpoints on the roots of the delay—often measured in decades—that it takes for conventional medicine to adopt new lifesaving approaches. Often, by the time a new test becomes mainstream, there's already a new and even better test that, again, won't be adopted for far too long. For the purposes of this book, it is important to recognize this reality and realize that by taking an active role in your own health, you can be on the leading—rather than the lagging—edge of our rapidly accelerating health knowledge.

EXAMPLES OF NEW TESTS YOU PROBABLY HAVEN'T HAD (BUT NEED)

Consider ultrafast CT coronary artery calcium (CAC) scans of the heart. This technology has been available for over a decade and a half, and, even though multiple studies have shown that it is effective at detecting coronary artery disease very early in its course—at a time that effective preventive treatment can still be done—the American Heart Association (AHA) still does not recommend its use for primary screening. Even without the AHA's official blessing, however, more and more cardiologists have finally started to use CAC cardiac screening in their practices. It has just about reached a tipping point and will probably become mainstream in the next few years— just in time to be out of date and for a new paradigm shift to occur.

Another example is the use of NTX urine testing to tell whether a woman is at risk of developing bone loss years before osteoporosis is diagnosed. NTX testing is currently *not recommended* by mainstream practitioners for routine screening purposes. It is approved only for monitoring patients who have already been diagnosed with osteoporosis to see how well their medication is working. We suggest you use this simple, noninvasive test to help determine whether you are a candidate for osteoporosis before you begin to lose bone and have to try to play catch up.

An inexpensive genomics test, the *APO E* genomics test, which we strongly advise, can tell whether you carry a gene that triples your risk for Alzheimer's disease. Most physicians either don't know that this test exists or, if they do, still don't suggest it to their patients—even those with a strong family history of the disease—because genomics testing is simply *too new* to them and not yet recommended by their specialty board. Eventually, all these tests and others like them will enter the mainstream. But, before they do, several years will pass and many people who could have benefited from early detection will be diagnosed too late in the course of their disease for full recovery.

THE IMPORTANCE OF BEING THOROUGH

Another problem is that most physicians do not do a thorough job of screening their patients. Preventive medicine has never been a priority for most physicians, and it's hard to get a lot done when the face-to-face time of an office call between a physician and patient averages 11 minutes. When was the last time your family physician checked your body fat or measured how many pushups or situps you could do? We know that declining hormones are a critical component of the aging process. Has your physician ever recommended a check of your testosterone, DHEA, estrogen, or progesterone levels? Tests such as these are critical components of a thorough health evaluation. Yet, because of pressures placed by health insurers, doctors are often not thorough and take shortcuts.

This isn't the case for every profession. It is instructive to compare how a pilot prepares before takeoff in an airplane to how a doctor examines a patient. A pilot *always* uses a checklist and "preflights" his airplane along with his copilot. It might be their 10,000th flight, but they still use the checklist. "Radio," "Check." "Fuel tank," "Check." "Tire pressure," "Check." They systematically go through every system of the airplane before takeoff. This is important because people's lives are at stake and they don't want to make a single mistake—ever. When a doctor examines a patient, use of the checklist is optional. Sometimes they use one, sometimes they don't. Sometimes they check everything on the list, sometimes they don't. People's lives are at stake here too, so what's the difference? When you come right down to it, *it's because the pilot is on the plane along with his passengers.* If he forgets to check something and the plane goes down, he dies too. If a doctor makes a mistake and forgets to check something, only the patient suffers. This may sound cynical, but if the doctor's life were on the line along with the patient's, we can guarantee you they'd be more thorough.

Another factor that greatly influences conventional medicine is the influence of the billions of marketing dollars behind pharmaceutical drugs. Please note that we're not at all opposed to prescription drugs per se and, in fact, elsewhere in this book we recommend a number of drugs that should be considered by you and your physician for certain conditions. But there are many natural supplements that are also lifesavers that do not have the same marketing muscle behind them. One reason is that our patent system makes it impossible to patent a naturally occurring substance. Without patent protection it doesn't make business sense to spend millions—or billions—of dollars promoting a natural substance that a competitor can market as well. Another reason is that the manufacturer of a new drug now incurs nearly a billion dollars of costs before obtaining an approved FDA claim of efficacy. Since synthetic drugs can be patented, the business model for going through the FDA process can be justified, but for a natural supplement that anyone can manufacture it doesn't make good business sense to spend these enormous sums to obtain an FDA-approved claim that anyone can use. That's why all of the natural supplements in your natural foods store say something like "take two a day" but give no indication of what they are good for. However, the first amendment on free speech is still in place, so we are not going to hesitate to share this information with you. Our recommendation is that you apply the best that *both* conventional and alternative medicine have to offer to stay on the cutting edge of our increasing understanding of biology.

When you go to see a lawyer or an accountant, you can usually get more done and help these specialists do a better job of assisting you if you arrive at your appointment with a game plan for what you'd like to accomplish—rather than relying on them to come up with all of the ideas on their own. There is no reason that you shouldn't apply the same logic to your encounters with your healthcare providers. Don't be afraid to tell your doctor what you want. If your doctor doesn't like your taking an active role, get another doctor! Your relationship with your physician should ideally be a collaboration between knowledgeable partners and not regarded as an association between an expert who knows almost everything and a layperson who knows next to nothing.

In addition, thanks to the Internet and the increasing transparency of most disciplines—including medicine—it's now possible for you to perform some of the tests mentioned below at home without involving your doctor at all. But whether you are tested through your healthcare provider or do some of these tests on your own, the choice is yours. To be able to take full advantage of the breakthroughs in biomedical technologies coming in the decades ahead, you must be the primary advocate for your own health. Remember, when it comes to your healthcare—you're the only one on the plane.

Reader: So, Ray and Terry of the future, what's prevention and early detection like in the years ahead?

Ray2023: They're actually quite a bit different from your day. The 2020s are proving to be the decade when early detection has really come into its own. Unlike your day, when most diseases weren't diagnosed until after symptoms appeared, this state of affairs is much less common. We detect disease much earlier today by detecting it at the level of cells.

Terry2034: Today in 2034, it is at the level of molecules.

Ray2023: The two leading causes of death in 2009 were heart disease and cancer. The reason they killed so many was that most people didn't realize they had these problems until it was often too late. Most people didn't know they had any heart disease at all until the day they had a heart attack, and a third of first heart attacks were fatal. The majority of cancers had already metastasized by the time they were found—meaning they were virtually incurable.

Reader: People no longer have heart attacks or die from cancer?

Terry2023: Not exactly, but they are becoming much rarer, as we've discussed in the heart disease and cancer chapters. It's been several years since any of my patients have had a heart attack. Heart surgeons are needing to be retrained in new surgical specialties like replacement organ transplantation since there is so little need for cardiac bypass surgery. The overwhelming majority of cancers are discovered at a very early stage, and we have much more effective means to treat cancer now.

Reader: What exactly has changed to make these major killers so much less of a problem?

Ray2023: It's largely due to early detection thanks to advances in scanning technologies. As regards heart disease, we now have scanners able to detect even the tiniest build-ups of plaque in the arteries. With such early detection and the powerful gene therapies we have available today, heart disease is almost completely treatable. Similar scanning technologies along with blood markers make it possible for today's doctors to detect most types of cancer at very early stages, long before they've had a chance to spread.

Terry2023: In addition, thanks to new, effective, and far less toxic and invasive cancer therapies, a diagnosis of cancer is no longer feared nearly as much as it was in your time or regarded as an almost automatic death sentence. In the cancer chapter, we talked about how insights into cancer stem cells, which were formative in your day, have now transformed the treatment of cancer.

Ray2023: There's also been a big change in hospitals and medical centers today. Almost no one is treated in big hospitals anymore, and most medical therapies and procedures that used to be done in the hospital are now done at specialized outpatient centers. At the cancer centers, minimally invasive surgical techniques and highly focused beams of radiation lead to nearly painless removal or destruction of cancerous tissue. A visit to a cancer center today is like a visit to the dentist in your day. And, by the way, we go to the dentist far less often these days, as almost everyone has been vaccinated against cavities and gum disease and highly effective toothwashes have replaced brushing and flossing.

Ray2034: Early detection of disease was indeed perfected in the 2020s but has given way to much more powerful forms of preventive medicine in the 2030s. Although we have even better scanners and blood tests for early detection, we don't need to use them as much anymore. Thanks to programmable nanobots circulating in our bloodstreams, we are able to prevent much of the damage from disease processes from happening in the first place. There's relatively little disease to detect anymore.

Terry2034: People are actually running out of things to die from.

Reader: It sounds like I should target living to 2023.

Ray2034: Yes, but if you can get to 2023, the biotech advances will help you get to 2034.

Terry2023: And accomplishing that is the point of our *TRANSCEND* program.

9

RELAXATION

"If you are distressed by anything external, the pain is not due to the thing itself, but to your estimate of it; and this you have the power to revoke at any moment."

—MARCUS AURELIUS

"What, me worry?"

—ALFRED E. NEUMAN

There's no way around it, life is stressful. From major threats like war, natural disaster, and accidents to the more mundane demands of your job or a troubled personal relationship, there's always something to worry about. The best things in life may be free, but they can be stressful as well: falling in love, the birth of a baby, a promotion at work. And while it may appear that many of the causes of your stress are external and beyond your control, you actually hold the keys to how stressed you become, to the betterment or detriment of your health.

For most of human existence, life was tenuous at best, with the ever-present risk of attack from animal predators or other humans. Our ancestors survived to pass their genes to the next generation because bodily systems evolved to help them deal effectively with such life-or-death situations by either fighting or running for their lives.

This "fight-or-flight" mechanism remains intact in modern humans, always at the ready should you ever be faced with threat to life or limb. It's an involuntary, highly tuned response and another example of our outdated genetic heritage. The instant you perceive that you are at risk, a series of internal processes is set in motion by your *amygdala*, the part of your brain where danger is perceived. It tells your pituitary gland to release ACTH (adrenocorticotropin hormone), which triggers the release of the stress hormone cortisol from the adrenal glands. This makes more energy available to your muscles and brain to strengthen your physical and mental capabilities.

The adrenal glands also produce adrenaline and noradrenaline. To free up internal resources for muscle and mind, these hormones suppress your digestive, immune, and reproductive systems, which aren't needed to address the immediate crisis (and won't do you much good in the long run if you don't make it out alive). Your blood pressure, blood sugar, and cholesterol levels rise to fuel the action, as do fibrinogen levels to speed blood clotting in case of injury. Your heart rate and respiration also go up for added energy, and your pupils dilate to better see the threat or a means of escape.

When circumstances demand that fight or flight actually take place, the resulting burst of physical exertion uses up this excess energy as intended, and your stress hormone levels gradually return to normal when the danger is past. In modern times, however, where life-or-death situations are relatively rare, your body's fight-or-flight mechanism still remains ready for action and, when not called upon to get you through a real emergency, will react nonetheless to the everyday stressors of life. From snarled traffic or a rebellious teenager in the house, to an upcoming job interview or the falling Dow Jones, constant stimulation of the fight-or-flight response without the expected physical release can lead to a range of serious ailments.

The *Yale University School of Medicine Heart Book* states, "When the work to be done is mental, the hormones and fats that have been mobilized for action are not used up. The unnecessarily high heart rate and blood pressure set up a condition of increased turbulence in the bloodstream, which in turn increases the tension on the walls of the arteries." Numerous studies have confirmed that chronic stress can contribute to decreased immune system function, gastrointestinal disorders, type 2 diabetes, cancer, rheumatoid arthritis, heart disease, and stroke. Stress with no constructive outlet can lead to anxiety, depression, difficulty concentrating, insomnia, and substance abuse, as well as compulsive eating, gambling, and sexual activity. Perhaps most important, chronic stress accelerates the aging process.

HOW STRESSFUL IS IT?

Almost anything can trigger stress. One person's passion for rock climbing is another's fear-of-heights nightmare. Even universally acknowledged stressors such as losing a home in a hurricane can be experienced differently from one person to the next. It's been more than 40 years since researchers Thomas Holmes and Richard Rahe completed their classic and frequently cited study linking elevated stress levels to higher rates of illness. They asked 5,000 patients to rank 43 life events for relative stress and used that data to devise their Holmes-Rahe Social Readjustment Rating Scale. Participants assessed life events in hindsight, so the data are subjective, but the scale is still used today as an indicator of relative stress and risk of health breakdown. By adding up the scores for all the events you have experienced in the past year, you can get a sense of how susceptible you may be to getting sick. According to the Holmes-Rahe prediction model, a score of 150 or less per year indicates relatively low risk, while scoring higher than 300 indicates a serious health risk.

HOLMES-RAHE SOCIAL READJUSTMENT RATING SCALE

1. Death of a spouse	100
2. Divorce	73
3. Marital separation	65
4. Imprisonment	63
5. Death of a close family member	63
6. Personal injury or illness	53
7. Marriage	50
8. Fired from job	47
9. Marital reconciliation	45
10. Retirement	45
11. Change in health of family member	44
12. Pregnancy	40
13. Sexual difficulties	39
14. Gain a new family member	39
15. Business readjustment	39

16. Change in financial state	38
17. Death of a close friend	37
18. Change to different line of work	36
19. Change in frequency of arguments with spouse	35
20. Major mortgage or loan	31
21. Foreclosure of mortgage	30
22. Change in responsibilities at work	29
23. Child leaving home	29
24. Trouble with in-laws	29
25. Outstanding personal achievement	28
26. Spouse begins or stops work	26
27. Begin or end school	26
28. Change in living conditions	25
29. Revision of personal habits	24
30. Trouble with boss	23
31. Change in work hours or conditions	20
32. Change in residence	20
33. Change in school	20
34. Change in recreation	19
35. Change in religious activities	19
36. Change in social activities	18
37. Minor mortgage or loan	17
38. Change in sleeping habits	16
39. Change in number of family gatherings	15
40. Change in eating habits	15
41. Vacation	13
42. Holidays	12
43. Minor violation of law	11

Obviously, you can't avoid all stressful events, and even low levels of stress can take a toll over the long term. However, you can take concrete actions to prevent stress from getting the best of you.

GETTING A GRIP ON STRESS

What options do you have for managing your stress? Essentially, it all boils down to lifestyle and attitude. By lifestyle, we mean the day-to-day, minute-by-minute activities of your life. Even some of the most routine lifestyle choices will have a profound impact on your stress levels and on your health. By attitude, we mean your outlook on life. Your expectations are not just an idle reflection of the future; your perspective turns out to be a self-fulfilling prophecy. Numerous studies have demonstrated that people who are optimistic, trusting, easygoing, tolerant, and contented tend to be happier and healthier and to live longer. This is not to say that "slacker" is healthy and "workaholic" is not. Being competitive and ambitious, even being a workaholic, can be positive and healthy if your drive to achieve is approached positively and constructively. This means keeping the four Cs in mind when tackling the task at hand:

- Challenge—choosing a difficult objective worthy of your efforts

- Commitment—staying on task to meet the challenge

- Curiosity—cultivating an open mind and a desire to learn

- Creativity—using your imagination to innovate

Of course, the demands of our fast-paced, competitive 21st-century global economy can make cultivating and maintaining a positive attitude and healthy lifestyle difficult. Therefore, managing your stress, as with all other aspects of our *TRANSCEND* program, must be seen as an ongoing process, not something you squeeze in on the weekend between paying the bills and dropping off the kids at soccer practice. You will need to figure out what works best for you as you consider incorporating the following suggestions into your daily routine.

Balance is a good place to begin. Your life has three major poles: family/friends, work, and self. All are important to living a happy, fulfilling, and healthy life. But as the saying goes, "Nobody ever lay on their deathbed wishing they'd made more money or spent more time at the office." The same applies to the other poles. Being too self-centered or overly invested in family or friends can also leave you off balance. Assess your priorities on a regular basis, and work to leave time for all the major poles of your life: time to share love and leisure with family and friends, time to make your work challenging and fulfilling, and time for yourself to relax, learn, create, and grow.

Balancing your busy life requires effective time management. One surefire way to guarantee stress is to overcommit and overschedule. We realize this can be a challenge—after all, there are only so many hours in a day, and forgoing sleep will only exacerbate your level of stress. But with a plan, you can become more efficient and slowly shift how you allocate your hours to better reflect your priorities.

Knowing your priorities is the first step. Make a list of what you have to do each day. Then decide what you want to accomplish that day in addition to your regular routine. Over a week or two, track how much time you actually spend on those activities. Are you spending time on activities you hadn't originally considered? Are you wasting time on low priorities and unnecessary activities? Are there ways you could be more efficient or delegate? Consider how you can develop a schedule and overcome procrastination. Learn when and how to say no (something the authors of this book are still trying to learn how to do!). The rewards of having more time to spend doing what is important to you and of having time to be spontaneous are well worth the effort, and the reduction in stress will make you healthier and more energized.

Close interpersonal relationships provide substantial protection against the negative effects of stress. Human evolution is familial and tribal. Social connection nurtures us with love, intimacy, support, and meaning, whereas isolation fosters anxiety, depression, and illness.

Make sure there are people in your life with whom you can talk openly and without embarrassment about your dreams, doubts, hopes, and fears. This intimate sharing with someone you trust and who cares about you can lead to some of the most profound and meaningful times of your life. It is also a powerful stress reliever, as the simple act of articulating your thoughts and feelings can often bring new clarity to what may have been causing you stress. Close relationships require cultivation, whether with your spouse or partner, members of your family, or friends and colleagues. And if you don't have anyone you feel you can really talk with, then seek out a mental health professional, a teacher, a member of the clergy, or a social worker. Just don't bottle up what's on your mind and in your heart.

Of course, relationships are two-way, and "being there" for others is equally important to building and maintaining strong interpersonal bonds and to managing your stress. Being supportive of others, giving of your time and concern, and, perhaps most important, *listening*, all contribute to your own well-being. Truly listening requires opening your mind to the ideas, concerns, and feelings of others and, much of the time, keeping your mouth shut. Don't daydream while someone else is talking. Feeling empathy and compassion, expressing understanding, and offering encouragement are the keys.

We need a variety of close relationships in our lives to keep us happy and healthy, and all those relationships require attention to flourish. However, finding that "special someone" and forming a successful, long-term marriage or committed partnership is in a relationship class all its own. Surveys find that those who are lucky and skilled enough to have such a relationship in their lives are healthier and live longer than those who don't. It takes luck to cross paths with the right person at the right time, but it takes skill to make such an intimate relationship work through all the changes and challenges two people will face over a period of many years. If you are lucky enough to have found your *other half*, it is essential that you learn the skills and do the work necessary to keep the relationship strong, fresh, and growing.

Being careful about what you put into your body is also critical to managing your stress. Obviously, that means avoiding addictive drugs as outlined in the section "How NOT to Deal with Stress" later in this chapter. Other suggestions include:

- NO tobacco or illegal drugs at all

- Prescription benzodiazepines only for acute conditions and under the care of your physician; buspirone if needed for prolonged use

- Alcohol only in moderation (no more than one drink per day for women and men over 65; two drinks per day for men younger than 65)

- Caffeine only in moderation (primarily from green tea)

Not so obvious, however, is the fact that your diet also has a profound effect on your ability to cope with stress. An unhealthy diet damages your body, weakening your ability to respond to stress. On the other hand, following our *TRANSCEND* recommendations for a low-sugar, low-glycemic-load diet (see Chapter 11) will strengthen your resistance to stress. A healthy diet reduces your risk of obesity, cardiovascular disease, cancer, and other serious illness, thereby also helping you avoid the enormous stress that comes with those conditions. With a healthy diet, your body will also be better able to regulate hormone levels, thus improving your mood and sleep patterns. Getting an adequate amount of good-quality sleep is also very important to relieving stress and maintaining a balanced life.

Exercise can also be a powerful stress reliever. In addition to its vital role in improving overall health and stamina, aerobic exercise burns up some of those excess fats and sugars that get released through the fight-or-flight response. And in what has been termed "runner's high," a strenuous aerobic workout also releases endorphins, brain chemicals that counter stress and promote a sense of well-being.

Taking breaks from your routine also works wonders when it comes to stress management. However, many people are so stressed by job pressures that they don't even take all the vacation time and holidays that they are due. A 2003 survey of business executives by Management Recruiters International found that 47 percent of respondents did not use all their vacation days that year, while a 2004 survey by Expedia.com found that 35 percent of employees didn't use all their vacation time. Some people *never* take a vacation. If you are feeling so pressured by your work that you can't take time off, that's a good indication that you are ignoring your level of stress. And it doesn't have to be a vacation in the classic sense of spending a week in the Caribbean or on the ski slopes. Stay home and read for pleasure, play your piano, work on your car, have fun with the kids, plant a garden ... just periodically make some space in your life to fall into a completely different routine, where your time is spent doing what you love at your own pace and whim.

While getting away from your routine for a week or more at a time is essential and can be quite rejuvenating, you can also recharge your batteries

HOW TO MEDITATE

The following is a simple meditation technique based on both Eastern and Western traditions, including techniques developed at Harvard Medical School by Dr. Herbert Benson:

- *Find a quiet, comfortable environment where you feel safe and will not be interrupted by other people, the phone, or other distractions.*

- *Sit comfortably on a chair or a cushion on the floor, slowly close your eyes, and take two or three slow, deep breaths.*

- *Become aware of your breathing. Take a few slow, deep breaths in and out, letting your whole body relax on each out-breath.*

- *After a few minutes of relaxing in this way, let your breathing become natural and relaxed and, on each out-breath, begin repeating in your mind (not aloud) a simple two-syllable word or sound. A "soft" word or word pair with no consonants is best. A suggestion is "ah one."*

- *Let the sound repeat in your mind on each out-breath. This is the essence of the meditation process. This is not a concentration exercise; just let the sound happen in your mind without concentrating on it. The mental sound will become more subtle as you do this, eventually just becoming a feeling of the sound.*

- *If thoughts intrude (as they will) or you notice that you have stopped repeating the sound, that's all right. Simply bring your mind gently*

by fitting shorter breaks into your life. Take an afternoon off to visit a museum or go to a concert. Take a class one night a week to learn about something not related to your work but fascinating to you. Get a massage. In addition to the break in your routine, massage is a great way to relieve both emotional and physical stress. A 2004 meta-analysis of massage therapy by the American Psychological Association found that massage relieves pain, reduces anxiety and depression, and temporarily reduces blood pressure and heart rate. Human touch also releases endorphins and is a powerful method to induce a sense of relaxation.

Finally, inducing the "relaxation response" on a daily basis can help manage your stress. The term was first applied in the mid-1970s by Dr. Herbert *back to your breathing and begin again to let the "sound" repeat itself. Relax. Allow whatever happens to happen. This is not an exercise in focusing and disciplining the mind, but in letting your mind free.*

- *Let your thoughts come and go like clouds in the sky. Don't be distressed if a negative thought or worry occurs—just like a cloud, it will change its shape and dissipate. Just watch yourself experiencing your thoughts as if you were another person.*

- *Continue the process for 15 to 30 minutes (it is better to open your eyes briefly to check the time than to be wrenched out of your meditation by an alarm). Try for 15 minutes three or four times a day if under increased stress.*

- *When you decide it is time to finish your meditation, stop the mental sound from repeating and, with your eyes still closed, bring your thoughts back to your current time and location. Take a few deep breaths in and out and slowly open your eyes. Then, when you are ready, slowly stand and get on with your day, now feeling relaxed and refreshed.*

Practice this technique once or twice a day, preferably before eating or 2 hours after eating as the digestive process can reduce the benefits. Over time, you will find it will become easier to slip into a deep meditative state of mind.

Benson and his colleagues at the Harvard Medical School and Beth Israel Hospital to their discovery that certain practices, such as meditation and yoga, produce physical and emotional effects that are the opposite of the fight-or-flight response. They found that these techniques can reduce respiration and heart rate and lower blood pressure and sugar levels. When practiced regularly over the long term, inducing the relaxation response has the power to permanently reduce blood pressure and enhance digestion, sleep patterns, mental acuity, bloodflow, and mood. Numerous studies have since confirmed these results by using a variety of approaches and techniques to evoke the relaxation response.

Any of the methods described here will work. The key to attaining stress management benefits is to find a technique that you will actually use. Your chosen method will do you no good if you don't practice it. Learn, explore, experiment.

Various forms of meditation have been practiced for thousands of years in most religious traditions, including Judaism, Christianity, Islam, Buddhism, and Hinduism. However, you don't need to follow a religious tradition to benefit from meditation. Most meditation techniques share a common focus on the breath, on the body, or on a word or sound that is repeated silently or aloud. For stress management purposes, most practitioners recommend meditating for 15 to 30 minutes at a time either once or twice a day. The objective is not to concentrate on anything, rather to calm the mind, allowing thoughts to passively drift in and out, always gently moving your mind back to the breath or sound.

There are also techniques that combine meditation with physical movement. Yoga, for example, is a 26,000-year-old Indian tradition that uses a series of stretching movements and controlled breathing to invoke the relaxation response while also building muscle strength, balance, and flexibility. The same benefits can also be gained from tai chi, a series of slow, calming movements that evolved from ancient Chinese martial arts.

Similar in some ways to meditation is a method called visualization, or guided imagery. Instead of gently bringing the mind back to the breath or a sound, the idea here is to imagine the sights, sounds, tastes, smells, and feelings of being in a state of deep relaxation in a safe and tranquil environment. Your thoughts influence how you feel and act, so consciously imagining a situation can produce almost the same results as actually experiencing it.

A wealth of books, Web sites, and audio and video recordings are available to help you explore various meditation and visualization techniques and find the one that best fits your personality and lifestyle. You may, however, find that you will become more adept and avoid mistakes by learning directly from a qualified teacher. But don't be concerned about needing to become a master in a technique. The benefits of the relaxation response can be attained even at a beginner's level. If you do it on a regular basis, you will develop the mental technique that works for you.

Biofeedback is a technology-assisted method of evoking the relaxation response. Equipment is used to monitor vital signs such as heart rate and blood pressure to determine your level of stress while continuously providing auditory or visual feedback to guide you into a more relaxed state of mind. As you learn what that state of mind feels like and become more skilled at reaching it, you can more easily reach it without the aid of the equipment. Biofeedback is usually practiced in a clinic, though biofeedback machines can be purchased for home use.

Reader: I've tried several of the techniques you mention, but they haven't worked for me. I even took some prescription benzodiazepines for a while and eventually managed to get off of them. Isn't there anything else that is more powerful than vacations, massages, and biofeedback?

Terry: In my clinic, I often recommend EFT and use of the AlphaStim for my patients who are troubled with chronic stress.

Reader: I've never heard of these.

Terry: EFT stands for the Emotional Freedom Technique. It's a novel method of stress reduction, developed by Gary Craig in the 1990s as a simplification of another technique used by psychologists known as Thought Field Therapy. The nice thing about EFT is that it can be learned easily and practiced at home. EFT involves the use of tapping with the fingers on a series of acupressure points on the head and upper body while repeating certain statements at the same time. People can use this technique to reduce their stress levels and also to work on many other types of physical and emotional problems.

Reader: Sounds great, how can I learn more?

Ray and Terry: You might start by going to Gary Craig's Web site, www.emofree.com.

Ray: There's another great resource called the AlphaStim. Alpha waves are the type of brain waves associated with deep relaxation. When you meditate, for instance, you increase the alpha waves in your brain. The AlphaStim is a small battery-controlled device that has two cotton-covered electrodes that you attach to your ear lobes. Using this machine for 20 to 60 minutes a day will increase your brain's alpha waves and can help with chronic stress, depression, and insomnia. Some people even use the AlphaStim for pain relief.

Terry: With the use of EFT and the AlphaStim, I find that many of my patients are able to get effective control of their stress without resorting to any type of medications.

Reader: Okay, so what's coming in the future?

Ray2023: It's a lot easier to take a vacation. Today we have full-immersion visual-auditory virtual reality with images written directly to your retinas from your eyeglasses. You can travel instantly to a completely convincing tropical beach or take a nap on a cloud. Virtual reality may not solve all of your real-world problems, but it sure makes it easier to "get away."

Terry2034: Today, in 2034, we can go to full-immersion virtual reality environments that incorporate all of the senses using the nanobots in our nervous system. So now you can actually feel the warm, moist spray of the ocean as you walk on your virtual beach.

Reader: But you still have your real-world problems, I assume?

Ray2034: Well, there are always problems to be concerned with. But the nanobots are beginning to alleviate the mental sources of chronic stress. For example, they can detect and alleviate negative and repetitive thought patterns. We have some good models of how our brains create stressful situations and are beginning to master the means of reprogramming them.

We understand that managing your stress can require a lot of time and effort. If you aren't already employing some of the stress management tools outlined above, making them part of your life might seem daunting. But don't let yourself get stressed out by stress management! Confucius said that "a journey of a thousand miles begins with one step," so approach this one step at a time. Try to make it enjoyable. Each little change you can make to your lifestyle and attitude will have a positive impact on your health and longevity. Over time, these new habits will become as integral to your daily life as some of your stressful habits may be now. And over time, as each new practice is added to the previous ones, you will find yourself living a happier, healthier, and longer life.

How NOT to Deal with Stress

Some people react to stress by becoming angry. Ask yourself the following questions: Do you take things personally? Do you argue a lot? Do you resort to violence when under pressure? A great deal of research has demonstrated significantly increased risk of heart disease and higher mortality rates from heart attack among people who exhibit excessive levels of anger, cynicism, suspicion, and hostility. Dealing with life's ups and downs in negative ways such as these maintains your body's fight-or-flight mechanism in a nearly perpetual state of readiness. This is bad for your body and an unpleasant way to go through a stress-shortened life, not to mention increasing the stress levels of those around you.

Another all-too-common but counterproductive response to stress is compulsive eating. The pleasure of digging into a half-gallon of ice cream may temporarily take your mind off your troubles, but it does nothing to address the underlying causes of your stress or help you establish an effective stress management routine. In fact, habitual eating is likely to add more stress to your life from guilt and embarrassment. And, of course, stress-induced eating will result in your getting more stressed when you get on the scale.

Drug abuse is another widespread but extremely counterproductive response to chronic stress. The common thread in substance abuse is a temporary relief from feelings of stress at the price of greatly increasing stress in the long term, including possible addiction.

The most damaging drug overall may be nicotine because of the sheer number of users and its severe toxic effects. Tobacco is physically and psychologically addictive, seductively advertised, relatively inexpensive, and easy to acquire. As a response to stress, nicotine can perk you up as it boosts levels of adrenaline and other hormones. The ritual of smoking can provide a calming break from the pressures of life, and the oral gratification can trigger unconscious feelings of comfort associated with nursing as an infant. But the deadly effects of tobacco have been well documented for decades. Even today, after billions of dollars have been spent on antismoking education campaigns, the tobacco death toll still tops 440,000 people every year in the United States. Overwhelming evidence links tobacco use to cardiovascular disease, heart failure, cancers of the lung and larynx, and emphysema, among other diseases.

Alcohol, when used in careful moderation, has been shown to have health benefits and promotes longevity, but the key word here is moderation. It is all too easy to cross over the line at which any potential benefits are replaced by serious risks. Moderate alcohol consumption means up to one alcoholic drink per day for women and men older than 65 and up to two a day for men under 65. Consuming more than these amounts can raise the risk of heart disease and cancer and cause hypertension, brain damage, gastrointestinal disorders, and cirrhosis of the liver. Driving under the influence of alcohol is a factor in at least half of all traffic accidents. Overall, about 30 million Americans abuse alcohol, and there are more than 100,000 alcohol-related deaths in the United States annually.

As a remedy for chronic stress, excessive alcohol is clearly a catastrophe. Though limited consumption can provide temporary respite from stress by inducing mild feelings of relaxation and promoting interpersonal interaction and connectedness within social situations, alcohol is a physically and psychologically addictive depressant, and abuse can damage interpersonal relationships, sabotage careers, and deepen feelings of anxiety, depression, and isolation.

Caffeine may be the most commonly abused drug of all. Widely advertised and readily available, it is woven into the fabric of American life. Most adults consume some caffeine, and tens of millions abuse it. You may be among those who kick-start each morning with a cup of coffee and down additional cups at regular intervals throughout the day. You may also be consuming high levels of caffeine in soft drinks and, at more extreme concentrations, in high-octane energy drinks.

Small amounts of caffeine can be beneficial, and it does have the power to help you concentrate, stay alert, and overcome the fatigue associated with stress. But when consumed in excess, it can be quite addictive, and its downsides are considerable. These include heart arrhythmias, hypertension, headaches, and digestive disorders. When used to get you through the day after a poor night's sleep, caffeine can remain in your system that evening and ruin another night's sleep. More caffeine is needed to get you through the next day, leading to a worsening cycle of insomnia and caffeine abuse.

A study published in the journal *Archives of Internal Medicine* in 2002 found that drinking several cups of coffee can significantly raise adrenaline and blood pressure levels, exacerbating the effects of stress. And in 2000, the journal *Psychopharmacology* published a study showing that even small amounts of caffeine aggravated chronic anxiety and panic disorders in some participants. Of course, to anyone who has experienced a case of "coffee jitters," this would come as no surprise. You will feel much better and be healthier if you cut out all sources of caffeine except tea, especially green tea. With about one-fourth of the caffeine found in coffee, green tea is rich in powerful antioxidants as well as the amino acid L-theanine. A study published in 2007 in the journal *Biological Psychology* found that L-theanine actually reduced psychological and physiological stress responses.

When symptoms of stress become unbearable or interfere significantly with daily life, a common response from health professionals is to prescribe one of the drugs known as benzodiazepines. These include Valium (diazepam), Tranxene (chlorazepam), Xanax (alprazolam), Ativan (lorazepam), and Klonopin (clonazepam), among others. While these drugs can ease a patient through an acute stress-induced episode, they must be used with extreme caution because the risks are substantial. There is a serious danger of addiction, and extended use can bring about the complete opposite of the desired effect, such as a persistent state of heightened anxiety, worsened insomnia, and chronic depression. While these drugs do have a place in alleviating acute symptoms, such as those of a panic attack, they do not offer a safe or effective solution to chronic stress.

Buspar (buspirone), on the other hand, is a prescription drug used in the treatment of stress that is not only effective but lacks many of the problems associated with the benzodiazepines such as addiction potential or tolerance. For people who have been unable to adequately control their anxiety with lifestyle measures and nutritional supplementation alone, buspirone can provide a significant degree of benefit safely and inexpensively. If it's needed, talk with your doctor about a prescription.

GABA (gamma aminobutyric acid) is a supplement, which is also a naturally occurring neurotransmitter, that can act as a mild tranquilizer. You can find GABA in your natural food store. It is non–habit forming and is effective at relieving stress. For some it may cause drowsiness and, if so, should be taken at bedtime in a dose of 500 to 1,000 milligrams.

L-theanine, the calming ingredient in green tea, is also available as a supplement and, in doses of 50 to 200 milligrams, is effective at relieving stress and does not induce drowsiness.

Nicotine, alcohol, caffeine, and the benzodiazepines are all legal and widely available. Together, they pose the greatest health risk to the greatest number of people. Nearly half the adult population of the United States, about 75 million people, abuses one or more of these legal drugs. However, the use of illegal drugs is also an all too common, risky, and equally ineffective response to chronic stress.

Marijuana is the most commonly used illegal drug. According to the 2006 National Survey on Drug Use and Health (NSDUH), 40 percent of the U.S. population older than age 12 (97.8 million) has tried marijuana at least once and 10.3 percent (25.4 million) used it at least once during the year. Although often regarded as a relatively benign drug, marijuana use is associated with a number of detrimental effects on health. Frequent upper respiratory tract infections, increased heart rate, heart rhythm disturbances, and memory loss are common. Studies have shown that regular marijuana smokers scored lower on standardized tests of memory, attention, and learning even if they hadn't used the drug for over 24 hours. Marijuana smoke has many of the same types of carcinogens as tobacco smoke, and heavy users are at increased risk of lung cancer. Although many people say they get high in order to decrease stress, marijuana use is associated with increased levels of anxiety and panic attacks in many individuals. Marijuana is also illegal, although some states have decriminalized possession of small amounts. There were 1.9 million arrests for drug-related violations in 2006; 4.8 percent of these were for marijuana sale and 40 percent for possession, so the legal status of marijuana can certainly add to one's stress level.

Cocaine, whether in the form of powder or crack, is a particularly pernicious drug. Consumption stimulates the release of dopamine in the brain, which is experienced by the user as extremely pleasurable. While this sense of euphoria may seem like a great alternative to enduring the symptoms of chronic stress, the effects are very short-lived, and cocaine use actually worsens stress symptoms. As supplies of dopamine and other neurotransmitters are depleted in the brain by chronic cocaine exposure, the drug-induced pleasure rapidly fades and the user is left in an unpleasant mental and physical state. Consuming more cocaine releases more dopamine and briefly restores the euphoria, but as more dopamine is used up, even more cocaine is required to get back to that state. This can quickly lead to addiction, especially in genetically prone individuals. Eventually, continued use exhausts the brain's capacity for dopamine production and disrupts its ability to regulate dopamine and other neurotransmitters. It becomes difficult for a user to even think clearly, and ever larger doses of cocaine are required just to avoid the unbearable misery brought on by such low levels of critical brain chemicals. Cocaine addiction can also lead to states of paranoia, aggressiveness, and violence, even when the addict is not under the influence. Addiction to drugs such as cocaine is a well-known and insidious slippery slope.

When you consider the lengths people will go to, such as becoming addicted to harmful drugs, to relieve feelings of stress, it becomes clear why developing a constructive program of relaxation should be one of the most important goals of your *TRANSCEND* program.

EARLY DETECTION—STRESS MANAGEMENT

It has been over 50 years since the type A and type B personality types were first described by cardiologists Meyer Friedman and Ray Rosenman. Their initial concept was that the type A personality, the classic workaholic who is impatient, competitive, and often hostile, is associated with a higher risk of cardiovascular disease. The type B individual, on the other hand, being more relaxed, patient, and easygoing, was felt to be less susceptible to risk. The problem with these distinctions is that they rely on outward appearances. They don't tell what's going on inside the mind and body, where it really counts. Subsequent research has shown that even though many type A people appear harried and nervous on the surface, they are actually quite calm on the inside. In fact, type A people who work very hard, but have a passion for their work, live longer and are healthier. Working hard doesn't appear to be a problem as long as it's coupled with enjoyment, challenge, and passion. Only the type A subgroup characterized by hostility appears to be at higher cardiovascular risk. When confronted with a stressful situation, oftentimes type A people do not react with elevations of blood pressure or heart rate, factors that can increase cardiac risk over time. By the same token, many type B people, who appear on the surface to be calm and relaxed most of the time, will react to stressful situations with significant—and dangerous—increases in heart rate and blood pressure, all the while appearing laid-back and peaceful on the surface.

As a result, the connection between the type A personality (other than hostility) and cardiovascular risk has been largely discredited and replaced with assessment tools that look deeper—at how a person really reacts to stress on the inside rather than how they *seem* to react on the surface. A simple method of doing this is to subject a person to a controlled stress and then measure what happens to their blood pressure. A test you can administer yourself is the "cold pressor test," which produces good data as to how a person reacts to stress on the inside. In this test, which you can easily do at home with the help of a partner, you immerse one of your hands up to the wrist in a basin of ice water for 2 minutes. You check your blood pressure right before placing your hand in the cold water and then again immediately afterward. It is normal for a person to experience a small increase in blood pressure, up to 15 points, for either the systolic or diastolic readings. People who have increases in their blood pressure greater than this are referred to as "hot reactors," while people whose blood pressure does not rise more than 15 points are classified as "cold reactors."

Being a hot reactor means that every time you experience stress, your body reacts with increased blood pressure. This puts you at risk of developing hypertension or elevated blood pressure in the future as well as developing coronary heart disease. If you do the cold water pressor test and find that you are a hot reactor, you should give a higher priority to reducing your stress level using the suggestions in this chapter.

10

ASSESSMENT

"Beginning your program 2 years ago was the start of a magical, new part of our lives: healthier, fitter bodies, clearer, happier minds, and a truly close and exciting relationship. Never again will we have to troll bookstores for the next new diet and health book; we have the best already. **We intend to live this way forever!**"

SAVANNA (68) AND JACK (65), U.S. VIRGIN ISLANDS

The two pillars that will help you to live—and remain in optimal health—long enough to take full advantage of the phenomenal break-throughs that will occur in technology and medicine in the next couple of decades are *Prevention* and *Early Detection* of disease. These two pillars con-stitute the Assessment portion of the *TRANSCEND* program. Prevention is assessment of yourself and Early Detection is most often assessment done with the help of your healthcare providers, but in this chapter we will focus on self-assessment tests you can do on yourself.

The way in which preventive medicine is typically practiced today is rarely effective. Mainstream health recommendations are often so watered down or misleading as to be ineffective. In addition, people often defeat their own health programs by telling themselves, "It's too hard to reduce calories," "I don't have time to exercise," or "I'm addicted to sugar." We feel that people *are* willing to take the steps needed to stay healthy and extend their lives—particularly, when they know that doing so may enable them to live far, far longer. In this chapter, we will describe health-promoting changes that quickly become self-sustaining because of the feelings of well-being that they generate. We'll also describe a program for early detection of disease—how to get the type of medical care that is powerful enough to detect disease very early in its course—at a point where curing these conditions is relatively easy. It's like stepping away from the edge of a cliff before you fall off. Someday soon these interventions will be relatively automatic, but, today, it still requires some effort on your part.

PREVENTION

Prevention is the first of our two pillars and consists largely of things that you can do yourself—the lifestyle choices you make day in and day out. In some ways, prevention is even more important than early detection. After all, who wouldn't prefer to prevent a disease entirely rather than detect it—no matter how early?

Your keys to disease prevention are based primarily on the lifestyle choices you make—diet, exercise, sleep, care of the brain, and management of stress—all key components of the *TRANSCEND* program. Recent studies have confirmed how important the lifestyle choices you make can be on how long you live. The Female Nurses' Study has been tracking 77,782 female U.S. nurses (age range, 34 to 59) since 1980. Five lifestyle risk factors were found to profoundly influence mortality from all causes as well as cancer and cardiovascular mortality in particular. The overall risks were as follows:

- Smoking—linked to 28 percent of deaths
- Low physical activity—17 percent
- Overweight—14 percent
- Poor diet—13 percent
- Alcohol (either none or excess)—7 percent

Each of these risk factors increased a woman's risk of dying prematurely. These risk factors were independent of one another and were additive—the more unhealthful lifestyle choices, the greater the effect. Having all five of these risky behaviors increased the overall risk of death 43 percent more than having none of the five. Cardiovascular mortality increased more than eightfold and cancer mortality by a factor of 3.26. The lifestyle choices you make play a profound role on how long you will live. Remember that by living longer now, you will get to a point in time where the clock will tick much longer.

Yet, another of our major theses is that our genetic code, our biological software, is better suited to Paleolithic times than to the world of today. If we were to wait for our genes to adapt to modern times using natural selection, it would take many thousands of years. Biotech breakthroughs—which we have referred to as "Bridge Two"—are going to accomplish this in a matter of several decades. So in the meantime, we need to follow present-day or "Bridge One" recommendations through the use of proper diet, exercise, stress management, nutritional supplementation, prescription medications, and other therapies. Targeted use of these *TRANSCEND* interventions will enable us to reprogram—that is, to *transcend*—many aspects of our outdated biological "software" to make them more applicable to life in the Information Age.

EARLY DETECTION

Early detection is the second pillar that will enable you to live and remain healthy long enough to be able to take full advantage of the next stages in technological evolution. If you can't prevent a disease, at the very least, you want to detect it at the earliest possible stage, when effective treatment and full recovery are still possible.

Most of the testing used for Early Detection will be done by your doctor. There are many tests, however, that you can do at home on yourself and that you should get in the habit of doing regularly. These include breast self-exams for women and testicular self-exams for men, heart rate and blood pressure measurement, body fat, waist-to-hip ratio, and fitness testing. Let's look at how you can perform these assessment tests on yourself.

SELF-TESTING

Breast Self-Exams—Women

A review published in July 2008 of 388,000 women found that regular breast self-exams had no effect on breast cancer deaths. In fact, all it did was double the number of breast biopsies that were done. For breast self-exams to be useful, you need to know what you are looking for. By examining your breasts regularly, you will become aware of any abnormal changes, but you need to know what changes require follow-up with your doctor. Be alert for any of the following:

- A new lump or knot in the breast or armpit that doesn't resolve after your menstrual period—particularly one that is hard and immobile

- Dimpling or thickening of the skin of the breast such that it appears irregular, like the peel of an orange

- Discharge from the nipple that is dark or bloody or any discharge that occurs without squeezing

- Any changes in the size or shape of your breasts

- Inversion or thickening of the nipple

- Any redness or scaling of the skin

- Any lump that doesn't go away after 2 months

Testicular Self-Exams—Men

Testicular cancer can occur at any age but is the leading solid cancer in men between 20 and 34 years of age. If detected early, however, it is nearly 100 percent treatable. Untreated testicular cancer, on the other hand, can spread to lymph nodes and lungs. Men should get in the habit of performing testicular self-exams on a regular basis. Young men should do so about once a month.

- Perform an exam after a bath or shower when the scrotum is warm and the tissues relaxed

- Use one hand to support the testicle and the other to examine

- Roll the testicle between the thumb and index finger, looking for any type of lump, particularly one that is hard

- Note that normal testicle tissue has the consistency of a hard-boiled egg (testicular cancer is typically painless)

- Feel the back of the testicle and examine the ropy structure called the epididymis; the epididymis is normally somewhat tender and irregular

In addition, the vital signs of heart rate and blood pressure are two other tests you should monitor yourself periodically.

VITAL SIGNS

Most doctors check your vital signs each time you come in because they are just what their name implies—vital to life. The four vital signs are heart rate, blood pressure, breathing rate, and temperature. Your personal wellness program should include regular measurements of your heart rate and blood pressure, which you can easily do either at home or with your healthcare provider.

HEART RATE

It's amazing how many people don't know how to check their own heart rate or pulse, but we think it's important that everyone learn how to perform this basic skill. Checking your pulse is very simple to learn and can provide you with a wealth of information about your cardiac status. Chinese medicine practitioners check 12 different types of pulses in the wrists, but luckily all you have to learn is one, and this one is simple.

The easiest place to check your pulse is at the wrist. Simply place the index and middle fingers of your other hand on the inside of your wrist on the thumb side. Count the pulsations for 20 seconds and multiply by 3 and you have your heart rate. You might want to try it now. You may need to feel around to find it if you haven't done this before, but most people can find their wrist pulse with a little practice. A common mistake is pressing too hard, so just feel lightly. There are also inexpensive heart rate monitors that you can buy in the local drugstore or sports shop that are simple to use.

The more physically fit you are, the lower your resting heart rate. Lance Armstrong, for instance, has a resting heart rate of only 32 beats per minute. Men typically have slower resting heart rates than women, but in general, you want your resting heart rate to be less than 84. An optimal heart rate is less than 70. If your resting heart rate is greater than 100, you may either be very unfit or have an underlying medical problem, such as heart disease, a thyroid problem, or anemia, and you should see your doctor.

Blood Pressure

Blood pressure is easy to measure at home with inexpensive automated devices that are widely available. There are four categories of blood pressure:

< 120/80 mm Hg is optimal
120/80–130/85 mm Hg is normal
130/85–140/90 mm Hg is high normal
>140/90 mm Hg is high

About 40 percent of Americans have readings in the optimal range, 24 percent have normal readings, 13 percent have high normal readings, and 23 percent have high blood pressure. Most doctors recommend medication when readings are consistently higher than 140/90; however, there are health risks any time the blood pressure is above the optimal range of 120/80. According to a study by the National Heart, Lung, and Blood Institute published in 2008, the chance of having a heart attack or stroke increases dramatically the higher your blood pressure. The table below shows the risk of a cardiovascular event such as a heart attack or stroke over a 10-year period for women and men between 35 and 64 years of age:

TABLE 10-1: RISK FOR CARDIOVASCULAR EVENT OVER 10 YEARS

BLOOD PRESSURE RANGE	PERCENT CARDIOVASCULAR EVENT RISK (WOMEN)	PERCENT CARDIOVASCULAR EVENT RISK (MEN)
Optimal	1.9	5.8
Normal	2.8	7.6
High normal	4.4	10.1

This table shows that just having high normal blood pressure, which doesn't even require medication, increases heart attack and stroke risk 230 percent in women and 70 percent in men compared to optimal ranges. Weight loss, regular exercise, and reduced consumption of high-glycemic-index foods are simple and effective ways you can move your normal or high normal blood pressure readings closer to the optimal range.

For people whose blood pressure remains high normal or even in the lower part of the high range despite implementing the *TRANSCEND* program, we recommend a traditional Chinese medicine herbal formula known as Uncaria-6 made by Seven Forests, also known as Gou Teng Jiang Ya Pian. This inexpensive herbal formulation appears to be safe and effective in many cases and has few side effects. It is widely available from acupuncturists who also practice Chinese herbal medicine.

Body Composition

In addition to vital signs, it is important to know your body composition. Body composition is a measurement of how much of your body is made up of fatty tissue and how much is nonfat, which includes everything else, such as muscle, bone, and blood. You can get additional information about your body composition by checking your waist-to-hip ratio, which tells where the fat on your body is distributed. These two measurements are just as important as—if not more important than—your scale weight. Body composition and the waist-to-hip ratio are rarely measured by most physicians, so you may need to do these tests on yourself.

Body Fat

Body fat can be divided into two main types: *essential fat*, which is needed for survival and reproductive function, and *storage fat*, which serves as a reservoir of calories. This is another way that our genetic "software" is out of date. The "fat insulin receptor gene," which is an ancient gene, basically says to hold on to every calorie. That made sense thousands of years ago but does not make sense today. Men need a minimum of 2 to 5 percent essential fat, and women need 10 to 12 percent. In many ways, fat behaves like any other organ of the body and has essential roles. Essential fat is needed to cushion organs such as the heart, spleen, and intestines. Over half of the nonwater weight of the brain is fat, and, in women, fat is involved in regulation of the sex hormones such as testosterone and estrogen. In both sexes, fatty tissue secretes important hormones, such as leptin, resistin, adiponectin, interleukin-6, and tumor necrosis factor alpha, that help regulate numerous metabolic processes.

As a general rule, men should maintain their total body fat at 10 to 17 percent and women, 18 to 26 percent. Optimal percentages are at the lower ends of these ranges, so men might set an optimal goal of 10 to 12 percent and women 18 to 20 percent, although trained athletes will typically have even lower amounts than these. If you weigh 164 pounds (the average weight of an American woman in 2002) and your body fat is 34 percent, you are carrying 52 pounds of fat. To achieve a healthier 24 percent body fat, you would need to lose 22 pounds (you would then weigh 142, close to the average weight of American women in 1960).

The most accurate method of measuring body fat is in an underwater tank, but you can get a close approximation with handheld devices or weight scales that display body fat as well as weight. Some of these units are notoriously inaccurate, so it's a good idea to have your body fat tested at a doctor's office or gym with a more accurate device, such as an impedance meter or an underwater tank, and then compare this measurement to the one you obtained on your home device to make sure you're getting accurate readings.

Waist-to-Hip Ratio

Your health can be affected not only by how much body fat you have but by where it is distributed as well. Some people—mostly women—tend to store fat on their upper thighs and buttocks, creating the so-called pear shape. The downside to this type of body fat is mostly cosmetic because it is associated with less health risk than fat that is stored in the midsection of the body—the apple shape.

Weight stored in the midsection, above the beltline, the classic "beer belly," is more common in men and is technically called *central adiposity*. It is a hallmark of *metabolic syndrome*, a major risk factor for cardiovascular disease. Measuring your waist-to-hip ratio provides a simple screening test for metabolic syndrome. To perform this test, all you need is a paper or plastic tape measure. Measure the circumference of your abdomen at its widest place—typically right at the umbilicus or belly button. Just keep your abdomen relaxed—don't suck it in. Next, measure your hip circumference at its widest point. Now, divide your waist measurement by your hip measurement and compare it to the chart below.

TABLE 10-2: WAIST-TO-HIP RATIOS AND HEALTH RISKS

MEN	WOMEN	RANGE
≤0.95	≤0.80	Optimal
0.96–1.0	0.81–0.85	Average
>1.0	>0.85	High risk

EXERCISE AND FITNESS TESTING

The human body was designed to get regular exercise, and doing so is critical to maintaining optimal health. As discussed in Chapter 14, three forms of exercise are needed to maintain high level fitness: aerobic training, strength training, and flexibility training. You can do the following four simple tests at home to assess your current level of fitness. We suggest two tests to measure strength, one for flexibility and one for aerobics.

Upper-Body Strength

To measure your upper-body strength, men should perform standard "military" pushups with only toes and palms touching the ground. Women can modify the test by having their knees on the floor as well. Keeping the legs, hips, and back straight, let your body go down until your chest is about 4 inches from the floor, then return to your starting position. Do as many pushups as you can without pausing.

PUSHUPS

Core-Body Strength

The strength of your core, which includes your abdominal, back, and hip muscles, can be assessed by seeing how many half situps you can do. Lie with your back flat on the floor, knees bent at a right angle, and palms down. Keeping your palms on the floor, do a situp to 45 degrees, then return to the floor. Shoulders need to touch the floor at the end of each rep, but not necessarily your head. Do as many repetitions as you can in 60 seconds.

HALF SITUPS

FLEXIBILITY

An easy way to get a general idea of your overall flexibility is the simple toe-touch test. In this test, you stand upright with your arms at your side and, keeping your knees straight, slowly lean forward as far as you can without bouncing. Men should be able to reach the floor with their fingertips, and women should be able to put their palms flat on the ground.

AEROBIC CONDITIONING

If you have been performing regular aerobic exercise, please feel free to do this test. If you have not been exercising regularly, particularly if you are over 40 years of age, you should *not* perform this test on yourself. Instead, you need to talk with your doctor and arrange to have an exercise treadmill test done first before engaging in aerobic exercise or attempting this test on yourself.

To do this test of aerobic fitness, you will run or walk 1.5 miles as fast as you can. Use the odometer on your car to measure a 1.5-mile course or go around a standard ¼ mile track six times. Don't eat for 2 hours before the test. Warm up first, then complete the course as fast as you can and use the chart below to assess your aerobic fitness.

TABLE 10-3: TIME TO COMPLETE 1.5-MILE RUN

	MEN			WOMEN		
AGE	OPTIMAL	AVERAGE	POOR	OPTIMAL	AVERAGE	POOR
20	10m	11m 15s	12m 30s	12m 45s	14m 30s	16m 30s
40	10m	11m 15s	12m 30s	14m 30s	16m 45s	19m
60	13m	15m 15s	17m 45s	16m 30s	20m	24m
80	15m 30s	18m 30s	22m 30s	20m	24m 30s	31m

Terry2023: With all this discussion about exercise and fitness, you need to remind your readers to be careful! Last year I had a nasty spill on my mountain bike. I was going pretty fast down steep, rocky terrain when all of a sudden, my front wheel locked and I flew head over heels over the handlebars.

Reader: Aren't you kind of old to be riding a mountain bike so aggressively?

Terry2023: Hey, I'm only 76. Nowadays that's considered young.

Ray2034: People don't get hurt nearly as often doing these types of sports in the 2030s since most people prefer to do dangerous recreational activities these days in virtual reality. This past month I've been skydiving, mountain climbing, helicopter skiing, and drag racing—all in full immersion virtual reality incorporating all of the senses. I could feel the moist snowy air in my face as I skied down the mountain. And I really enjoyed them, knowing that I didn't have to worry about getting hurt or killed no matter what I did.

Terry2023: There is virtual reality today with eyeglasses that beam three-dimensional images to our retinas, but the tactile devices are not fully realistic yet. But I probably should have done the mountain biking in virtual reality anyway. I ripped a ligament and tore the joint capsule.

Reader: Did you need surgery?

Terry2023: No, I didn't, although in your day I would have. Fortunately, back in 2009, I made arrangements to collect some of my stem cells for future use. Stem cell procedures are very common today, and most injuries and degenerative processes are treated with stem cell therapies rather than surgery. But it's best if you can use your own stem cells to repair or regenerate your own tissues. There's now an even more advanced stem cell approach being adopted that turns your own skin cells into adult stem cells for these procedures. It also corrects any genetic defects in the cells before they're used. But for now I chose to use the more conservative approach using the stem cells I harvested and saved back in 2009.

Reader: So maybe I should save my own stem cells for future use?

Ray2023: It's not a bad idea. You can arrange to harvest some of your own stem cells right now and freeze them for your future use. Terry and I made arrangements to save some of our cells back in 2009.

Terry2023: We each saved about two dozen tiny vials of our more youthful stem cells. This is actually the fourth vial I've used so far. The first was to rejuvenate the cartilage in my left knee, which had been damaged many years ago from a ski accident. A few years ago, my knee had begun to show some early signs of arthritis, so my doctor and I used my previously saved and more youthful stem cells. After the injection, the cartilage in my knee grew back and the pain went away.

Reader: And what did you use the second vial for?

Terry2023: Actually, for cosmetic purposes.

Reader: A stem cell face-lift?

Terry2023: Not a full face-lift, but I used some of my stored stem cells to remove some of the deep lines and wrinkles on my face when I was in my midsixties. I had also started to lose a lot of hair—male pattern baldness—and I didn't like how gray I was turning. So, I tried stem cell therapy and used both the second and third vials for these cosmetic applications.

Reader: Looks like it worked well.

Terry2023: Thanks, you're very kind. And I still have about 20 more vials available for when I need them in the future.

Terry2034: And you will. I mean, I did.

11

NUTRITION

An article published in 2008 in the prestigious *Journal of the American College of Cardiology* suggests that damage to your arteries occurs after just one unhealthy meal (for example, cheeseburger, French fries, and large soft drink). On the other hand, your body begins to repair that damage following just one healthy meal (such as salmon, salad, and green tea).

For most of us, food is so much more than merely healthy or unhealthy fuel for the body. What we eat is intricately entwined with our cultural identities and family traditions. Sharing a meal with others can be a delight and helps strengthen interpersonal bonds. The varied flavors of a well-prepared dish can be a rich pleasure.

Food is also a major industry determined to influence what you eat through a constant barrage of clever marketing. And what you eat has a profound impact on your health; after all, you are what you eat (and also what you think, as we explained in Chapter 1)! Extensive research now confirms that poor nutrition is a major contributor to a host of serious conditions, including cancer, heart disease, stroke, hypertension, obesity, and type 2 diabetes. *Nutrition* is, of course, an essential element of our *TRANSCEND* pro-

WATER

All life depends on it. About 60 percent of your body is made up of water, and every system and organ requires water to function. Water carries nutrients to your cells and flushes waste and toxins out. While you might survive as long as a few weeks without food, you wouldn't last more than a few days without water.

Many health experts recommend that most people need to drink between 8 and 10 glasses of water a day. Others refine it to a half ounce of water per pound of body weight each day. (That's water, not sodas, coffee, beer, fruit juices, etc.) Several recent review papers, however, have failed to find substantial scientific justification for this recommendation. We still feel that even though recommendations for the amount of water have not been scientifically validated, it is still important that you drink the right type of water.

Specifically, tap water and bottled water often contain environmental toxins, so we believe it is important that you filter tap water and know the quality and source of bottled water. For instance, a 2008 survey of municipal water supplies in 24 major metropolitan areas revealed trace amounts of antibiotics, anticonvulsants, mood stabilizers, and sex hormones affecting the drinking water of 41 million Americans. Therefore, it is important that you filter the water from your tap before you drink it. Chemicals such as chlorine are also added to kill bacteria and other pathogens, but these chemicals are toxic to the body. Filtering tap water before you drink it is like removing the plastic grocery bag from your broccoli before you eat it. Both are necessary to protect your food or water before you consume them, but these protective materials still need to be removed before the food or water enters your mouth. Water filters are available in a wide variety of styles and price ranges, and there are filters that meet every budget. Also, when drinking bottled water, avoid soft plastic containers that are made with phthalates, a contaminant that can leach into the water and affect hormone-sensitive tissues such as breast tissue in women and prostate tissue in men.

gram. In this chapter, we'll describe how you can optimize your diet for health and longevity. We've also included a number of sample recipes that are simple, tasty, and healthy to help you implement our diet recommendations into your daily life.

GOOD AND BAD CALORIES

There are four sources of calories. The three you learned in school—carbohydrates, proteins, and fats—and the one you probably learned about after school or on weekends—alcohol. Debate has raged for years about how much of each of these types of calories should be consumed for optimal health and weight. Every year or so, a "new and improved" weight loss *cure* appears in the bookstores and rockets to the top of the bestseller list. Many of these diets advocate radical decreases in one type of calorie and greater emphasis on another. Millions have lost significant amounts of weight following the Atkins-type diet, which centers on drastic cuts in all carbohydrate foods but general acceptance of a wide variety of proteins and fats. Very low fat diets such as Pritikin are just the opposite and call for dramatic reductions in dietary fat as their mainstay. There is no question that people who follow Atkins will lose weight and show improvement in some lab tests, such as lower triglycerides (blood fats) and higher high-density lipoprotein (HDL) or good cholesterol, while very low fat diets have been shown to reduce cholesterol deposits from within the heart arteries.

But there are health risks associated with following the Atkins diet over the long term, such as eating the wrong kinds of fat, strain on the kidneys caused by consuming excessive amounts of protein (especially animal protein), and increased risk of osteoporosis—and some concern has been raised that eating so much animal-based food could increase cancer risk. Although very low fat diets have many virtues, they can be deficient in calcium, vitamin B_{12}, and essential fats for growing children, pregnant or nursing mothers, or people who are physically very active, so supplementing with these nutrients is important when on very low fat diets. To their credit, however, both diet programs have made changes in response to recent nutrition research: The Atkins diet has begun restricting saturated fat, and most very low fat diets have recommended that minimum amounts of omega-3 fats be consumed (or that EPA-DHA supplements be used).

Rates of long-term adherence are low for both of these diets. After a few months, people on Atkins get bored with eating so few carbohydrates (yes, people do get tired of just eating bacon and hamburgers) and the very low fat folks develop cravings for fat.

Although Dean Ornish is often associated with the very low fat school of thought, his Spectrum program is actually quite flexible. Like our *TRANSCEND* program, he recommends a personalized program that matches your specific health circumstances. For patients with significant heart disease, his "reversal" program, which includes a very low fat diet, was the first to scientifically prove that lifestyle changes can actually reverse the buildup of plaque in arteries. For others, Dr. Ornish recommends a program similar to the one in this book that includes a range of healthy foods such as vegetables, fruits, legumes, and unrefined carbohydrates. Dr. Ornish was actually one of the first to recommend omega-3 fatty acids, which he has been doing for the past 25 years.

It's clear that most people eat too many of their calories in the form of car-bohydrates—sugary foods and refined and high-glycemic carbohydrate "white" foods such as white bread, white rice, white pasta, and white potatoes in particular—there is value in eating a wide variety of foods. We feel that *all* of the sources of calories have value, and that it makes sense to eat a variety of foods from each type. Rather than saying that carbs are bad and fats are good or vice versa, it's more important to realize that carbohydrates, fats, pro-teins—and even alcohols—are not good or bad as a category. Instead, under-stand that there are good and bad varieties of carbohydrates, proteins, fats, and alcohols. Optimal health is easiest to achieve by making most of your choices from the healthy varieties of these food types.

CARBOHYDRATES

Throughout most of the long course of human evolution, refined carbohy-drates, such as refined flour products and white sugar, did not exist, while complex carbohydrates or starches such as potatoes, cereals, and grains made up a relatively small portion of our ancestors' daily fare. Our bodies still function much as they did tens of thousands of years ago, so eating a diet lower in carbohydrates and higher in certain kinds of protein and fats is the most favorable to good health. But since the time humans first learned to harvest and grind grains, refined carbohydrates have become "staples" in our diet. Bread, pasta, white rice, cookies, cake, candies, soft drinks . . . the typical modern diet is now so heavily weighted toward carbohydrates, and in particular, refined sugars and starches, that it's literally killing us. In fact, the principal cause of obesity, type 2 diabetes, and metabolic syndrome is overconsumption of these kinds of carbohydrates.

The most basic carbohydrates are simple sugars (monosaccharides) such as fructose found in fruit, and glucose, the critical blood sugar that provides energy to your cells. Simple sugars combine to form disaccharides (meaning "double sugars"), such as table sugar (sucrose), and polysaccharides (meaning "many sugars"), such as amylose (starch) and cellulose (the woody, indigestible structural material in plants).

The faster a carbohydrate is digested and absorbed, the faster it releases glucose into your bloodstream. Refined-carbohydrate foods are digested and absorbed rapidly, thus, they raise your blood sugar very quickly whenever you eat them. This lies at the heart of why a diet rich in refined sugars and starches is so unhealthy. A simple monosaccharide such as glucose can be absorbed by the lining of the small intestine directly where it immediately enters the bloodstream. Disaccharides, such as sucrose (table sugar) and lactose (from milk products), can't be absorbed directly, but are easily broken down by enzymes (sucrase and lactase, respectively) into simple sugars, which are then quickly absorbed. Even amylose, the main carbohydrate found in grains and starchy vegetables such as potatoes, is easily broken down by the enzyme amylase in saliva and pancreatic fluids into simple sugars and is quickly absorbed into the bloodstream.

Insulin: A Double-Edged Sword

The rapid digestion and absorption of refined sugars and starches triggers a rapid rise in your blood glucose level, which signals your pancreas to excrete insulin into your bloodstream. Insulin is a pancreatic hormone that moves glucose into your cells to either be burned immediately for energy or signals the conversion of glucose into glycogen and then to triglyceride (fat) to be stored for later use. Any time you eat a sugary food, insulin levels spike, which can sometimes result in a sharp drop in blood sugar a few hours later. The resulting hypoglycemia (low blood sugar) then triggers cravings for more foods high in refined sugars and starches, which leads to another spike in insulin and so on.

When this spike-and-plunge cycle continues over time, your cells lose their sensitivity to insulin and ever-increasing amounts of insulin are needed to move glucose into the cells. This is known as *insulin resistance* and is the primary cause of *the metabolic syndrome,* a serious condition associated with increased blood pressure, higher risk of coronary artery disease, and acceleration of the aging processes.

Insulin resistance can also lead to *type 2 diabetes,* which is a breakdown in the body's ability to process carbohydrates. In type 2 diabetes, the pancreas can become so overworked that it eventually fails and can't produce enough insulin to lower blood sugar levels. This crippling disease causes problems with blood circulation, dramatically increases the risk of heart attacks, and is associated with a 15-year decrease in life expectancy. Largely because of excessive consumption of simple carbohydrates and sugary foods, the number of people with type 2 diabetes in the United States has increased tenfold in the past 35 years, and as of 2008, type 2 diabetes affected over 21 million Americans. In just the 10 years between 1997 and 2007, the incidence of type 2 diabetes nearly doubled in the United States, from 4.8 to 9.1 per 1,000 people.

Excess sugar in the blood also contributes to the formation of AGEs (advanced glycation end products). These sticky conglomerations of sugar and protein molecules interfere with the work of your enzymes and accelerate aging processes. Some cases of high blood pressure (due to stiffening of the blood vessel walls from glycation) and age spots are indicators of AGE production at work.

High blood glucose levels inhibit your immune system and interfere with vitamin C's crucial role in combating infection and building body tissues. High blood sugar stimulates growth of a broad range of pathological cells, including fungal infections such as candida (yeast) and cancer, and boosts adrenaline production, triggering the fight-or-flight stress reflex.

Consumption of refined sugars and starches is also the primary cause of weight gain and obesity. Insulin brings your blood glucose level under control by converting excess sugar into fat and moving it into fat cells for storage. The more sugary, starchy carbohydrates you eat, the more weight you gain. It's that simple. And the craving for more sugary, starchy carbohydrates caused by the cyclical drops in blood glucose level only exacerbates the problem.

Clinical research, including a study published in 2001 in the journal *Nutrition Reviews*, demonstrates that refined sugars and starches are addictive and that eliminating them from your diet can cause withdrawal symptoms. But eliminating such simple carbohydrates from your diet is the single most important thing you can do to reduce risk of metabolic syndrome and type 2 diabetes. It is also the best thing you can do to lose excess weight and maintain a healthy body weight. If you completely cut out these foods, the addictive cravings will dissipate after a week or two. If you just cut back, the addiction and craving will remain.

Glycemic Index, Glycemic Load . . . and How to Keep Your Blood from Turning into "Pink Cream"

Not all carbohydrate foods are unhealthy. In fact, many are highly desirable, containing crucial vitamins, minerals, fiber, and other valuable nutrients. To help you determine whether a particular food falls into the "good carbs" group, you need to know its *glycemic load*, and that is determined, in part, by its *glycemic index*. This may sound complicated, but let us explain.

The speed at which a particular food is converted into glucose in the blood, and thus how fast it boosts your insulin level, is called its glycemic index. Because glucose is the primary blood sugar and needs no processing to enter the bloodstream, scientists have assigned glucose a glycemic index of 100. The glycemic index for any particular food is determined as a function of how fast it turns into sugar in the bloodstream as compared to glucose itself. For example, corn flakes have a relatively high glycemic index of 92, very close to glucose, while peanuts are quite low at 14. Two slices of white bread have a glycemic index of 73, while two slices of whole-grain bread have a glycemic index of 55. This is because the fiber content of the whole-grain bread slows the digestive process and the speed at which the carbohydrates are converted to glucose, so it has a lower glycemic index. Foods made from whole grain are also better for you because they still contain all the nutrients that the refining or milling process strips away. Similarly, a piece of fruit with all its fiber and other nutrients intact has a lower glycemic index and is better for you than a glass of juice squeezed from that fruit.

Any time you eat a high-glycemic food, there is a sudden rush of sugar or glucose into your bloodstream. The body has minimal ability to store glucose, so the only options are: (a) burn it as fuel immediately or (b) turn it into triglycerides and store it as fat. Unless you happen to be running a marathon or doing some other type of extreme physical exertion, you don't burn very much sugar at any one time, so most of the sugar in high-glycemic foods you eat is converted into triglycerides and eventually stored as fat. If you consume a high-glycemic food such as a candy bar or sugary soft drink and have a sample of your blood drawn 60 minutes later—unless you're doing something physically very strenuous and burning all the sugar as soon as you eat it—you'll find your blood sample filled with triglycerides or fat particles. There will be so much fat floating in your bloodstream, in fact, that it will look like *pink cream.* The same thing happens any time you eat any high-glycemic food. Drink a glass of orange juice, and 60 minutes later, your blood looks like pink cream; a sugary breakfast cereal with skim milk and a banana—same thing. A bowl of white spaghetti, a sandwich on white bread, white potatoes—your blood changes from red to pink because it has so much fat in it. The pink color lasts only a few hours (unless you have diabetes) as the fat particles gradually are cleared from your bloodstream and enter your fat cells—adding to your weight and waistline. If you have undiagnosed and, thus, untreated type 2 diabetes, which is true of 6 million Americans, or are one of the 57 million with prediabetes (also called metabolic syndrome), the fat remains in your bloodstream for many hours, or in more severe cases, your blood simply stays pink—that is, filled with fat—all the time.

In general, foods with a lower glycemic index are better for you than those with a higher glycemic index, but that's not always the case. That's why a food's *glycemic load* is a much more useful indicator. The glycemic load for a serving of food represents the approximate amount of insulin needed to process it based on *both* the speed at which it is converted to glucose (its glycemic index) *and* the total amount of sugar it has, a factor determined by the total amount of carbohydrates in the serving. For example, a serving of green peas with a glycemic index of 75 *appears* to be worse than white bread with a glycemic index of 73. But when you factor in the total number of grams of carbohydrates contained in a serving of peas compared with that in a serving of bread, the actual "load" on your system—and the amount of insulin needed—is much lower with the peas. Glycemic load is calculated by multiplying the number of grams of carbohydrates in the serving by the glycemic index of that food. Thus, the glycemic load for two slices of white bread is 28 grams (total carbs) × 73 percent (glycemic index) = 20. A serving of peas with 7.5 grams of carbs and a glycemic index of 75 would have a much lower glycemic load of only 5.5. The amount of insulin released into your bloodstream is directly related to the glycemic load of the meal, so a serving of bread creates a much greater overall impact on insulin levels than a serving of green peas.

The following table will help you determine which foods are to be avoided and which can be included in a healthy diet. Unrefined, high-fiber, low–glycemic load foods are best, and your mother was right when she told you to eat your vegetables—nonstarchy vegetables are not only very low in glycemic load but are loaded with many valuable nutrients, such as vitamins and phytochemicals.

TABLE 11-1: GLYCEMIC INDEX AND GLYCEMIC LOAD

FOOD	GLYCEMIC INDEX (GLUCOSE = 100)	SERVING SIZE	CARBOHYDRATE PER SERVING (GRAMS)	GLYCEMIC LOAD PER SERVING
Dates, dried	103	2 oz	40	42
Russet potato (baked)	85	1 medium	30	26
Cornflakes	81	1 cup	26	21
Jelly beans	78	1 oz	28	22
Puffed rice cakes	78	3 cakes	21	17
Doughnut	76	1 medium	23	17
Soda crackers	74	4 crackers	17	12
White bread	73	1 large slice	14	10
Table sugar (sucrose)	68	2 tsp	10	7
Pancake	67	6" diameter	58	39
White rice (boiled)	64	1 cup	36	23
Brown rice (boiled)	55	1 cup	33	18
Spaghetti, white; boiled	38	1 cup	48	18
Spaghetti, whole wheat; boiled	37	1 cup	42	16
Rye, pumpernickel bread	41	1 large slice	12	5
Orange, raw	42	1 medium	11	5
Pear, raw	38	1 medium	11	4
Apple, raw	38	1 medium	15	6
All-Bran cereal	38	1 cup	23	9
Skim milk	32	8 fl oz	13	4
Lentils, dried; boiled	29	1 cup	18	5
Kidney beans, dried; boiled	28	1 cup	25	7
Pearled barley; boiled	25	1 cup	42	11
Cashew nuts	22	1 oz	13	3
Peanuts	14	1 oz	6	1

[Taken from the Linus Pauling Institute at Oregon State University]
http://lpi.oregonstate.edu/infocenter/foods/grains/gigl.html
Jane Higdon © 2009 LPI, used with permission

Recommendations for Carbohydrate Consumption

Our recommendations regarding dietary carbohydrate consumption are based on your current state of health. For most people, we recommend that carbohydrates make up 33 to 40 percent of total calories. This may represent a significant reduction for some, as many people consume 60 percent or more of their daily calories in the form of carbohydrates. We also recommend that the majority be of the low-glycemic variety.

Some people need to cut their carbohydrate consumption even more sharply for a period of time to help them lose weight and give the cells of their pancreas a rest and the other cells of the body a break from the constant barrage of so much insulin. These people need to follow what we call the *Low-Carbohydrate Corrective Diet* for a period of months (or sometimes a year or more) to help rectify imbalances. Then, once optimal body weight, blood pressure, blood sugar, and insulin levels have been restored, they can move to the *Moderate-Carbohydrate Diet.* Some people with type 2 diabetes, intractable weight problems, or carbohydrate addiction may need to follow the *Low-Carbohydrate Corrective Diet* permanently. To determine your optimal carbohydrate consumption level, first decide whether you need to begin with the Low-Carbohydrate Corrective Diet or the Moderate-Carbohydrate Diet and then follow the corresponding recommendations.

The Low-Carbohydrate Corrective Diet should be followed by people with health conditions caused or exacerbated by a high-glycemic-load diet and whose lives may depend on breaking their carbohydrate addiction. These include:

- People trying to lose weight. Eliminating high-glycemic-load foods is critical to successfully reaching and maintaining your optimal weight (see Chapter 13 for additional recommendations for weight reduction).

- People with metabolic syndrome. The metabolic syndrome affects about a third of the adult U.S. population and results from insulin resistance brought on by consumption of a high-glycemic-load diet.

- People with type 2 diabetes. Untreated metabolic syndrome often evolves into this crippling disease, which affects about 9 percent of the U.S. population, with nearly one-fourth of those currently undiagnosed.

• People with elevated risk of heart disease. A high-glycemic-load diet boosts cholesterol and other lipid (fat) levels, raising your risk of heart disease.

• Carbohydrate addicts. It is natural to crave sweets and other high-glycemic carbohydrate foods. Since these foods are either immediately available for energy or are easily converted to fat, eating them conferred a powerful survival advantage to our cavemen and cavewomen ancestors. Our Stone Age genes ensure that we still jump at the chance to eat sugary foods whenever we can. For our cave-dwelling ancestors, this wasn't very often, but for us, sugary foods are available 24/7. Yet, we have found that it is very easy to break the addiction to sugary foods. Simply eliminating them from the diet—as you will do as part of the Low-Carbohydrate Corrective Diet—will result in a rapid decrease in sugar cravings within a matter of days.

To begin reversing the damage caused by high-glycemic-load carbohydrates, here are our recommendations for the Low-Carbohydrate Corrective Diet:

• Eat no more than 20 percent of your total calories in the form of carbohydrates (see Table 11-2).

• Cut all high-glycemic-load foods from your diet, including anything that contains sugar or refined starch (breads, pastries, pasta, candy, soft drinks, etc.) and all high-starch vegetables (potatoes, corn, etc.). In recent years, many new low-carbohydrate breads, desserts, and pastas have come on the market, making it easier to eat the kinds of foods you enjoy without the high glycemic load of refined sugars and starches.

• Steer clear of grains and fruit juices.

• Eat very small amounts of low-glycemic-load fruits (e.g., melons, berries).

• Eat moderate amounts of legumes (peas, beans, lentils, peanuts, etc.) and nuts (walnuts, cashews, almonds, etc.).

TABLE 11-2: MAINTENANCE CALORIE LEVEL AND RECOMMENDED CARBOHYDRATE LEVEL FOR THE LOW-CARBOHYDRATE CORRECTIVE DIET

WEIGHT (POUNDS)	SEDENTARY LIFESTYLE		MODERATELY ACTIVE LIFESTYLE		VERY ACTIVE LIFESTYLE	
	TOTAL CALORIES	CARB GRAMS	TOTAL CALORIES	CARB GRAMS	TOTAL CALORIES	CARB GRAMS
90	1,170	59	1,350	68	1,620	81
100	1,300	65	1,500	75	1,800	90
110	1,430	72	1,650	83	1,980	99
120	1,560	78	1,800	90	2,160	108
130	1,690	85	1,950	98	2,340	117
140	1,820	91	2,100	105	2,520	126
150	1,950	98	2,250	113	2,700	135
160	2,080	104	2,400	120	2,880	144
170	2,210	111	2,550	128	3,060	153
180	2,340	117	2,700	135	3,240	162
190	2,470	124	2,850	143	3,420	171
200	2,600	130	3,000	150	3,600	180
210	2,730	137	3,150	158	3,780	189
220	2,860	143	3,300	165	3,960	198
230	2,990	150	3,450	173	4,140	207
240	3,120	156	3,600	180	4,320	216

• Eat unlimited amounts of low-starch aboveground vegetables (cabbage, Brussels sprouts, broccoli, kale, mustard greens, Swiss chard, collard greens, spinach, lettuce, peppers of all colors, Chinese cabbage, bok choy, snow peas, celery, cauliflower, zucchini, cucumbers, etc.), preferably fresh and lightly cooked.

• Speak with your health provider about using a starch blocker (a medication or supplement designed to deactivate the enzyme amylase that digests starch). We recommend starch blockers in addition to reducing carbohydrates, not as a substitute for reducing carbohydrates.

Table 11-2 offers guidelines for calorie and carbohydrate consumption for the Low-Carbohydrate Corrective Diet. To establish your limits, first choose your activity level (sedentary, moderately active, or very active) and your current weight. The corresponding number in the "Total calories" column represents the maximum number of calories you can consume each day and still maintain your current weight. (To lose weight, we recommend that you work within the calorie limits set for your *desired* weight.)

The Moderate-Carbohydrate Diet is for everyone who is not on the Low-Carbohydrate Corrective Diet, as well as for people in the Low-Carbohydrate group after they have come back into balance and corrected their weight problems, blood pressure, and blood sugar levels. People with intractable weight problems, carbohydrate addiction, or type 2 diabetes may need to stay on the Low-Carbohydrate Corrective Diet. People who do the Low-Carbohydrate Corrective Diet temporarily and then switch to the Moderate-Carbohydrate Diet should also stay at the lower end of the carbohydrate range (i.e., 33 percent of total calories as carbohydrates).

Our recommendations for the Moderate-Carbohydrate Diet include:

• Drastically limit your consumption of high-glycemic-load foods, including anything that contains sugar or refined starch (breads, pastries, pasta, candy, soft drinks, etc.) as well as high-starch vegetables (potatoes, rice, corn, etc.). In recent years, many new low-carbohydrate breads, desserts, and pastas have come on the market, making it easier to eat the kinds of foods you enjoy without the high glycemic load of refined sugars and starches.

• Eat moderate amounts of whole grains.

• Eat small amounts of low-glycemic-load fruits.

• Eat moderate amounts of legumes (peas, beans, lentils, peanuts, etc.) and nuts (walnuts, cashews, almonds, etc.).

• Eat lots of low-starch aboveground vegetables (see examples above).

• Consider using a starch blocker.

TABLE 11-3: MAINTENANCE CALORIE LEVEL AND RECOMMENDED CARBOHYDRATE LEVEL FOR THE MODERATE-CARBOHYDRATE DIET

WEIGHT (POUNDS)	SEDENTARY LIFESTYLE			MODERATELY ACTIVE LIFESTYLE			VERY ACTIVE LIFESTYLE		
	TOTAL CALORIES	CARB GRAMS		TOTAL CALORIES	CARB GRAMS		TOTAL CALORIES	CARB GRAMS	
		33%	40%		33%	40%		33%	40%
90	1,170	97	117	1,350	111	135	1,620	134	162
100	1,300	107	130	1,500	124	150	1,800	149	180
110	1,430	118	143	1,650	136	165	1,980	163	198
120	1,560	129	156	1,800	149	180	2,160	178	216
130	1,690	139	169	1,950	161	195	2,340	193	234
140	1,820	150	182	2,100	173	210	2,520	208	252
150	1,950	161	195	2,250	186	225	2,700	223	270
160	2,080	172	208	2,400	198	240	2,880	238	288
170	2,210	182	221	2,550	210	255	3,060	252	306
180	2,340	193	234	2,700	223	270	3,240	267	324
190	2,470	204	247	2,850	235	285	3,420	282	342
200	2,600	215	260	3,000	248	300	3,600	297	360
210	2,730	225	273	3,150	260	315	3,780	312	378
220	2,860	236	286	3,300	272	330	3,960	327	396
230	2,990	247	299	3,450	285	345	4,140	342	414
240	3,120	257	312	3,600	297	360	4,320	356	432

A Few Words about Sweeteners

Sugar is bad for you!

Well, maybe a few more words are called for because this point can't be stressed enough. We feel sugar is so bad that we call it the "White Satan." There is overwhelming clinical evidence linking consumption of high-glycemic-load carbohydrates to increased risk of type 2 diabetes, cardiovascular disease, cancer, obesity, AGEs, and the metabolic syndrome. And without doubt, refined sugar in its many forms (cane sugar, beet sugar, high-fructose corn syrup, etc.) tops the list of most dangerous high-glycemic-load carbohydrates.

At the beginning of the 20th century, Americans averaged about 5 pounds of sugar each year. Today, estimated annual consumption of sugar in the United States is more than 150 pounds for every man, woman, and child. Much of this sugar is consumed in the form of soft drinks, averaging 53 gallons per person per year. Of course, these averages include a lot of people who consume little or no refined sugars, which leaves many others putting away enormous quantities indeed. And all that soda pop is turning these people's blood into pink cream. And then there are all the other prepared foods that are loaded with refined sugars, from the obvious breakfast cereals, cakes, cookies, and candies to not-so-obvious items such as frozen microwavable entrees—more pink cream. The alarming rise in obesity rates in recent years, including among children, is directly related to this glut of sugar in our diet.

Despite overwhelming evidence of the terrible health consequences of our national sugar addiction, public health authorities and most physicians in the United States have yet to fully raise the alarm. This is due in part to the power of the sugar industry, which, just as was seen in the past with the tobacco industry, has managed to thwart efforts by government health agencies to issue any official warnings about sugar consumption. Even on an international level, it wasn't until 2003 that the World Health Organization included recommendations in its health guidelines that people cut total calories consumed in the form of refined sugars to less than 10 percent—an amount that is still far too high. However, there's no need for you to wait for an official warning to take the initiative in your own life and eliminate or greatly reduce your consumption of sugar.

"But," you may be saying, "I love my sweets. What about sugar substitutes?" In our experience, the best course is to train your tastebuds to be happy without the regular stimulation of sweet foods, but if that's not possible, here are our recommendations:

- **Saccharin** was linked to bladder cancer in laboratory animals, and despite recent efforts by producers to have warnings to that effect removed from labels, we feel the evidence is still persuasive enough to recommend against its use.

• **Aspartame** (Nutrasweet and Equal) is also to be avoided because of a host of health concerns raised by several studies, such as one reported in 1991 in the journal *Pharmacology and Toxicology* linking it to significant imbalance of amino acids and neurotransmitters in the brain, including lowered serotonin levels.

• **Acesulfame-K** (Sunett, acesulfame potassium) is not recommended in light of studies linking it to possible genetic damage and stimulation of insulin production.

• **Sucralose** (Splenda) has yet to be definitively linked to any specific health risks. It is a modified form of sugar that passes through the digestive tract without being absorbed into the bloodstream, so it appears to be a better alternative than saccharin, aspartame, or acesulfame-K, but we advise caution until more research is done.

• **Stevia** is the only calorie-free sugar substitute that we do recommend—in small quantities. Stevia, derived from a South American shrub (Yerba dulce) is more than 100 times sweeter than table sugar. It has been used for at least 1,500 years for culinary and medicinal purposes in South America, and is widely used in Japan today. Stevia is not available as a sweetener in the United States, Canada, or the European Community; rather, it is marketed as a nutritional supplement. It is not "generally regarded as safe" by the FDA as animal experiments have shown that high doses can lead to reduced sperm production and could cause infertility in rodents. Research in humans thus far has not linked moderate stevia consumption with any harmful effects. In fact, numerous studies have demonstrated important health benefits, including lowered blood pressure and blood sugar levels, increased energy and mental activity, reduced tobacco and alcohol cravings, and antibacterial effects in the mouth. We feel the best course is to get used to not sweetening foods and beverages but feel that low or moderate amounts of stevia are acceptable.

FATS

A change has occurred in our thinking over the years relating to the role of dietary fats. We have discussed why we need to eat carbohydrate foods with a low glycemic load, such as nonstarchy vegetables. While these foods are nutrient-rich, they are relatively low in calories. Consequently, the bulk of our calories must come from protein and fat, especially if you are on the Low-Carbohydrate Corrective Diet. Yet, excessive protein consumption is associated with negative health consequences as well. Eating too much protein contributes to bone loss and places a strain on the kidneys. This is especially true with regard to animal protein. In addition, the maximum amount of protein that can be utilized at any one meal is about 30 grams (1 ounce). [Please note that you would need to eat 98 grams (3.5 ounces) of skinless chicken breast or salmon to get 28.5 grams (1 ounce) of protein.] Any excess beyond this is changed to (and stored as) fat. Therefore, if we are to lower our carbohydrates (to keep insulin low) and we shouldn't consume excessive protein, the only alternative is to eat more fat.

There are several advantages to dietary fat. Unlike carbohydrates or proteins, fats have a glycemic index of zero, so they don't raise insulin levels at all. But we need to realize that just as there are good and bad carbohydrates, there are good and bad fats; we need to emphasize these good fats. There are also dangers to eating excessive fat, which is why we recommend that you lower calories altogether as we discuss in Chapter 13, especially if you are on the Low-Carbohydrate Corrective Diet.

Fats are also essential to life. Without two fatty acids, the omega-3 fatty acid (alpha-linolenic acid) and the omega-6 fatty acid (linoleic acid), you would die. In addition, our ability to store excess food as fat is an evolutionary development that allowed our proto-human ancestors to survive periods of famine. But all those millions of years of evolution didn't prepare us for the typical modern diet and sedentary lifestyle so common to the developed world that has contributed greatly to the current epidemic of obesity, cardiovascular disease, diabetes, and other degenerative diseases.

Part of the problem is simply the excessive consumption and the wrong type of fats in the typical modern diet. Fats contain 9 calories per gram, compared with 4 calories per gram for carbohydrates and protein, and any calories you consume but don't burn will be stored in your body's own fat cells. Just one meal of a cheeseburger and fries at one popular fast-food restaurant contains 900 calories and 44 grams of fat. But the problem is more complex than just consuming too much fat. Your fat-related health risks are also determined by the particular types of fats that you consume. The typical Western diet contains a lot of very unhealthy fats, while even healthy fats are too often consumed in an unhealthy balance.

Like carbohydrates, all fat molecules are built primarily from carbon, hydrogen, and oxygen atoms. The number of atoms of each element and the ways those atoms are bonded to each other determine whether a fat is *saturated* or *unsaturated*. In unsaturated fats, one or more pairs of carbon bonding sites are still open, that is, they are not saturated or filled up with hydrogen atoms. These open bonding sites are available to interact biochemically with other molecules in your body in a range of processes critical to the function of your cells, organs, and systems. These open bonding sites also allow the fat molecules to bend, so unsaturated fats are flexible.

Saturated fats, on the other hand, are completely saturated with hydrogen atoms, leaving no carbon bonds available, so there is little these fats can do biochemically. Lacking open bonding sites, they are rigid, which is beneficial because you need some of this rigidity to maintain the structure of your cell membranes. Although essential in small quantities, if consumed in excess (which the typical modern diet does in spades), saturated fat will raise cholesterol levels and can end up being stored in your fat cells or clogging your arteries with cholesterol, raising your risk of hypertension, heart disease, type 2 diabetes, stroke, and more. Seventeen of the 44 grams of fat in the cheeseburger and fries meal above are of the saturated variety, exceeding the recommended daily upper limit for saturated fat for a 180-pound man.

Unsaturated Fats

Unsaturated fats, which are liquid at room or body temperature, are generally considered "good" fats, but they come in several varieties and must be consumed in the right balance for optimal health. Unsaturated fats can be *monounsaturated,* having one open pair of carbon bonds, or *polyunsaturated,* having two or more open pairs. Two crucial groups of polyunsaturated fats are the omega-3 and omega-6 fats. Omega-3 fats can be found in foods such as fish, walnuts, and flaxseeds, while omega-6 fats are found primarily in liquid vegetable oils.

Humans evolved eating a diet in which the ratio between omega-3 and omega-6 fats was close to 1:1. Today, however, it is estimated that the ratio in the typical Western diet is more like 25:1 in favor of the omega-6 fats! The consequences of this imbalance can be severe. Even though we need *some* omega-6 fats to live, they promote inflammation in the body, which contributes to a number of degenerative processes and chronic diseases (see Chapter 5). Omega-3 fats, on the other hand, counter inflammation. Since inflammation is at the heart of so many disease processes, maintaining a healthy balance between omega-3 and omega-6 fats is critical to good health and longevity.

ALA (alpha-linolenic acid), an omega-3 fat, is one of the two *essential* fatty acids (EFAs), meaning it is required for life. Since ALA is not produced in the human body, it must be consumed in foods or supplements. ALA reduces inflammation and blood pressure, improves tissue oxygenation, accelerates healing and muscle recovery from exertion, and aids in stress management. (Even though ALA is essential, moderation is the key as recent research has shown that consumption of excessive amounts of ALA may increase risk of prostate cancer.)

EPA (eicosapentaenoic acid) and **DHA** (docosahexaenoic acid) are also omega-3 fats and are derived from ALA. When consumed in sufficient quantity, EPA and DHA help lower blood pressure, decrease risk of atherosclerosis and heart disease, cut triglyceride and cholesterol levels, counteract the damaging effects of saturated fat and trans fatty acids, and slow the growth and spread of cancerous cells. With age, conversion of ALA to EPA and DHA typically diminishes, so supplementation with EPA and DHA is usually required to achieve optimal balance (see Chapter 12 for dosage recommendations). This can be considered one of our more important supplement recommendations.

LA (linoleic acid), an omega-6 fat, is the other fatty acid essential to life, but it is also not created in your body. However, because LA can lead to inflammation, it is important to balance consumption of LA with adequate amounts of omega-3 fats. The typical modern diet contains disproportionate amounts of LA from vegetable oils, including safflower, sunflower, soy, and sesame, so we recommend that you limit your consumption of such LA-rich oils.

AA (arachidonic acid) is found primarily in red meat, egg yolks, and other animal products. While the body requires some AA to make vital prostaglandins, most people can produce sufficient amounts. Arachidonic acid promotes inflammation, however, and excessive red meat consumption leads to high levels of AA and inflammation, which can lead to heart disease and other degenerative conditions (see Chapter 2).

Monounsaturated fatty acids have just one pair of open carbon bonds available for biochemical reactions in the body. This single open bonding site means that monounsaturated fats combine qualities of saturated fats, such as rigidity, with the fluidity of polyunsaturated fats. This makes monounsaturated fats highly beneficial. A healthful monounsaturated fatty acid is oleic acid (**OA**), found in olives and olive oil (preferably extra virgin), and it should constitute an important portion of your daily fat consumption.

Oleic acid is similar to ALA in its anti-inflammatory properties. It also helps prevent atherosclerosis by keeping arteries flexible. In addition to olives and olive oil, OA is also found in avocados and nuts (peanuts, pecans, cashews, filberts, and macadamias, among others). The Nurses' Health Study showed that women who consumed 1 ounce of nuts a day had dramatically lower rates of cancer and heart disease. While meats and animal products such as butter contain very small amounts of OA, the high levels of AA and other unhealthy fats also found in those foods make them a very poor choice for dietary OA.

POA (palmitoleic acid) is another monounsaturated fatty acid, but it is best avoided because it raises cholesterol levels. POA can be found in coconut and palm oils, which are used in many prepackaged baked goods and snack foods like cookies and chips, and is a primary ingredient in commercial non-dairy coffee creamers.

Saturated Fats

Saturated fats can be bad for you if eaten in excess, and a diet rich in saturated fat can contribute to metabolic syndrome, type 2 diabetes, obesity, and atherosclerosis. Typically solid at room or body temperature, saturated fats are sticky and can cause red blood cells to clump together, inhibiting their ability to carry oxygen to the cells. Saturated fats can also cause blood platelets to stick together and form blood clots that can cause a heart attack or stroke. The conventional Western diet contains too much saturated fat, primarily in meats (beef, pork, lamb), whole-milk dairy products (butter, cheese, milk, ice cream), and poultry skin, as well as in some plant products such as coconuts, coconut oil, palm oil, and palm kernel oil. Reducing consumption of saturated fat improves the health of most people.

Yet, saturated fats are not all bad. Being more rigid than unsaturated fats, they provide support to your cell membranes and serve as precursors to a variety of hormones and hormonelike substances. A middle-of-the-road approach to saturated fat consumption is to consume about 3 to 7 percent of your daily calories as saturated fat.

Trans Fats—Unsafe in Any Amount

Commercial food manufacturers have created a group of fats in the laboratory that are more beneficial to their bottom line than to your health. Known as trans fatty acids, these fats are designed to provide taste, "mouth feel," and cooking qualities that are similar to naturally occurring fats while being far less expensive to make and having a much longer shelf life. The production process, called hydrogenation, uses heat to infuse hydrogen into unsaturated liquid vegetable oils to create something that is very similar to saturated fats but far worse for you. The production process strips away any essential fatty acids, vitamins, minerals, or other useful nutrients, while toxic chemicals used in processing and created by the intense heat can remain.

The trans fatty acids found in margarine are one example, as are any kind of "hydrogenated" vegetable oil often found in fast foods, such as French fries, and in manufactured baked goods, chips, and other snack products. Consuming these faux fats inhibits the body's detoxification processes, decreases testosterone levels, and increases the risk of heart disease, metabolic syndrome, and type 2 diabetes. Trans fats are far more damaging than saturated fats because they raise "bad" low-density lipoprotein (LDL) cholesterol levels *and* lower "good" HDL cholesterol. Health risks associated with trans fats have been so well substantiated that their use has been banned in restaurants in New York City, Philadelphia, and Seattle. Legislation banning trans fats in the entire state of California was passed in 2008 and will take effect in 2010.

As of 2006, FDA estimates were that the average daily intake of trans fat in the U.S. population was approximately 5.8 grams per day, or about 2.5 percent of calories. This may seem like very little, but there is no RDA for trans

Nutrition Facts

Serving Size 1/2 cup (about 82g)
Servings Per Container 8

Amount Per Serving

Calories 135   Calories from Fat 65

	% Daily Value*
Total Fat 7g	**11%**
Saturated Fat 3g	**15%**
Trans Fat 0g	
Cholesterol 55mg	**18%**
Sodium 40mg	**2%**
Total Carbohydrate 17g	**6%**
Dietary Fiber 1g	**4%**
Sugars 14g	
Protein 3g	

Vitamin A 10% • Vitamin C 0%

Calcium 10% • Iron 6%

*Percent Daily Values are based on a 2,000 calorie diet. Your daily values may be higher or lower depending on your calorie needs:

		Calories:	2,000	2,500
Total Fat	Less than		65g	80g
Saturated Fat	Less than		20g	25g
Cholesterol	Less than		300mg	300 mg
Sodium	Less than		2,400mg	2,400mg
Total Carbohydrate			300g	375g
Dietary Fiber			25g	30g

Calories per gram:
    Fat 9 • Carbohydrate 4 • Protein 4

fats and the optimal amount for you to eat is zero. The FDA now requires food manufacturers to list the trans fat content of foods on a separate line as part of the "Nutrition Facts" panel.

Yet, just because the label reads 0g of trans fat does not mean that it doesn't contain any. According to the FDA, "Food manufacturers are allowed to list amounts of trans fat with less than 0.5 gram ($\frac{1}{2}$ g) as 0 (zero) on the Nutrition Facts panel. As a result, consumers may see a few products that list 0 gram trans fat on the label, while the ingredient list will have 'shortening' or 'partially hydrogenated vegetable oil' on it." This means that even though food manufacturers must list trans fat content, they can manipulate the data by changing the serving size. For example, by lowering the serving size from four cookies containing 1.8 grams of trans fat, an amount that would need to be listed on the label, to a serving size of one cookie, which at 0.45 gram would be less than 0.5 gram, the information would be rounded off and listed on the label as zero. Therefore, you need to read the ingredients list as well, and if it includes any type of *partially hydrogenated* oil (soybean, cottonseed, canola, corn, sunflower, etc.), this product contains trans fats—and even though the Nutrition Facts panel says it contains 0 grams of trans fat—it does have trans fats and should be avoided. *Don't* consume any of these dangerous fats!

Cholesterol

Cholesterol is not a fat, but we include it here because the manner in which your body processes it is quite similar. While most discussion about cholesterol tends to focus on lowering its levels in the body because of the health risks, you actually need a certain amount of cholesterol to live. Your cell membranes require it to function properly, and it provides raw material for production of critical hormones, including estrogen, progesterone, testosterone, and cortisone.

The bad press, however, is warranted. An elevated level of cholesterol in the blood is a well-documented risk factor for atherosclerosis, which occurs when waxy cholesterol deposits (plaque) form along the inner surfaces of blood vessel walls (see Chapter 2 for a more thorough description of this process).

While consumption of cholesterol-rich foods can contribute to elevated levels in the blood, as much as 75 percent of the cholesterol in your body is produced by your body, primarily in the liver. And the leading cause of over-production of cholesterol by the body is overconsumption of calories, especially in the form of high-glycemic-load carbohydrates, excessive saturated fat, and hydrogenated trans fats. Chronic stress is an additional contributor to high cholesterol levels because the constant demand for the stress hormone cortisol triggers production of more cholesterol to provide the raw materials for cortisol production.

To reduce your blood cholesterol levels, it is critical that you strictly limit your consumption of high-glycemic-load carbohydrates, eat moderate amounts of saturated fat, and avoid trans fats completely. You should also limit consumption of foods high in cholesterol, such as fatty red meat, shellfish, whole milk products, full-fat cheese, butter, and whole egg yolks. And, finally, you want to take measures to control and reduce the stress in your life as much as possible.

Summary of Fats and Cholesterol

Given the risks to your health and longevity associated with eating either too much fat or the wrong kind of fat, it is essential that you become very selective regarding the amount and composition of fats in your diet.

- Most of your fat calories should come from healthy omega-3 sources, such as fish and fish oil, as well as extra-virgin olive oil, avocado, and nuts.

- Limit consumption of saturated fats to 3 to 7 percent of total calories.

- Avoid whole milk products such as butter, cream, ice cream, and cheese. You don't want to consume skim milk or zero-fat dairy products either because these still contain lactose or milk sugar. Since skim milk has sugar, but no fat to slow its absorption, this turns it into a high-glycemic-load food, which should be avoided. Low-fat (1 to 2 percent) dairy products are preferred.

- Avoid fatty meats such as fatty beef, pork, and lamb. In addition to saturated fat and cholesterol, these meats can also contain high concentrations of pesticides, hormones, antibiotics, and other chemicals used in industrial agriculture. Small amounts of lean beef or buffalo (which is a very-low-fat red meat) are acceptable.

• Limit poultry consumption to small servings of white meat chicken or turkey (skin removed), preferably from "free-range" birds that have been raised without antibiotics or hormones.

• Cut all hydrogenated trans fats from your diet by avoiding margarine, shortening, manufactured baked products, chips, and other fried snacks.

• Consume most of your fat calories from foods that are rich in healthy fats, including:

○ Fish that is high in EPA and DHA, such as salmon (wild salmon is higher in EPA and DHA and lower in mercury than farm-raised), trout, and sardines.

○ Raw nuts, such as walnuts, almonds, peanuts, pecans, and pistachios, and seeds, such as flaxseeds, pumpkin seeds, and sunflower seeds. The health benefits of nuts have been so well established that the FDA now allows some nuts and nut-based foods to carry the claim: "Eating a diet that includes 1 ounce of nuts daily can reduce your risk of heart disease." Be careful, however, to avoid nuts and seeds roasted in unhealthy vegetable oils or with sugary coatings, and stick with nut butters ground from 100 percent whole nuts with no added sugars, flavorings, or hydrogenated trans fats.

○ Extra-virgin olive oil (avoid other commercial cooking and salad oils).

○ Low-glycemic-load vegetables with small amounts of healthy fat.

○ Tofu.

• Avoid deep-frying. Instead, sauté or stir-fry your foods, and only use extra-virgin olive oil. To prevent heat from causing toxic chemicals and free radicals to form in the oil and food, be careful to keep the oil at a moderate temperature (under 180°F; smoking oil is too hot). You can put a little bit of water into the pan before adding extra-virgin olive oil to moderate the temperature.

• Supplement with EPA and DHA (see Chapter 12 for dosage recommendations).

• Limit cholesterol consumption to no more than 1,400 milligrams per week (700 milligrams per week if you are at high risk of cardiovascular disease).

PROTEIN

Life in all its incredibly diverse forms is built from proteins. All proteins are assembled from amino acids in a remarkable and complex process known as biosynthesis. Your DNA, the genetic code that makes you uniquely who you are, contains all the information needed to link various amino acids together into all the different proteins required to build the cells and structures that make up all the organs and systems of your body.

Your body requires a constant supply of these amino acid building blocks to live. And while your body can produce its own supply of many amino acids, eight are known as essential amino acids because you must obtain them from your food. Consuming *complete* protein from sources such as meat, milk, and other animal products provides you with all eight essential amino acids. But you don't need to consume complete proteins as long as you eat all eight essential amino acids over the course of each day, not necessarily at every meal.

According to the National Institutes of Health and the American Heart Association, most healthy adults can do well on a minimum of 50 to 60 grams of protein each day. Daily needs for sedentary adults are 0.36 grams of protein per pound of body weight, but for people who exercise regularly, these needs increase by 10 grams per 60 minutes of exercise. This means that an active 130-pound woman who averages 60 minutes of exercise per day would need a minimum of 52 grams of protein a day, while a 180-pound man who averages 60 minutes a day would need at least 75 grams. To meet these requirements, you can use the following ballpark approximations. Count an average of:

- 1 gram of protein for each serving of fruit or vegetables
- 5 grams per egg or handful of nuts
- 10 grams per cup of milk
- 15 grams per cup of beans
- 25 grams for every 3- to 4-ounce serving of meat, fish, or poultry

The above daily protein requirements are minimal amounts, but you don't want to consume disproportionate amounts of protein—excess protein has been associated with dehydration and may increase the risk of gout, kidney stones, osteoporosis, and possibly some types of cancer.

We believe that you can eat more protein if it is from vegetable sources rather than animal sources; however, specific guidelines for optimal amounts or upper limits of vegetable protein have not yet been established.

You only need to consume about 1 gram of each of the essential amino acids a day, so a single small serving of animal protein containing all eight essential amino acids each day should be sufficient. However, most meat, poultry, and dairy products also contain relatively high amounts of saturated fat and cholesterol. Therefore, serving sizes of most red meat and poultry should be limited to 3- to 4-ounce servings. Red meat should be limited to low-fat cuts, and white meat from naturally fed poultry with the skin removed is also acceptable. Fish is a good source of animal protein, especially salmon, which is also high in healthy omega-3 fats. Because wild salmon is richer in beneficial omega-3 fat and contains lesser amounts of contaminants, it is preferable to farm-raised salmon when possible. Dairy products should be of the reduced-fat variety.

Vegetable proteins are an even better choice. Not only do you avoid the saturated fat and cholesterol found in animal proteins, you also reduce your exposure to the hormones, pesticides, antibiotics, and other agricultural chemicals found concentrated in meat and dairy products. And given the pressing need to address environmental concerns, vegetable proteins have the added benefit of being produced about 20 times more efficiently than meat proteins.

Even if you follow a vegetarian or vegan diet, you should have no problem getting all the protein you need if you eat a variety of protein-rich foods that contain all eight essential amino acids among them. Nuts, legumes (peas, lentils, beans, etc.), and whole grains are all healthy sources of vegetable protein. Soybeans are a particularly good protein source, being the only complete vegetable protein containing all eight essential amino acids. Soy also contains valuable vitamins, minerals, phytochemicals, and fiber. In addition to traditional foods, such as tofu, soy protein can now be found in a wide range of products, including soy milk, soy burgers, and soy pastas.

A number of health benefits have also been attributed to consumption of soy products, such as a slight lowering of cholesterol levels. Soy may also be beneficial in helping prevent cardiovascular disease, some hormonally sensitive cancers, and menopausal osteoporosis. Yet, moderation must also be applied to soy. Excessive consumption of soy products has been found to decrease absorption of iron, zinc, and possibly other minerals; can adversely affect thyroid function; and may decrease fertility in men.

Summary of Protein

For people in both carbohydrate groups, we also recommend that you:

- Consume most of your protein from vegetable sources such as soy products (tofu, soy burgers, tempeh, miso, etc.), along with other legumes such as beans and lentils, nuts, and whole grains.

- Limit consumption of animal protein to fish, 3- to 4-ounce servings of lean meat such as white meat chicken and turkey, and egg whites or egg substitutes (which are 99 percent egg white).

- Avoid fatty meat, whole eggs, and whole milk products that contain high levels of saturated fats and cholesterol.

ALCOHOL

In addition to the three major sources of calories—carbohydrates, fats, and proteins—there is a fourth distinctly different type of calorie—alcohol. The debate that has raged over the benefits and risks of the other food groups pales in comparison to the controversy over alcohol. Some religions condemn drinking and prohibit touching even a drop, while others use it as part of their holiest ceremonies. Until recently, the conventional wisdom held that there were essentially no health benefits associated with drinking alcohol. Moderate alcohol consumption was thought of as a negative—but tolerable—health risk, while excessive consumption was extremely harmful. Recent studies, however, have shown that moderate drinking does appear to provide some health benefits. It now appears that people who drink in moderation are both healthier and live longer than people who don't drink at all.

That being said, we won't go so far as to suggest you start drinking for your health if you currently do not! But, *if* you do drink, we offer the following advice and guidelines. It appears that moderation is the key; moderate drinking has been defined for us by the U.S. Department of Agriculture and the Dietary Guidelines for Americans very precisely as:

- Up to two alcoholic drinks per day for men under 65
- Up to one alcoholic drink per day for men over 65 and for women

Please note that these are daily figures, and not cumulative. Younger men can have up to two drinks a day and women and older men up to one a day—on any given day. Not drinking at all during the week and then having between 7 and 14 drinks on Saturday night is not a good idea!

In addition, an alcoholic drink is defined as either:

- 12 ounces of beer
- 5 ounces of wine
- 1.5 ounces of distilled spirits

By following these provisos for moderate drinking, it is possible to lower your risk of heart disease, stroke, and dying of heart attack. You also lower your risk of diabetes and gallstones. But alcohol is no panacea. While overall mortality and death from cardiovascular disease appears to be lowered by moderate alcohol consumption, cancer risk is not. Risk of several cancers increases with even modest alcohol. Alcohol consumption is strongly associated with cancers of the mouth, pharynx, and esophagus. A 2007 Kaiser Permanente study of over 70,000 women found that drinking one to two alcoholic drinks a day increased breast cancer risk 10 percent. Three or more drinks a day increased their risk 30 percent. A study from Northwestern University found that patients who drank at all were diagnosed with colorectal cancer 5.2 years earlier on average.

In addition, because of the addictive nature of alcohol, it is far too easy for some people to cross over the line and drink more than moderate amounts. Excessive alcohol consumption is associated with a number of serious health problems in addition to cancer, such as cirrhosis of the liver, heart muscle damage, and pancreatitis. Chronic alcoholism is a leading cause of traffic accidents and fatalities, and more than 16,000 people die each year as a result of motor vehicle accidents in which alcohol played a part. Alcohol increases rates of criminal activity and is felt to play a role in 25 percent of all violent crimes. Over 14 million Americans meet the criteria for alcohol abuse or alcoholism, and if you have had problems with alcohol in the past, or have a strong family history of alcoholism, the slope is probably too slippery for you to drink. Other contraindications to drinking include liver disease, use of certain prescription or over-the-counter medications, and history of ulcer disease.

If you have no contraindications and are able to drink in moderation, there are relative benefits in drinking certain types of alcohol for different groups. Red wine contains a phytonutrient known as resveratrol, which has been shown to increase longevity in animal experiments. Although it doesn't contain enough resveratrol to confer the types of benefits seen in the animal experiments, red wine is thought to help explain the "French paradox," where people in southern France eat diets very high in saturated fat, yet have a low incidence of coronary heart disease. The drier the red wine, the more beneficial flavonoids it contains, so cabernets are the best, followed by syrahs and pinot noirs. Red wine also contains tannins, which inhibit absorption of iron. Although a small effect, this is beneficial for men and postmenopausal women who have difficulty ridding their bodies of excess iron. Premenopausal women lose iron each month through menstruation, so they might consider drinking white wine, which doesn't interfere with iron absorption.

Historically, beer has been thought less beneficial than wine, but some recent studies suggest just the opposite. Beer drinking increases vitamin B_6 levels 30 percent, and vitamin B_6 is helpful in lowering levels of homocysteine (see Chapter 5). Wine does not contain B_6, and homocysteine levels rise after drinking wine or distilled spirits. The Harvard Alumni Health Study found that wine or beer consumption had no effect on prostate cancer risk. Hard liquor consumption, on the other hand, increased risk 61 to 67 percent. In summary, it appears that moderate alcohol consumption may provide some modest health benefits (although certain health risks as well) and is acceptable as long as neither medical contraindication nor tendency towards abuse exists. So don't overdo it!

BRINGING IT ALL TOGETHER

We know the thought of putting our nutritional recommendations into practice may seem a bit overwhelming if your current diet is quite different, but don't let that stop you from taking the first step. One small step after another will lead over time to a healthier diet and a healthier you. Clinical evidence makes it quite clear that the benefits will be well worth your efforts.

To help you visualize our dietary recommendations, we have included our *TRANSCEND* Food Pyramid as well as a number of sample recipes. You're probably familiar with the original food pyramid published in 1992 by the U.S. Department of Agriculture (USDA) and widely promoted in most public schools and in many health facilities. Unfortunately, many of its recommendations were dead wrong. It failed to distinguish between high- and low-glycemic-load carbohydrates. It also made no distinction between healthy and unhealthy fats or between protein sources high in saturated fats and cholesterol (e.g., red meats, dairy products) and much healthier options (e.g., legumes, nuts, fish).

A team at Harvard Medical School published a revised food pyramid in 2002 that improved on the USDA version by moving high-glycemic-load carbohydrates from the bottom of the pyramid (eat a lot) to the top (eat sparingly) and instead split the bottom of the pyramid between vegetable oils and whole grains. We disagree, however, with its emphasis on grains as well as the use of vegetable oils (other than extra-virgin olive oil). Though better than refined starches, whole grains are higher in glycemic load than vegetables, as are fruits, which they place at the same level as vegetables. The Harvard Food Pyramid makes no distinction between healthy oils, such as extra-virgin olive oil, which is high in oleic acid, and other commercially processed vegetable oils that are high in proinflammatory omega-6 fats and trans fatty acids. It also doesn't distinguish between fish, which is high in omega-3 fats, and less desirable sources of proteins such as poultry and eggs. And it includes high-fat dairy products, which we believe should be avoided altogether.

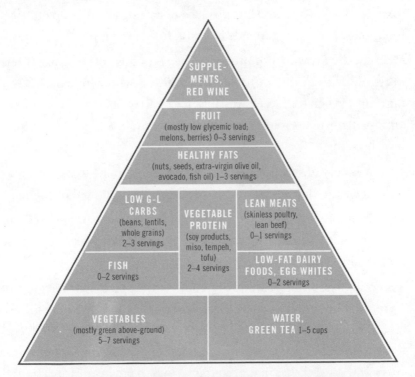

THE *TRANSCEND* FOOD PYRAMID

Our nutritional recommendations are reflected in the *TRANSCEND* Food Pyramid, which emphasizes low-glycemic-load carbohydrates, low-fat proteins, and healthy fats.

As you assess your current diet and devise plans to improve it as needed, we offer these additional recommendations:

• Eat organic whole foods whenever possible. "Organic" applies to foods grown in soil that is rich in natural nutrients and without the use of chemical fertilizers, pesticides, and herbicides so prevalent in conventional farming. "Whole" means foods that are complete, that have not been refined or processed in ways that remove the valuable vitamins, minerals, protein, phytochemicals, and fiber found in food as it comes from the field. Animal proteins are designated "organically raised" when the animal is fed a strict diet of organic feed with no hormones or antibiotics. (If organic produce is unavailable, you can reduce your exposure to toxins by soaking nonorganic produce in a solution of ¼ cup of 3 percent food-grade hydrogen peroxide in a sink full of water for 20 minutes.)

• Eat locally grown fresh produce whenever possible because it retains most of its vitamins, minerals, enzymes, phytochemicals, and other nutrients. When fresh is unavailable, frozen is the best alternative because almost all nutrients remain. Avoid canned produce as most nutrients will have been lost.

• Eat a variety of vegetables and fruits of many colors to please your palate and ensure that you consume a full spectrum of vital nutrients.

• Don't overcook your foods. Vegetables lose nutrients when overcooked. In addition, baking or frying foods to the point of browning (caramelizing) produces acrylamide, a potent carcinogen.

• Spice up your diet. A study of people with type 2 diabetes found that 1 gram of cinnamon a day (¼ teaspoon twice daily) significantly lowered blood sugar, fatty acids in the blood (triglycerides), "bad" LDL cholesterol, and total cholesterol. And curcumin (turmeric), a spice used in curry and other dishes, has been found to be a potent anti-inflammatory that is useful against cancer, Alzheimer's disease, and other conditions.

• Use vinegar in your diet. A study reported in *Diabetes Care* in 2004 found that vinegar dramatically reduces the rise in blood sugar and insulin after eating. Include vegetables pickled in vinegar with your meals and use salad and vegetable dressings made with vinegar and extra-virgin olive oil (make your own or use commercial dressings that do not contain added sugars or preservatives).

• Drink mostly water, green tea, and freshly squeezed vegetable juice (celery, cucumber, and fennel are ideal; smaller amounts are good of red and green leaf lettuce, romaine lettuce, endive, escarole, spinach, parsley, and kale; carrots and beets are high in sugar, so use in moderation).

• Start the day with a good breakfast to avoid fatigue and low blood sugar levels. And spreading several smaller meals and snacks out over the day is preferable to eating one or two large meals as this puts less strain on your digestive system and decreases insulin spikes that lead to insulin resistance and carbohydrate cravings. Make sure those frequent small meals and snacks are healthy, for example, low-starch vegetables or a small serving of fruit.

• Read food labels:

  ○ Fiber is counted as part of total carbohydrates and total calories but, since fiber is not digested, you can delete the number of fiber grams from the total carbohydrate count and delete 4 calories per fiber gram from total calories when calculating your consumption.

  ○ Examine the ingredients list to determine whether carbohydrates are high-glycemic load or low-glycemic load and whether fats are healthy or unhealthy.

  ○ Don't be fooled by "healthy"-sounding terms. For example, "High in Polyunsaturated Fat" could mean healthy anti-inflammatory omega-3 fats or unhealthy, inflammation-causing omega-6 fats. "Monounsaturated" could mean healthy, anti-inflammatory oleic acid or unhealthy, cholesterol-raising palmitoleic acid. "No Cholesterol" does you no good if the product contains cholesterol-raising saturated fats or trans fatty acids. And "Cold Pressed" refers only to the way the oil was extracted, not whether any other potentially harmful steps were used in preparation.

○ Avoid products with additives, especially those you don't recognize.

To help translate our dietary recommendations into more practical terms, we have included a number of sample recipes from the Kurzweil and Grossman kitchens. Each of these menu items incorporates our basic principles of healthy eating. Of equal importance, we've tried every one of them, and we hope you'll agree they're not only healthy—they're delicious as well. Bon appétit!

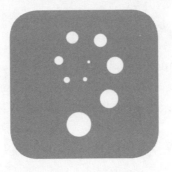

THE

TRANSCEND

RECIPES

The following recipes from Ray's and Terry's kitchens—with contributions from our health-conscious friends and colleagues—are designed to delight your palate while making it easy to follow our nutritional guidelines. We've selected these recipes because they're fresh and different and also easy to prepare.

Our goal in these recipes is to demonstrate that eating the *TRANSCEND* way is an enjoyable and satisfying experience. We've also tried to keep these recipes simple so they fit into today's busy lifestyles.

Sample 7-Day Menu

Monday

B Israeli Breakfast, Pome-Grapefruit Shake

L Autumnal Squash Soup, Scallop and Blue Cheese Salad

D Corn Chowder, Steeped Tofu Slices, Brussels Sauté, Lemony Caraway Carrots

Tuesday

B Fiesta Omelet, whole-grain sprouted bread

L Oven-Steamed Chicken and Leeks, Aegean Salad

D Miso Soup, Green Tea Poached Salmon, Carrot Salad, Garlicky Kale, Basic Japanese Pickles

Wednesday

B Triple Berry Cooler

L Israeli Salad, Mediterranean Baked Cod, endive salad with Springy White Wine Vinaigrette

D Spicy Thai Stir-Fry, Spring Salad, Black Beans with Cilantro, Frozen Fruit

Thursday

B Japanese Breakfast

L Cabbage Soup, Ginger Turkey Burgers, Shirataki with Sautéed Vegetables

D Vegetarian Chili Verde, Cucumber Salad, Vegetarian Stuffed Tomatoes, spinach greens with Light and Creamy Dressing

Friday

B Chocolate Avocado Shake

L Thai Chicken Wrap, green salad with Sin-Free Stevia Salad Dressing

D Quick Vegataki Soup, Steamed Snapper with Orange Sauce, Sautéed Quinoa Pilaf, Cool Melon Salad

Saturday

B Waldorf Salad with Light and Creamy Dressing, Guiltless Egg White Omelet
L Simply Springtime Lentil Soup, Indian Salad
D Hoisin Chicken, Roasted Vegetables with Thyme, Garlicky Kale

Sunday

B Tofu Scramble, Avocado Salad
L Nutty Salmon, Cauliflower with Indian Spices
D Provençal-Style Chicken, Herbed Zucchini, green salad with Springy White Wine Vinaigrette

Snacks

Roasted Red Pepper Hummus
Whipped Strawberry-Banana Mousse
Various Smoothies (Antioxidant Blast/Patriotic/Chocolate Mint/Smoothie and Fruity/Peach Delight/Chocolate Cherry/Piña Colada/Cool As a Cucumber Health)

BREAKFAST

ISRAELI BREAKFAST

Traditional Western diets often focus on unhealthy breakfast foods, but incorporating vegetables, healthy fats, and proteins into the first meal starts the day off right. In Israel, breakfast contains a lot of fresh vegetables.

> 1 ounce feta cheese
>
> 2 tablespoons hummus
>
> ¼ red bell pepper, sliced
>
> ¼ green bell pepper, sliced
>
> ¼ cucumber, peeled and sliced
>
> 3 ounces cooked salmon

Arrange feta, hummus, bell peppers, cucumber, and salmon on a serving plate.

1 SERVING

NUTRITION INFORMATION PER SERVING:
Calories: 286
Fat: 12 g (38% calories from fat)
Protein: 40 g (56% calories from protein)
Carbohydrates (net): 4 g (6% calories from carbohydrates)

FIESTA OMELET

Egg white protein is a complete protein, which means it provides all eight essential amino acids. Egg whites or egg substitutes are an excellent way to add protein to your diet without adding lots of calories, fat, and cholesterol. Plus, you can tuck fresh garden produce inside and enjoy the harvest!

1 medium-heat chile pepper, such as the poblano, seeds removed and chopped

2 small shallots, minced

$\frac{1}{2}$ cup chopped cherry tomatoes

1 cup egg whites or egg substitute

Spray a nonstick skillet with cooking spray and heat over medium-high heat. Cook pepper for 2 minutes. Add shallots, and cook another 1 to 2 minutes, until softened. Add tomatoes and cook while stirring for 1 minute until slightly soft. Remove vegetables to a bowl.

Return skillet to heat. Add egg whites or substitute and cook for 1 minute, until slightly set. Add vegetables to one side of the egg layer. When the top of the eggs begins to set, flip the egg-only side over the vegetables. Cover and continue to cook for 1 minute, or until no runny egg whites remain and omelet is nicely puffed.

2 SERVINGS

NUTRITION INFORMATION PER SERVING:
Calories: 82
Fat: 3 g (4% calories from fat)
Protein: 14 g (68% calories from protein)
Carbohydrates (net): 5 g (28% calories from carbohydrates)

JAPANESE BREAKFAST

In addition to being healthy and delicious, a Japanese breakfast is easy to prepare if you have leftovers from a Japanese style supper the day before. This combination results in a nearly perfect distribution of calories across protein, carbs and fats and is an ideal morning meal.

1 cup Miso Soup (prepared the day before or made fresh, page 253)

¼ cup steamed brown rice

1 sheet nori seaweed, cut into strips

1 ounce natto

4 ounces Green Tea Poached Salmon (page 270)

5–10 Basic Japanese Pickles (page 292)

Heat miso soup in a medium saucepan over medium heat. Be careful not to bring to a boil. High temperatures destroy miso's healthy bacteria. Serve in a bowl. Arrange the rice, nori, natto, and salmon on a plate and serve.

1 SERVING

NUTRITION INFORMATION PER SERVING:
Calories: 412
Fat: 15 g (32% calories from fat)
Protein: 37 g (34% calories from protein)
Carbohydrates (net): 25 g (34% calories from carbohydrates)

GUILTLESS EGG WHITE OMELET

Mushrooms add texture and flavor to enhance the taste of this egg white omelet. They are a good source of selenium, which assists in the body's natural detoxification process.

¼ cup fat-free milk

10 small mushrooms, sliced

1 small onion, finely chopped

1 small potato, baked or microwaved until just tender, then cubed

2 teaspoons herbes de Provence

8 egg whites

In a medium sauté pan, bring the milk to a boil. Add the mushroom slices and onion to the milk. The liquid volume will increase as the mushrooms cook. Reduce the amount of liquid to 3 tablespoons. The mushrooms will have darkened, but not nearly as much as when cooked in butter or fat.

On low heat (the heat retained in the pan will probably be enough), toss the cubed potato and herbes de Provence with the mushroom mixture. This will absorb the remaining liquid.

In a 10-inch skillet, either nonstick or sprayed with cooking spray, cook the egg whites (previously colored with turmeric or saffron, if desired) on medium-high. Coat the bottom of the pan with the eggs to make a flat disk.

Arrange the vegetables in the middle of the disk and fold the two side flaps over. Serve immediately.

2 SERVINGS

NUTRITION INFORMATION PER SERVING:

Calories: 222

Fat: 1 g (3% calories from fat)

Protein: 19 g (32% calories from protein)

Carbohydrates (net): 34 g (65% calories from carbohydrates)

TOFU SCRAMBLE

Tofu is a versatile plant protein that adapts to the flavors of most dishes. Due to its texture, soft tofu works best in this dish. Flaxseed is a good vegetarian source of healthy omega-3 fats. For a twist, ingredients such as chopped onion, pepper, tomatoes, or scallions can be sautéed in the same pan before adding the egg mixture.

> 4 egg whites
>
> 4 ounces soft tofu
>
> 2 tablespoons grated cheese
>
> 1 tablespoon golden flaxseed (optional)

Combine the egg whites and tofu in a blender and pulse briefly for 3 to 5 seconds (be careful not to overblend).

Coat a medium nonstick skillet with cooking spray and place over medium heat. Add egg and tofu blend and scramble for 5 minutes. Top with cheese and flaxseed, if desired.

2 SERVINGS

NUTRITION INFORMATION PER SERVING:
Calories: 83
Fat: 3 g (31% calories from fat)
Protein: 13 g (64% calories from protein)
Carbohydrates (net): 1 g (5% calories from carbohydrates)

SOUPS

AUTUMNAL SQUASH SOUP

A healthy diet doesn't mean you have to eliminate creamy soups. Soy milk and a little stevia rescue this traditional favorite and reinvent it as a tasty recipe featuring beta-carotene-rich squash.

1 tablespoon extra-virgin olive oil

1 quart vegetable broth, divided

1 large onion, chopped

2 large leeks (white and pale green parts only), chopped

2 cups chopped butternut squash

1 large potato, peeled and chopped

1 cup finely chopped carrots (about 3 medium)

1 Granny Smith apple, peeled, cored, and sliced ¼" thick

⅓ cup dry white wine (optional)

½ cup unsweetened soy milk

¼ teaspoon freshly ground nutmeg

1 packet stevia

Sea salt and pepper to taste

Combine the olive oil and 1½ tablespoons of broth in a large pot over medium heat. Add the onion and leeks and cook for 5 minutes or until onion is translucent. Add the squash, potato, carrots, apple, and remaining vegetable broth. Bring to a boil, then reduce heat to medium-low, cover, and simmer until the vegetables are soft, about 20 minutes.

Puree the soup in a blender or food processor, working in batches as necessary, or use an immersion blender to puree the soup right in the pot (be careful to avoid splatters). Return soup to the pot and stir in wine, if using, and soy milk. Add the nutmeg and season to taste with stevia, salt, and pepper. Simmer gently for at least 5 minutes, and up to 2 hours for best flavor.

8 SERVINGS

NUTRITION INFORMATION PER SERVING:
Calories: 109
Fat: 2 g (18% calories from fat)
Protein: 3 g (10% calories from protein)
Carbohydrates (net): 18 g (66% calories from carbohydrates)

CORN CHOWDER

Although corn is high in carbohydrates, variety is the key to enjoying your food, and this creamy low-fat chowder makes the perfect occasional accent to a spicy meal. For a fresher taste, use 1 pound of frozen corn with 1 cup of water or, if in season, fresh corn kernels cut off the cob. For a chunky texture, reserve one-third of the corn and add it to the soup after blending the other ingredients.

1 clove garlic
$1/2$ cup coarsely chopped Vidalia onion or 4 scallions, chopped
1 can (16 ounces) corn
$1/3$ cup fat-free milk
$1/3$ cup nonfat dry-milk powder
Sea salt

Pulse the garlic and onion in a food processor until finely chopped (if using scallions, reserve until the end). With the motor running, add the corn with its liquid, milk, and dry-milk powder.

Transfer mixture into a medium saucepan and heat gently over medium heat until warmed through. Stir in scallions, if using. Season to taste with salt.

2 SERVINGS

NUTRITION INFORMATION PER SERVING:
Calories: 248
Fat: 2 g (8% calories from fat)
Protein: 14 g (20% calories from protein)
Carbohydrates (net): 44 g (72% calories from carbohydrates)

MISO SOUP

This soup can be enjoyed any time of the day. See also Japanese Breakfast (page 248). Dashi powder is available at most supermarkets and natural foods stores. Miso soup should never be boiled, as the high temperature would kill the healthy bacteria contained in the miso.

4 cups water

2 teaspoons dashi powder

¼ cup red, white, or brown miso

1–2 long strips wakame seaweed, finely crumbled

1–2 scallions, finely chopped

2–4 ounces tofu, chopped into small cubes

Heat the water in a medium saucepan over medium-high heat. Stir in the dashi powder and bring to a low boil. Reduce heat to low and add the miso, wakame, scallions, and tofu. Simmer for 10 to 15 minutes, until flavors combine.

4 SERVINGS

NUTRITION INFORMATION PER SERVING:
Calories: 36
Fat: 1 g (25% calories from fat)
Protein: 2 g (22% calories from protein)
Carbohydrates (net): 4 g (53% calories from carbohydrates)

CABBAGE SOUP

Eating cabbage is associated with lower risks of certain kinds of cancer. Lycopene, a phytonutrient found in tomatoes, is a strong antioxidant that also helps lower cancer risk. The apples provide a layer of sweetness.

    4 teaspoons extra-virgin olive oil
    2 ribs celery, chopped
    1 medium onion, chopped
    1 leek, sliced into thin rounds
    2 quarts reduced-sodium vegetable broth
    ½ large head cabbage, shredded
    2 tart apples, peeled and chopped
    1 can (28 ounces) diced tomatoes with juices
    1 can (10.75 ounces) condensed tomato soup
    Salt and pepper
    Lemon juice

Heat the olive oil in a large pot over medium-high heat. Add the celery, onion, and leek and sauté for 5 to 7 minutes. Add the broth and simmer for 5 to 10 minutes until onion is translucent. Stir in cabbage, apples, tomatoes, and soup, plus 3 soup cans of water. Reduce heat to low and simmer for 1 to 1½ hours. Just before serving, season to taste with salt, pepper, and lemon juice.

10 SERVINGS

NUTRITION INFORMATION PER SERVING:
Calories: 129
Fat: 3 g (19% calories from fat)
Protein: 5 g (12% calories from protein)
Carbohydrates (net): 21 g (69% calories from carbohydrates)

VEGETARIAN CHILI VERDE

This hearty chili makes a great one-pot meal. Beans and legumes provide healthy vegetable proteins and complex carbohydrates.

1 tablespoon extra-virgin olive oil

1 small onion, chopped

1½ cups reduced-sodium vegetable broth

2 cans (4 ounces each) chopped green chilies

1½ teaspoons garlic powder

1 teaspoon ground cumin

1 teaspoon dried oregano

1 teaspoon dried cilantro

¼ teaspoon ground red pepper

1 can (15 ounces) cannellini beans

1 can (15 ounces) pink kidney beans

Heat the olive oil in a large nonstick pot over medium-high heat. Add the onion and cook for 5 minutes or until soft. Add the vegetable broth, chilies, garlic powder, cumin, oregano, cilantro, and red pepper. Reduce heat to low and simmer for 15 minutes to allow flavors to combine. Stir in the beans with their canning liquid and simmer for 8 minutes longer or until heated through.

6 SERVINGS

NUTRITION INFORMATION PER SERVING:
Calories: 209
Fat: 3 g (13% calories from fat)
Protein: 13 g (21% calories from protein)
Carbohydrates (net): 25 g (66% calories from carbohydrates)

QUICK VEGATAKI SOUP

Shirataki is an excellent low-carb noodle substitute made from a water-soluble fiber substance called glucomannan, which is derived from the Konnyaku root. Available online in its simplest state from Miracle Noodle or made with tofu and sold at natural foods stores, shirataki contains zero net carbohydrates and very few calories. Preparation: Some people may find the natural smell of the shirataki noticeable. To reduce the aroma, increase rinsing time and/or parboiling time (try using a double boiler) for up to 10 minutes each.

1 package (8 ounces) shirataki spaghetti-style noodles, coarsely chopped

2 teaspoons extra-virgin olive oil

4 ribs celery, finely chopped

1 medium cipollini onion, finely chopped

1 bunch kale, stems removed and leaves coarsely chopped

1 can (28 ounces) diced tomatoes

6 cups reduced-sodium vegetable broth

$\frac{1}{2}$ teaspoon garlic powder

Basil and oregano

Salt and pepper

Rinse the noodles for 5 to 10 minutes in cold running water and drain thoroughly.

Heat the olive oil in a large pot over medium-high heat. Add the celery and onion and cook for 3 to 4 minutes until vegetables begin to soften. Add the kale and tomatoes and bring to a low boil. Stir in the broth and garlic powder and reduce heat to medium. Cook for 20 minutes, stirring occasionally, until kale is tender. Add noodles and heat for another 5 minutes until warmed through. Season to taste with basil, oregano, salt, and pepper.

8 SERVINGS

NUTRITION INFORMATION PER SERVING:

Calories: 77

Fat: 2 g (20% calories from fat)

Protein: 4 g (18% calories from protein)

Carbohydrates (net): 11 g (63% calories from carbohydrates)

SIMPLY SPRINGTIME LENTIL SOUP

Red chard and red lentils make for a vibrantly colorful fiber- and protein-rich soup,
perfect for the early spring.

> 1 tablespoon extra-virgin olive oil
> $^2/_3$ cup finely chopped celery (1 or 2 ribs)
> $^1/_3$ cup finely chopped leek (white part only)
> $^1/_2$ cup finely chopped white onion
> $1^1/_2$ tablespoons finely chopped shallot
> 2 teaspoons minced garlic
> 2 quarts reduced-sodium vegetable broth
> 4 cups chopped red chard leaves, stems removed
> $1^1/_4$ cups dry red lentils

Heat the olive oil in a medium saucepan over medium heat. Add the celery, leek, onion,
shallot, and garlic and cook for 5 to 10 minutes until onions are translucent.

Add the vegetable broth, chard, and lentils, raise heat to medium-high, and bring to a boil.
Reduce heat to low and simmer for 30 minutes until lentils are just tender, but not soft
(do not overcook the lentils).

8 SERVINGS

NUTRITION INFORMATION PER SERVING:
Calories: 94
Fat: 2 g (18% calories from fat)
Protein: 7 g (25% calories from protein)
Carbohydrates (net): 9 g (57% calories from carbohydrates)

SEAFOOD POZOLE

Brighten this nutritious, protein-rich take on traditional Mexican stew by serving it with lime wedges.

   2 teaspoons extra-virgin olive oil

   1 small onion, thinly sliced

   2 teaspoons ground cumin

   1 can (15 ounces) yellow hominy, rinsed and drained

   3 cups reduced-sodium vegetable or seafood broth

   2 cups or 1 can (16 ounces) diced tomatoes and juice

   1 can (4 ounces) chopped green chilies

   ¾ pound cod fillet, cut into ¾" pieces

   1 lime, cut in 6 wedges

   Salsa or hot-pepper sauce (optional)

Heat the oil in a large skillet. Add the onion and cook, stirring occasionally, about 5 minutes, until tender. Add the cumin during the last minute and stir. Add the hominy, broth, tomatoes and their juice, and chilies. Cover the skillet and bring to a boil; reduce heat and simmer for 5 minutes.

Add the fish and continue to simmer, stirring gently occasionally, until fish is opaque through (cut to test), 2 to 4 minutes. Ladle into soup bowls to serve, passing lime wedges and salsa or hot-pepper sauce, if desired, separately.

6 SERVINGS

NUTRITION INFORMATION PER SERVING:
Calories: 165
Fat: 3 g (18% calories from fat)
Protein: 17 g (42% calories from protein)
Carbohydrates (net): 10 g (40% calories from carbohydrates)

SALADS AND DRESSINGS

SCALLOP AND BLUE CHEESE SALAD

Spinach greens are packed with vitamins and nutrients and provide a healthy base for salad proteins. In this dish, scallops and blue cheese are drizzled with a warm, tangy dressing.

- 3 tablespoons fresh lemon juice
- 3 tablespoons white wine vinegar
- 2 tablespoons + 2 teaspoons extra-virgin olive oil, divided
- ¼ cup chopped fresh parsley
- 2 teaspoons minced garlic
- 1 tablespoon Dijon mustard
- 2 tablespoons water
- 1 tablespoon minced shallot
- 1 package (7 ounces) fresh spinach
- 4 ounces blue cheese, crumbled
- 1 pound sea scallops

In a small glass or stainless steel pan over medium heat, combine the lemon juice, vinegar, 2 tablespoons of the olive oil, parsley, garlic, mustard, water, and shallot. Whisk until smooth. When mixture begins to simmer, reduce heat and keep warm.

Divide the spinach among 4 plates and sprinkle with cheese.

Rinse the scallops and pat dry; halve large sea scallops horizontally if necessary. Heat 1 teaspoon of the olive oil in a large skillet over high heat. When hot, add half the scallops and cook, stirring often, until opaque through center, 1 to 2 minutes. Lift scallops from pan and set aside. Repeat with remaining scallops and oil. Arrange scallops on spinach and drizzle with warm dressing.

4 SERVINGS

NUTRITION INFORMATION PER SERVING:
Calories: 328
Fat: 19 g (51% calories from fat)
Protein: 34 g (41% calories from protein)
Carbohydrates (net): 4 g (8% calories from carbohydrates)

CUCUMBER SALAD

The springtime flavor of dill complements the fresh, cool taste of cucumber in this lively salad. Dressing with soy yogurt conserves calories and adds protein to the dish.

2 medium cucumbers, peeled, seeded, and sliced paper-thin

½ teaspoon salt

⅓ cup plain soy yogurt

2 tablespoons lemon juice

Stevia

1 tablespoon finely chopped fresh dill, or 1 teaspoon dried dill

Paprika (optional)

Spread the cucumbers in a single layer in a shallow dish. Sprinkle with the salt and set aside at room temperature for 20 minutes.

Meanwhile, in a small serving bowl, combine the yogurt and lemon juice. Add stevia to taste and stir until well combined.

Squeeze the cucumber slices gently to remove excess liquid and pat dry with paper towels. Stir the cucumber slices into the yogurt mixture to coat evenly.

Cover and refrigerate for 2 hours. Sprinkle with dill and paprika, if desired, before serving.

4 SERVINGS

NUTRITION INFORMATION PER SERVING:

Calories: 56

Fat: 1 g (20% calories from fat)

Protein: 3 g (19% calories from protein)

Carbohydrates (net): 7 g (61% calories from carbohydrates)

AEGEAN SALAD

Incorporating leafy greens into your diet doesn't mean your food has to taste bland and boring. Arugula is rich in nutrients and adds a touch of spice to mixed greens.

  1 head romaine lettuce, torn into bite-size pieces
  1 bunch arugula
  1 cucumber, peeled and chopped
  2 tomatoes, chopped
  1 bell pepper, chopped
  3 tablespoons extra-virgin olive oil
  $\frac{1}{2}$ teaspoon lemon zest
  3 tablespoons lemon juice
  6 sprigs spearmint
  2 cloves garlic
  16 kalamata olives
  $\frac{1}{8}$ cup crumbled feta cheese

In a large salad bowl, combine the romaine, arugula, cucumber, tomatoes, and bell pepper. In a food processor, whirl the olive oil, lemon zest, lemon juice, mint, and garlic. Pour the olive oil mixture over the salad and top with olives and cheese.

4 SERVINGS

NUTRITION INFORMATION PER SERVING:
Calories: 132
Fat: 9 g (63% calories from fat)
Protein: 4 g (8% calories from protein)
Carbohydrates (net): 6 g (29% calories from carbohydrates)

CARROT SALAD

Fresh carrots are sweetened with stevia, a sugar substitute that has been studied for its health benefits, including its ability to help regulate blood sugar levels. Red cabbage may help prevent certain cancers and adds both color and fiber to this dish.

> 1 packet stevia
> 2 tablespoons extra-virgin olive oil
> 1/2 teaspoon Dijon mustard
> 4 large carrots, shredded
> 2 ribs celery, chopped
> 1/2 cup shredded red cabbage
> 4 finely chopped scallions
> 1/8 cup raisins
> 1/4 cup finely chopped parsley
> 3 or 4 sprigs spearmint, finely chopped
> 1/8 cup currants (optional)

In a small bowl, whisk together the stevia, olive oil, and mustard. In a large serving bowl, combine the carrots, celery, cabbage, scallions, raisins, parsley, and mint. Pour stevia mixture over the carrot mixture and toss to coat. If desired, add currants for a sweeter salad.

6 SERVINGS

NUTRITION INFORMATION PER SERVING:
Calories: 80
Fat: 5 g (52% calories from fat)
Protein: 1 g (4% calories from protein)
Carbohydrates (net): 8 g (44% calories from carbohydrates)

ISRAELI SALAD

Israel has a very low incidence of obesity. It's no coincidence that fresh vegetables and salads are a major part of every meal. Israelis eat this salad with breakfast, lunch, and supper. Try adding a chopped sprig of fresh mint or crumble in ¼ cup of feta cheese for variety.

 3 cucumbers, finely chopped
 3 tomatoes, finely chopped
 1 green bell pepper, finely chopped
 3 scallions, finely chopped (optional)
 1 tablespoon extra-virgin olive oil
 2 tablespoons fresh lemon juice
 Sea salt
 Ground black pepper

In a large bowl, combine the cucumbers, tomatoes, bell pepper, and scallions, if using. Just before serving, toss with the olive oil and lemon juice. Season to taste with salt and pepper.

4 SERVINGS

NUTRITION INFORMATION PER SERVING:
Calories: 79
Fat: 4 g (43% calories from fat)
Protein: 2 g (9% calories from protein)
Carbohydrates (net): 8 g (48% calories from carbohydrates)

SPRING SALAD

Limited intake of low-fat dairy is part of our recommended diet. Fat-free dairy products are not healthy because they contain sugar but no fats to slow its absorption. Low-fat cottage cheese is the best option. Individual servings of this colorful salad may be served in a cupped leaf of iceberg or Boston lettuce.

2 medium-size cucumbers, peeled and sliced

2 medium tomatoes, chopped into ½" pieces

4 scallions, sliced

6 radishes, sliced

12 ounces low-fat cottage cheese (1% fat)

12 ounces plain soy yogurt

¼ teaspoon sea salt

Freshly ground black pepper

In a large mixing bowl, combine the cucumbers, tomatoes, scallions, radishes, cottage cheese, yogurt, and salt. Season to taste with pepper. Toss and divide into equal portions.

6 SERVINGS

NUTRITION INFORMATION PER SERVING:

Calories: 99

Fat: 2 g (19% calories from fat)

Protein: 10 g (42% calories from protein)

Carbohydrates (net): 8 g (39% calories from carbohydrates)

WALDORF SALAD WITH LIGHT AND CREAMY DRESSING

For a change of pace, try this healthy take on the traditional Waldorf salad. You'll never miss the mayonnaise. The fruity flavors also work as a breakfast salad.

Light and Creamy Dressing

½ cup plain unsweetened soy yogurt

2 tablespoons fresh lemon juice

2 teaspoons minced garlic

1 tablespoon minced shallot

2 teaspoons finely chopped parsley

½ teaspoon lemon zest

Dash sea salt

Stevia to taste

In a large bowl, whisk together the yogurt, lemon juice, garlic, shallot, parsley, and lemon zest. Add a dash of salt and season to taste with stevia.

Waldorf Salad

3 tart apples, such as Granny Smith, chopped

1 rib celery, chopped

2 cups halved seedless grapes

¼ cup currants or raisins

1 head romaine lettuce, torn into bite-size pieces

¾ cup walnuts, chopped and toasted

In the same bowl in which the dressing was prepared, mix in the apples, celery, grapes, and currants or raisins. Toss to coat.

Lay the lettuce on a large plate. Just before serving, arrange the apple mixture over the lettuce and top with the walnuts.

8 SERVINGS

NUTRITION INFORMATION PER SERVING:

Calories: 172

Fat: 8 g (39% calories from fat)

Protein: 3 g (7% calories from protein)

Carbohydrates (net): 22 g (54% calories from carbohydrates)

INDIAN SALAD

Simple ingredients can take on a unique flavor by adding a few Indian spices. Reduced-fat Greek yogurt is a smart dairy choice and can be used in any creamy salad in place of regular yogurt. It lends salads an extra-creamy texture while reducing calories and increasing protein.

    1 cucumber, peeled, seeded, and finely chopped
    1 teaspoon chopped onion
    ¼ teaspoon sea salt
    2 teaspoons ground cumin
    1 teaspoon ground coriander
    2 medium tomatoes, chopped
    1 red bell pepper, chopped
    ¾ cup low-fat (2%) Greek yogurt
    ¼ cup chopped parsley

In a small bowl, combine the cucumber, onion, and salt and let sit at room temperature for 30 minutes. Transfer to a strainer and squeeze to remove excess water.

Meanwhile, in a small dry skillet over medium-high heat, toast the cumin and coriander until fragrant.

In a large bowl, combine the cucumber mixture, spice mixture, tomatoes, pepper, yogurt, and parsley. Mix well.

2 SERVINGS

NUTRITION INFORMATION PER SERVING:
Calories: 115
Fat: 3 g (21% calories from fat)
Protein: 10 g (33% calories from protein)
Carbohydrates (net): 11 g (46% calories from carbohydrates)

AVOCADO SALAD

Avocados contain high levels of monounsaturated fats, vitamins, and fiber. They have been shown to help improve cholesterol levels, and the good fats they contain benefit the brain, joints, and heart. The anti-inflammatory oleic acid in avocados helps arteries remain flexible.

1 package (5 ounces) mixed baby spring greens

1 avocado, sliced

1 cucumber, peeled and chopped

1 green bell pepper, chopped

2 carrots, chopped

2 tomatoes, chopped

4 ounces Cheddar cheese, shredded (optional)

Place greens in a large bowl and top with avocado, cucumber, pepper, carrots, tomatoes, and cheese, if desired.

6 SERVINGS

NUTRITION INFORMATION PER SERVING:
Calories: 84
Fat: 5 g (53% calories from fat)
Protein: 3 g (12% calories from protein)
Carbohydrates (net): 4 g (35% calories from carbohydrates)

SOUTHWESTERN CHICKEN SALAD

Cilantro is a natural chelating agent, meaning it is able to bind to and remove some toxic heavy metals such as lead, mercury, or arsenic from the body. Look for a mayonnaise that is free of trans fats. Some brands also include plant sterols to help manage cholesterol levels.

4 boneless, skinless chicken breasts (6 ounces each)

1 cup chopped cilantro, with stems reserved for poaching liquid (optional)

½ cup low-fat (2%) Greek yogurt

3 tablespoons nonhydrogenated light mayonnaise

1 tablespoon lime juice

1 teaspoon chili powder

Sea salt

Ground black pepper

½ cup corn kernels, thawed if frozen

1 can (14.5 ounces) black beans, rinsed and drained

1 head romaine lettuce, torn into bite-size pieces

Place the chicken in a large pot and cover with 1½ quarts water. Add the cilantro stems, if desired. Bring to a gentle boil over medium-high heat. Reduce heat to low and simmer for 10 minutes. Remove from heat. Let chicken rest in poaching liquid for 20 minutes. Remove from poaching liquid and, when cool enough to handle, pull into shreds, removing any tough or fatty pieces.

In a large bowl, combine the yogurt, mayonnaise, lime juice, and chili powder. Season to taste with salt and pepper. Add the chicken, corn, and beans to the yogurt mixture and toss to combine. Serve on top of lettuce.

8 SERVINGS

NUTRITION INFORMATION PER SERVING:
Calories: 227
Fat: 6 g (23% calories from fat)
Protein: 32 g (60% calories from protein)
Carbohydrates (net): 10 g (17% calories from carbohydrates)

MEDITERRANEAN SALMON SALAD

Wild salmon is preferable to farmed salmon because it does not contain antibiotics, pesticides, dyes, and other toxic substances. Canned salmon is convenient for this zesty combination of Mediterranean flavors. This recipe doubles easily.

> 1 can (7.5 ounces) wild salmon, drained and flaked
> 1 teaspoon oregano or Italian seasoning
> 8 kalamata olives, pitted and halved
> 1 tablespoon minced shallot
> Freshly ground pepper
> $\frac{1}{2}$ lemon
> $1\frac{1}{2}$ teaspoons extra-virgin olive oil
> 2 tablespoons red wine vinegar
> $\frac{1}{4}$ cup crumbled feta cheese
> Rice crackers or sprouted grain bread (optional)

In a large bowl, combine the salmon, oregano or Italian seasoning, olives, shallot, and pepper to taste. Toss thoroughly. Squeeze the lemon over the mixture and add the oil and vinegar. Top with the cheese and serve with rice crackers or sprouted grain bread if desired.

2 SERVINGS

NUTRITION INFORMATION PER SERVING:
Calories: 253
Fat: 17 g (60% calories from fat)
Protein: 21 g (32% calories from protein)
Carbohydrates (net): 4 g (8% calories from carbohydrates)

SPRINGY WHITE WINE VINAIGRETTE

Making your own salad dressing ensures that you do not consume the unnecessary preservatives and sweeteners found in store-bought dressings. Extra virgin olive oil and vinegar both offer significant health benefits—use high-quality products for best results.

1 cup extra-virgin olive oil

¾ cup aged white wine vinegar

2 tablespoons chopped flat leaf parsley

1½ tablespoons dried basil

1 tablespoon Dijon mustard

½ tablespoon minced shallot

2 cloves garlic, minced

In a food processor, combine the oil, vinegar, parsley, basil, mustard, shallot, and garlic and pulse several times to blend well. Store tightly sealed in the refrigerator for up to 2 weeks.

16 SERVINGS

NUTRITION INFORMATION PER SERVING:
Calories: 124
Fat: 14 g (97% calories from fat)
Protein: 0 g (0% calories from protein)
Carbohydrates (net): 1 g (3% calories from carbohydrates)

SIN-FREE STEVIA SALAD DRESSING

Extra-virgin olive oil provides omega-3 fats and is the only oil that should be used in salad dressings. To sweeten dressings naturally, use a little stevia. Adding herbs such as oregano enhances the flavor and supplies antioxidant power.

> 1 cup red wine vinegar
> ¾ cup extra-virgin olive oil
> 1 packet stevia
> 2 tablespoons lemon juice
> Herbs as desired

In a large bowl, combine the vinegar, oil, stevia, lemon juice, and herbs.

16 SERVINGS

NUTRITION INFORMATION PER SERVING:
Calories: 93
Fat: 10 g (96% calories from fat)
Protein: 0 g (0% calories from protein)
Carbohydrates (net): 1 g (3% calories from carbohydrates)

FISH

GREEN TEA POACHED SALMON

Poaching chicken or fish in tea is a good way to add flavor without adding any fat or calories. This method infuses the protein with taste by cooking it submerged in the tea.

2 skinless wild salmon fillets (4 ounces each)
1 teaspoon garlic powder
Sea salt
Freshly ground black pepper
2 cups organic green tea

Sprinkle the salmon fillets with the garlic powder and season to taste with salt and pepper. Arrange fillets in a medium skillet and pour in just enough tea to cover salmon. Bring to a simmer over medium heat. Cook for 7 to 9 minutes or until fish is no longer opaque.

2 SERVINGS

NUTRITION INFORMATION PER SERVING:
Calories: 195
Fat: 10 g (46% calories from fat)
Protein: 24 g (50% calories from protein)
Carbohydrates (net): 1 g (4% calories from carbohydrates)

MEDITERRANEAN BAKED COD

Cod takes on the distinctive flavor of capers and olives in this Italian-inspired dish. For a steamed option, place the fish in a steamer basket over boiling water for 5 minutes. Add the remaining ingredients on top of the fish and steam for an additional 3 to 4 minutes.

> 4 cod fillets (6 ounces each)
> 1 can (14.5 ounces) Italian-style stewed tomatoes
> 1 tablespoon drained capers
> 16 pitted ripe black olives
> 2 tablespoons grated Parmesan cheese
> 1 teaspoon herbes de Provence

Preheat the oven to 375°F. Rinse the cod and pat dry. Lightly coat a large baking dish with olive oil cooking spray and arrange fillets in dish, folding thin tail ends under to give fillets even thickness.

Pour the tomatoes and their juices over fish. Sprinkle with the capers and olives, followed by the cheese and herbs.

Bake uncovered for 20 minutes or until the fish flakes easily with a fork.

4 SERVINGS

NUTRITION INFORMATION PER SERVING:
Calories: 239
Fat: 5 g (17% calories from fat)
Protein: 40 g (71% calories from protein)
Carbohydrates (net): 6 g (12% calories from carbohydrates)

STEAMED SNAPPER WITH ORANGE SAUCE

Citrus adds a tang to this tender snapper. For the best steamed fish, use a thick, firm variety such as black sea bass or mahi mahi, which has lower contaminant levels than striped bass.

$\frac{1}{4}$ cup orange juice

3 tablespoons rice vinegar

2 tablespoons light soy sauce

$1\frac{1}{2}$ teaspoons sesame oil

$\frac{3}{4}$ teaspoon fresh grated ginger

$\frac{3}{4}$ teaspoon orange zest

4 snapper fillets (5 ounces each)

1 pound baby bella mushrooms

4 green onions, cut in 1" lengths

$\frac{1}{2}$ pound green or wax beans, cut in 1" lengths

1 large carrot, sliced

In a small bowl, combine the orange juice, vinegar, soy sauce, oil, ginger, and zest. Set aside for 30 minutes so the flavors combine.

Rinse the fish and pat dry. Arrange on a steamer tray and set over a pan with 2 to 3 inches boiling water. Cover and steam for 1 to 2 minutes.

Top with the vegetables and steam until the fish is opaque through the center and the vegetables are crisp-tender, 3 to 5 minutes longer.

Transfer the vegetables and fish to individual plates and pour the sauce over each portion.

4 SERVINGS

NUTRITION INFORMATION PER SERVING:
Calories: 309
Fat: 6 g (17% calories from fat)
Protein: 39 g (51% calories from protein)
Carbohydrates (net): 22 g (32% calories from carbohydrates)

NUTTY SALMON

Salmon is chock-full of healthy omega-3 fats and protein. The almonds in this dish add texture and even more healthy fats.

> 1 tablespoon spicy brown or Dijon mustard
>
> 2 tablespoons lemon juice
>
> 2 tablespoons finely chopped almonds or pecans
>
> 1 tablespoon dried bread crumbs
>
> 2 teaspoons garlic powder
>
> 1 tablespoon chopped fresh parsley
>
> 2 skinless wild salmon fillets (4 ounces each)
>
> Freshly ground black pepper

Preheat the oven to 350°F. Coat a 13" × 9" baking dish with olive oil cooking spray.

In a small bowl, mix the mustard and lemon juice. On a small plate, combine the nuts, bread crumbs, garlic powder, and parsley.

Season the salmon to taste with pepper and place in the mustard mixture. Turn to coat. Transfer fillets to the plate and press one side into the nut mixture.

Arrange fillets nut side up in the baking dish and bake for 15 to 20 minutes or until the fish flakes easily with a fork.

2 SERVINGS

NUTRITION INFORMATION PER SERVING:
Calories: 273
Fat: 16 g (50% calories from fat)
Protein: 26 g (38% calories from protein)
Carbohydrates (net): 6 g (12% calories from carbohydrates)

QUICK GARLIC TILAPIA

An exceedingly easy solution for a healthy dinner, tilapia is an outstanding and relatively inexpensive way to incorporate protein from fish into your diet.

> 2 tilapia fillets (4 ounces each)
> Salt and pepper
> 3 tablespoons white wine
> 1 teaspoon chopped garlic
> 1 plum tomato, chopped
> 3 teaspoons lime juice
> 1 teaspoon soy sauce

Season the tilapia to taste with salt and pepper. In a skillet on medium-high heat, heat the wine and garlic. Add the tilapia and cover. Cook for 8 minutes.

In a small bowl, combine the tomato, lime juice, and soy sauce and cover the fish with the topping. Cook for 2 more minutes and serve immediately. Be careful not to overcook because the tomatoes will become mushy.

2 SERVINGS

NUTRITION INFORMATION PER SERVING:
Calories: 176
Fat: 3 g (16% calories from fat)
Protein: 30 g (67% calories from protein)
Carbohydrates (net): 5 g (8% calories from carbohydrates)
Alcohol: (9% calories from alcohol)

CAJUN SALMON FILLETS

Bell peppers are low in calories and carbohydrates, making them an excellent option for many recipes. Salmon takes on a Cajun twist in this baked alternative.

4 wild skinless salmon fillets (4 ounces each)

2 teaspoons Worcestershire sauce

1 teaspoon lemon juice

1 teaspoon Cajun seasoning

1 small green bell pepper, seeded and chopped

1 small red bell pepper, seeded and chopped

Preheat the oven to 375°F. Rinse the fish and pat dry.

Arrange the fillets in a 13" × 9" baking dish. Rub the top of each fillet with Worcestershire sauce and lemon juice, then rub with the Cajun seasoning.

Sprinkle the peppers over the fish and bake uncovered for 10 to 12 minutes or until the fish flakes easily with a fork.

4 SERVINGS

NUTRITION INFORMATION PER SERVING:
Calories: 206
Fat: 10 g (45% calories from fat)
Protein: 25 g (47% calories from protein)
Carbohydrates (net): 3 g (8% calories from carbohydrates)

SEARED BALSAMIC SCALLOPS

A small amount of flour is all that's needed to make a satisfying coating for these flavorful scallops. Balsamic reduction provides culinary flair and dense flavor.

¼ cup balsamic vinegar

¼ cup red wine vinegar

¾ pound sea scallops

Salt and pepper

¼ cup flour

1 tablespoon extra-virgin olive oil

Chives, chopped, for garnish

In a small glass or stainless steel saucepan over low heat, combine the balsamic and red wine vinegars. Simmer until the mixture reaches a syrupy consistency. Set aside.

Rinse the scallops and pat dry; halve large scallops horizontally if necessary. Season to taste with salt and pepper. Dredge scallops in flour and shake off excess (scallops should be very lightly coated). Discard the flour.

Heat the oil in a large skillet over medium-high heat. Arrange the scallops in a single layer and cook until golden brown. Turn scallops and sear the other side until golden.

Drizzle plates with balsamic syrup. Place the seared scallops on top of the syrup and sprinkle with chives.

2 SERVINGS

NUTRITION INFORMATION PER SERVING:
Calories: 310
Fat: 9 g (26% calories from fat)
Protein: 40 g (52% calories from protein)
Carbohydrates (net): 11 g (22% calories from carbohydrates)

FRENCH BAKED COD

Cod is a versatile fish, transformed here by the unique essence of tarragon. Baking it in a marinade preserves the moisture and infuses the seafood with flavor.

>   2 tablespoons rice vinegar or white wine vinegar
>
>   2 tablespoons extra-virgin olive oil
>
>   1 tablespoon Dijon mustard
>
>   ½ teaspoon dried tarragon or 1 teaspoon chopped fresh tarragon
>
>   ⅛ teaspoon pepper
>
>   4 cod fillets (6 ounces each)

Preheat the oven to 450°F.

In a large bowl, whisk together the vinegar, oil, mustard, tarragon, and pepper. Rinse the fish and pat dry. Add fish to mustard mixture and turn pieces to coat. Refrigerate for 15 minutes.

Arrange the fish, overlapping thin areas, in a shallow baking pan and drizzle with any remaining marinade. Bake, uncovered, for 7 to 10 minutes, depending on thickness, or until fish flakes easily with a fork.

4 SERVINGS

NUTRITION INFORMATION PER SERVING:
Calories: 243
Fat: 8 g (31% calories from fat)
Protein: 39 g (68% calories from protein)
Carbohydrates (net): 0.3 g (1% calories from carbohydrates)

POULTRY

OVEN-STEAMED CHICKEN AND LEEKS

Steaming food in packets shortens the cooking time, keeps the ingredients moist and flavorful, and helps maintain the nutrient content of the food. It's convenient to make these packets ahead of time to pop in the oven for an easy, healthy dinner.

    2 boneless, skinless chicken breast halves (8 ounces each)
    4 teaspoons coarse Dijon mustard
    2 leeks (white part only), cut into matchstick-size pieces
    2 medium carrots, cut into matchstick-size pieces
    Fresh thyme
    Freshly ground black pepper

Preheat the oven to 400°F.

Slice the chicken breasts in half lengthwise to make 4 equal portions. Place one piece of chicken in the center of a 1-foot length of aluminum foil. Top with 1 teaspoon of the mustard and one-quarter of the leeks and carrots. Sprinkle with thyme and pepper. Purse the foil into a packet and crimp the edges to seal. Repeat with the remaining ingredients.

Arrange the packets on a baking sheet and bake for 12 minutes, or until a thermometer inserted in the thickest portion registers 165°F.

4 SERVINGS

NUTRITION INFORMATION PER SERVING:
Calories: 205
Fat: 4 g (17% calories from fat)
Protein: 33 g (69% calories from protein)
Carbohydrates (net): 6 g (14% calories from carbohydrates)

SPICY THAI STIR-FRY

This stir-fry is equally well suited for chicken or tofu. Use a natural peanut butter or grind your own to avoid extra oils. Low-fat canned coconut milk and jars of cut lemongrass are available in most specialty food aisles.

½ cup natural peanut butter

⅔ cup low-fat coconut milk

½ cup reduced-sodium vegetable broth

2 teaspoons curry powder

1 teaspoon red-pepper flakes

1 tablespoon soy sauce

1 tablespoon extra-virgin olive oil

½ pound boneless, skinless chicken breasts or tofu, sliced

Salt and pepper

1 pound fresh green beans, cut into 2" pieces, blanched

1 cup chopped broccoli florets

2 large red bell peppers, cut into strips

1–2 cups shredded carrots

2 tablespoons finely chopped fresh ginger

4–5 cloves garlic, minced

¼ cup sliced lemongrass

1 bunch green onions, chopped

3 tablespoons grated dry coconut

In a small bowl, whisk together the peanut butter, coconut milk, vegetable broth, curry powder, pepper flakes, and soy sauce. Set aside.

Heat a wok or large nonstick skillet over medium-high heat for 1 minute, and then add the olive oil. When the oil is hot, add the chicken or tofu. Cook, stirring frequently, for 1 minute or until the chicken begins to brown. Season to taste with salt and pepper. Add the beans, broccoli, bell peppers, carrots, ginger, and garlic. Cook for 3 to 5 minutes or until the peppers just begin to soften. Add the reserved peanut butter mixture, lemongrass, onions, and coconut. Cook 1 minute longer or until heated through.

6 SERVINGS

NUTRITION INFORMATION PER SERVING:
Calories: 378
Fat: 23 g (52% calories from fat)
Protein: 31 g (33% calories from protein)
Carbohydrates (net): 11 g (15% calories from carbohydrates)

GINGER TURKEY BURGERS

Fresh ginger provides the optimum flavor because the oils are released when the dish is being made. Mushrooms add texture to the ground turkey and lend their own meatiness to these patties. These can be served over shirataki noodles or spinach greens for a low-carb delicacy.

1¼ pounds extra-lean ground turkey

½ medium onion, finely chopped

3 ounces shiitake mushrooms, finely chopped

2" piece of ginger, peeled and finely chopped

1 tablespoon minced garlic

2 teaspoons reduced-sodium soy sauce

In a large bowl, combine the turkey, onion, mushroom, ginger, garlic, and soy sauce. With clean hands, gently mix until thoroughly combined. Form into 6 patties.

Heat a nonstick grill pan over medium-high heat. Cook patties for 5 minutes on each side, or until a thermometer inserted in the center of each patty registers 165°F.

6 SERVINGS

NUTRITION INFORMATION PER SERVING:
Calories: 236
Fat: 12 g (47% calories from fat)
Protein: 26 g (47% calories from protein)
Carbohydrates (net): 4 g (6% calories from carbohydrates)

THAI CHICKEN WRAPS

Cooling spearmint takes the spicy edge off, and lightly cooking the vegetables ensures that they retain their nutritional content and crispness.

⅓ cup natural creamy peanut butter	½ teaspoon sea salt
4 tablespoons low-sodium soy sauce	½ teaspoon pepper
4 tablespoons water	1 package (12 ounces) cole slaw mix
2 tablespoons extra-virgin olive oil, divided	1 bunch Broccolini, finely chopped
3 teaspoons minced garlic, divided	1 medium red onion, sliced into thin strips
1 packet stevia	1 medium red pepper, sliced into thin strips
¾ pound boneless, skinless chicken breasts, sliced into thin strips	2 teaspoons grated fresh ginger
	8 low-carb tortillas (10–12")
	Spearmint

In a small saucepan over medium heat, combine the peanut butter, soy sauce, water, 1 tablespoon of the olive oil, 2 teaspoons of the garlic, and the stevia. Stir frequently until smooth. Reduce heat and keep warm.

Heat the remaining 1 tablespoon of olive oil in a large skillet over medium-high heat. Add the chicken and remaining 1 teaspoon of garlic and cook, stirring frequently, for 2 to 4 minutes, until the chicken is cooked through. Add the salt and pepper, stir, and place in a bowl.

Return the skillet to the heat and add the slaw mix, Broccolini, onion, pepper, and ginger. Cook, stirring frequently, for 3 minutes, or until the vegetables are crisp-tender.

To assemble wraps, spread each tortilla with 1 tablespoon peanut sauce. Top with one-eighth of the chicken and vegetables and roll up, burrito-style. Serve with fresh mint sprigs and the remaining peanut sauce on the side.

8 SERVINGS

NUTRITION INFORMATION PER SERVING:
Calories: 393
Fat: 15 g (34% calories from fat)
Protein: 26 g (26% calories from protein)
Carbohydrates (net): 17 g (40% calories from carbohydrates)

HOISIN CHICKEN

A little of this delicious marinade goes a long way. You get all of the great flavor without all of the unnecessary calories. Occasional use of bone-in chicken (remove the skin to avoid unnecessary fats) is acceptable for variety and to maximize the flavor of the poultry.

3 pounds skinless chicken breasts and drumsticks
2 cloves garlic, chopped
2 tablespoons reduced-sodium soy sauce
2 tablespoons sherry
$1\frac{1}{2}$ tablespoons hoisin sauce
$\frac{1}{4}$ teaspoon hot-pepper sauce
$\frac{1}{2}$ teaspoon salt

In a large glass container, combine the chicken, garlic, soy sauce, sherry, hoisin sauce, hot-pepper sauce, and salt. Refrigerate for a few hours or overnight.

Preheat the oven to 350°F. Place the chicken pieces and marinade on a large baking pan and cover with aluminum foil. Bake for 60 minutes or until a thermometer inserted in the thickest portion registers 165°F.

8 SERVINGS

NUTRITION INFORMATION PER SERVING:
Calories: 323
Fat: 11 g (32% calories from fat)
Protein: 50 g (66% calories from protein)
Carbohydrates (net): 2 g (2% calories from carbohydrates)

PROVENÇAL-STYLE CHICKEN

Pounding the chicken breast creates a delicate cutlet that takes little time to cook by stovetop. Keep in mind that browning foods produces acrylamide, a potent carcinogen, so be careful not to overcook.

    1 pound boneless, skinless chicken breasts
    2 tablespoons extra-virgin olive oil
    4 cloves garlic, minced
    5 tomatoes, chopped
    4 ounces crimini mushrooms, chopped
    ¼ cup chopped fresh parsley
    2 tablespoons fresh thyme
    ⅓ cup white wine

Pound the chicken breasts to ½" thickness between two pieces of waxed paper.

Heat the oil in a large skillet over medium-high heat. Add the garlic and cook for 1 minute or until fragrant. Add the chicken and sear for 1 minute on each side. Add the tomatoes, mushrooms, parsley, thyme, and wine and cook for 10 minutes longer or until chicken is cooked through.

4 SERVINGS

NUTRITION INFORMATION PER SERVING:
Calories: 313
Fat: 11 g (32% calories from fat)
Protein: 37 g (49% calories from protein)
Carbohydrates (net): 10 g (14% calories from carbohydrates)

CHICKEN IN A POT

Slow cookers are a convenient way to prepare one-pot meals without extra fat or fuss. This dish features turmeric, an exceptionally healthy curry spice being studied for its anti-inflammatory and anticancer properties.

1 pint Brussels sprouts

3 pounds boneless, skinless chicken breasts, cubed

16 baby carrots or 4 large carrots, sliced

1 cup cubed turnips or rutabaga

1 cup cubed parsnips

1 cup sliced leeks (white part only)

12 small white onions, peeled

½ teaspoon dried thyme or 2 sprigs fresh thyme

1 bay leaf

1 tablespoon grated fresh ginger

1 teaspoon turmeric

2 whole cloves

2 cups reduced-fat chicken broth or water

½ cup white wine

2 tablespoons Worcestershire sauce

In a slow cooker, combine all ingredients except the Worcestershire sauce. Cook for 4 hours on high or 8 hours on low.

Add Worcestershire sauce. *Voilà!*

6 SERVINGS

NUTRITION INFORMATION PER SERVING:

Calories: 448

Fat: 7 g (15% calories from fat)

Protein: 67 g (63% calories from protein)

Carbohydrates (net): 15 g (16% calories from carbohydrates)

VEGETARIAN

SHIRATAKI WITH SAUTÉED VEGETABLES

A noodle dish with this few calories would be impossible without shirataki. The carbohydrates in this recipe are almost exclusively low-glycemic carbs from vegetables. The extra virgin olive oil contributes healthy fats, which come to a higher percentage because the caloric level is so low. This is an excellent dish to use in a caloric restriction diet.

   1 package (8 ounces) shirataki spaghetti-style noodles
   ½ bunch asparagus, chopped
   2 teaspoons extra-virgin olive oil
   4 tablespoons sliced shallots
   2 cloves garlic, minced
   1 large tomato, chopped
   1 tablespoon chopped fresh oregano
   Ground black pepper

Rinse the shirataki noodles for 5 to 10 minutes in cold running water. Bring 2 to 3 quarts of water to a low boil. Add the shirataki and cook for 3 minutes, stirring occasionally. Drain noodles and set aside.

Steam the asparagus for 2 minutes until just slightly softened. Heat the olive oil in a large nonstick skillet over medium-high heat. Add the shallots and garlic and cook for 2 minutes, until fragrant.

Add the tomato, asparagus, and oregano. Season to taste with black pepper. Cook for 2 to 3 minutes, stirring occasionally, until tomato begins to soften. Remove from heat. Add noodles to the pan and toss to combine.

2 SERVINGS

NUTRITION INFORMATION PER SERVING:
Calories: 115
Fat: 5 g (42% calories from fat)
Protein: 5 g (11% calories from protein)
Carbohydrates (net): 10 g (47% calories from carbohydrates)

STEEPED TOFU SLICES

Tofu is a well-balanced source of protein for vegetarians as soy contains all eight essential amino acids. Each ounce of tofu contains about 8 milligrams of soy isoflavones, which are thought to help prevent heart disease and lower cholesterol.

2 tablespoons soy sauce

2 tablespoons vegetable broth

½ tablespoon chopped green onion tops

1 teaspoon finely chopped lemongrass

1 teaspoon minced garlic

½ tablespoon extra-virgin olive oil

Pinch freshly ground black pepper

Pinch stevia

1 package (10.5 ounces) extra-firm lite tofu, cut into 10 pieces

In a shallow dish, combine the soy sauce, broth, onion, lemongrass, garlic, and oil. Season to taste with black pepper and stevia. Arrange tofu slices in soy sauce mixture. Cover and refrigerate overnight. Can be served using marinade as a sauce.

10 SERVINGS (1 OUNCE EACH)

NUTRITION INFORMATION PER SERVING:
Calories: 20
Fat: 1 g (40% calories from fat)
Protein: 2 g (45% calories from protein)
Carbohydrates (net): 1 g (15% calories from carbohydrates)

VEGETARIAN STUFFED TOMATOES

Tomatoes, a good source of lycopene and vitamins, stuffed with a healthy grain or legume, reemerge as a hearty side dish. Lycopene is fat-soluble and therefore is more readily absorbed when consumed with high-fat foods such as olive oil or cheese.

    4 large ripe tomatoes
    ½ recipe Sautéed Quinoa Pilaf (page 290)
    2 tablespoons extra-virgin olive oil, divided
    2 tablespoons grated pecorino cheese

Preheat the oven to 375°F. Core the tomatoes and remove the seeded centers. Stuff each tomato with 1 scoop of quinoa pilaf.

Apply 1 tablespoon of the oil to a metal baking sheet. Place the tomatoes on the sheet and drizzle with the remaining 1 tablespoon of oil. Sprinkle with the cheese. Roast for 20 to 25 minutes, until browned.

4 SERVINGS

NUTRITION INFORMATION PER SERVING:
Calories: 250
Fat: 14 g (48% calories from fat)
Protein: 9 g (12% calories from protein)
Carbohydrates (net): 20 g (40% calories from carbohydrates)

SAUTÉED QUINOA PILAF

Quinoa is a gluten-free ancient whole grain that can be used in moderation. Vegetarians will benefit from its complete protein (it contains all 8 essential amino acids). It has a pleasant nutty flavor and is a good source of magnesium, which helps relax blood vessels and minimizes the occurrence of headaches, as well as promotes a healthy heart.

    1 cup quinoa, rinsed and drained
    2 teaspoons extra-virgin olive oil
    2 cloves garlic
    1 red bell pepper, finely chopped
    2 baby bok choy, chopped (about 4 cups)
    ½ cup slivered almonds (skins on)
    Sea salt and black pepper
    1 bunch scallions, finely chopped

Bring 2 cups of water to a boil in a medium saucepan over medium-high heat. Add the quinoa. Cover and simmer for 15 to 20 minutes, until the liquid is absorbed.

Heat the olive oil in a wok or large frying pan on medium-high heat. Use the side of a chef's knife to smash the garlic cloves on a cutting board before adding them to the pan to flavor the oil. Add the bell pepper and cook for 2 to 3 minutes, stirring constantly, until it begins to soften. Add the bok choy and almonds. Cook for 1 minute longer or until the bok choy brightens in color. Add the quinoa and stir to combine. Season to taste with salt and pepper and add the scallions just before serving.

8 SERVINGS (1 CUP EACH)

NUTRITION INFORMATION PER SERVING:
Calories: 147
Fat: 6 g (35% calories from fat)
Protein: 6 g (15% calories from protein)
Carbohydrates (net): 15 g (50% calories from carbohydrates)

SIDES AND SNACKS

BRUSSELS SAUTÉ

Brussels sprouts, cruciferous vegetables like the cabbages they resemble, contain a wealth of vitamins A and C, as well as plenty of beta-carotene. Best of all, research indicates that they can help prevent colon cancer. For a seasonal adaptation, try substituting parboiled winter squash for the summer squash.

>   2 teaspoons extra-virgin olive oil
>   4 cloves garlic, minced
>   ½ pound fresh Brussels sprouts, trimmed and halved
>   1 large yellow summer squash, cut into ½" slices
>   1 large tomato, chopped

Heat the oil in a large nonstick pan over medium-high heat. Add the garlic and cook for 1 minute or until fragrant. Add the Brussels sprouts and cook for 2 to 3 minutes until they begin to soften. Add the squash and cook for 2 to 3 minutes longer or until it begins to soften. Add the tomato and cook for 1 to 2 minutes or until the Brussels sprouts and squash are tender and the tomato is lightly softened.

4 SERVINGS

NUTRITION INFORMATION PER SERVING:
Calories: 72
Fat: 3 g (41% calories from fat)
Protein: 3 g (10% calories from protein)
Carbohydrates (net): 7 g (49% calories from carbohydrates)

LEMONY CARAWAY CARROTS

Given this short list of ingredients, the big flavor may surprise you. The acidity of lemon balances the naturally sweet carrots and caraway seeds.

2 cups baby carrots (about 12 ounces)

1 tablespoon caraway seeds

Juice from 1 lemon (1–2 tablespoons)

1 teaspoon lemon zest

Place the carrots in a pot and add water to cover. Add the caraway seeds and cook over medium-high heat for 10 minutes, or until carrots are crisp-tender.

Drain well and return to the pot (some seeds will stay on the carrots, but you can add back as many as desired from the strainer). Add the lemon juice and zest and toss to coat.

4 SERVINGS

NUTRITION INFORMATION PER SERVING:

Calories: 32

Fat: 1 g (3% calories from fat)

Protein: 1 g (5% calories from protein)

Carbohydrates (net): 6 g (92% calories from carbohydrates)

GARLICKY KALE

Kale is loaded with vitamin K, which helps protect the body from cancer. If you are on blood thinners, it is important to try to maintain a stable vitamin K intake because changes in consumption affect the results of your dosage. Tangy lemon brightens the earthiness of the kale and naturally helps minimize the browning (oxidation) of the leaves.

2 tablespoons extra-virgin olive oil

3–4 cloves garlic, minced

1 bunch kale, stemmed and coarsely chopped

Juice of 1 lemon

$\frac{1}{4}$ cup water

Heat the oil in a large nonstick skillet over medium-high heat. Add the garlic and cook for 1 minute or until fragrant. Add the kale and cook, stirring constantly, for 2 minutes, tossing as it wilts. Stir in the lemon juice. Add the water and cover. Cook for 2 to 3 minutes until softened. Do not overcook (the kale should remain dark green).

2 SERVINGS

NUTRITION INFORMATION PER SERVING:

Calories: 207

Fat: 15 g (62% calories from fat)

Protein: 5 g (6% calories from protein)

Carbohydrates (net): 14 g (32% calories from carbohydrates)

BASIC JAPANESE PICKLES

Pickling is an easy way to preserve raw foods and a healthy way to include vinegar in your diet. Research shows that vinegar significantly reduces the rise in blood sugar and insulin after eating, and we recommend eating pickled vegetables with meals to take advantage of this effect. You can pickle a variety of firm vegetables such as cucumbers, cabbage, and celery using this simple method. Add a little miso to create a more complex flavor.

> 1 daikon radish, sliced into $\frac{1}{2}$"-thick pieces
> $\frac{1}{2}$ medium carrot, shredded
> 1 piece fresh ginger (about 2"), peeled and sliced
> $\frac{1}{2}$ teaspoon table salt
> $\frac{1}{4}$ cup rice vinegar
> 1 teaspoon soy sauce
> Stevia to taste (1 packet)
> $\frac{1}{4}$ teaspoon pickling salt (no iodine)

Sprinkle the radish, carrot, and ginger with table salt and let stand for 20 minutes. Squeeze out the excess water.

In a bowl, combine the vinegar, soy sauce, stevia, and pickling salt and mix well. Add the vegetables and toss to mix. Cover and refrigerate for at least 8 hours. In a tightly sealed container, these pickles can be stored for 1 to 2 weeks.

2 SERVINGS

NUTRITION INFORMATION PER SERVING:
Calories: 48
Fat: 0 g (4% calories from fat)
Protein: 1 g (9% calories from protein)
Carbohydrates (net): 7 g (87% calories from carbohydrates)

ROASTED VEGETABLES WITH THYME

The musky flavor of thyme and nutmeg complement these wonderful roasted vegetables. For variety, experiment with different types of squash, and try adding some jalapeños for an extra kick. These veggies are also suitable to assemble on skewers and grill.

  2 cups chopped butternut squash

  2 teaspoons extra-virgin olive oil

  2 tablespoons cider vinegar

  2 tablespoons lemon juice

  1 teaspoon dried thyme

  1 teaspoon garlic powder

  $\frac{1}{2}$ teaspoon ground black pepper

  $\frac{1}{8}$ teaspoon ground nutmeg

  Kosher salt

  4 large plum tomatoes, chopped

  2 bell peppers, chopped

  1 large onion, chopped

Preheat the oven to 400°F.

Place the squash chunks in a microwaveable bowl and add 1 to 2 teaspoons water. Cover and cook on high for 2 to 3 minutes or until very slightly softened.

Meanwhile, in a large bowl, combine the oil, vinegar, lemon juice, thyme, garlic powder, black pepper, nutmeg, and salt to taste. Add the squash, tomatoes, bell peppers, and onion. Toss gently to coat and let sit for at least 10 minutes.

Arrange the vegetable mixture in a metal roasting pan and bake in the lower third of the oven for 10 to 15 minutes or until browned and tender.

6 SERVINGS

NUTRITION INFORMATION PER SERVING:

Calories: 77

Fat: 3 g (29% calories from fat)

Protein: 2 g (6% calories from protein)

Carbohydrates (net): 12 g (65% calories from carbohydrates)

CAULIFLOWER WITH INDIAN SPICES

Turmeric and other fragrant Indian spices add kick to cauliflower—a healthy low-calorie cruciferous vegetable. Curcumin, an important component of turmeric, is a potent anti-inflammatory that combats cancer, Alzheimer's, and other conditions.

2 tablespoons extra-virgin olive oil, divided

1 small jalapeño chile pepper, finely chopped (discard seeds if you don't want too much heat)

1 piece fresh ginger (2"), peeled and finely chopped

2 teaspoons ground cumin

2 teaspoons ground coriander

½ teaspoon ground turmeric

½ teaspoon sea salt

2 small red-skinned potatoes, chopped into ¾" pieces

1 head cauliflower, cut into florets

⅔ cup water, plus more if needed

½ cup frozen peas

Heat 1 tablespoon of the oil in a large nonstick skillet over medium-high heat. Add the pepper, ginger, cumin, and coriander. Cook for 15 to 30 seconds, stirring constantly, until fragrant. Add the turmeric, salt, potatoes, and cauliflower. Toss to coat. Add the water, cover, and reduce the heat to a simmer.

Let the vegetable mixture simmer for 15 to 20 minutes until the cauliflower and potatoes are tender. (Add a little more water if the pan begins to dry out.)

Uncover and increase the heat to medium-high. Add the peas and the remaining 1 table-spoon of oil. Cook for 5 minutes or until the remaining liquid is gone and the vegetables are slightly browned.

6 SERVINGS

NUTRITION INFORMATION PER SERVING:
Calories: 126
Fat: 5 g (38% calories from fat)
Protein: 4 g (9% calories from protein)
Carbohydrates (net): 13 g (53% calories from carbohydrates)

HERBED ZUCCHINI

Steaming is the optimal method for cooking vegetables. You maintain crisp texture while preserving nutrient content. Oregano has the highest antioxidant content of any herb—a tablespoon of fresh oregano has the same antioxidant activity as an entire apple!

2 medium zucchini, sliced into $\frac{1}{2}$" pieces

1 tablespoon chopped fresh oregano

2 teaspoons extra-virgin olive oil

$\frac{1}{2}$ teaspoon garlic powder

Freshly ground black pepper

Pinch of sea salt

Bring 1 to 2" of water to a boil in a pan with a tight-fitting lid. Arrange the zucchini in a steamer basket and place over the boiling water. Cover and steam for 3 to 5 minutes until tender. (Do not overcook.)

Meanwhile, in a large bowl, combine the oregano, oil, garlic powder, pepper to taste, and salt. Transfer the zucchini to the bowl and toss to coat.

2 SERVINGS

NUTRITION INFORMATION PER SERVING:
Calories: 72
Fat: 5 g (57% calories from fat)
Protein: 1 g (4% calories from protein)
Carbohydrates (net): 5 g (39% calories from carbohydrates)

ROASTED RED PEPPER HUMMUS

Roasted red peppers are conveniently available precut in jars, but choose those packed in water over oil. Garbanzo beans (chickpeas) are full of fiber and contain molybdenum, which assists the body's detoxification process.

2 cans (16 ounces each) chickpeas, reserve ¼ cup liquid

1 cup roasted red peppers

Juice of ½ lemon

1½ tablespoons tahini

2 cloves garlic, crushed

½ teaspoon sea salt

2 tablespoons extra-virgin olive oil

In a blender or food processor, combine all ingredients except ¼ cup liquid. Pulse until smooth and blended, adding liquid through the feed tube until the ingredients reach the desired consistency.

Serve with your favorite raw veggies.

12 SERVINGS

NUTRITION INFORMATION PER SERVING:
Calories: 123
Fat: 4 g (28% calories from fat)
Protein: 4 g (12% calories from protein)
Carbohydrates (net): 14 g (60% calories from carbohydrates)

DESSERT

FROZEN FRUIT

Banana lends this fat-free dessert a sweet creaminess. Garnish with frozen grapes, berries, or pineapple to complete the presentation. For a variation, try ½ cup fruit juice (such as apple juice) or ½ cup of skim milk instead of the pineapple slices.

    1 large banana, peeled and sliced
    ¾ cup raspberries
    ½ bunch fresh spearmint
    1 cup sliced pineapple

Put the banana slices and raspberries in the freezer for at least 6 hours. Blend the frozen banana pieces, frozen raspberries, mint leaves, and cold (but not frozen) pineapple slices in a food processor until smooth.

4 SERVINGS

NUTRITION INFORMATION PER SERVING:
Calories: 65
Fat: 0 g (5% calories from fat)
Protein: 1 g (5% calories from protein)
Carbohydrates (net): 13 g (90% calories from carbohydrates)

COOL MELON SALAD

This cool and refreshing dessert features a splash of lime to keep the fruit crisp and fresh—lemon or orange juice would also work well. Garnish with a little plain soy yogurt for a creamy accent.

¼ medium honeydew melon, cut into chunks
¼ medium cantaloupe, cut into chunks
1 cup cubed watermelon
Juice of 1 lime
¼ teaspoon lime zest
Spearmint leaves

In a bowl, combine the honeydew, cantaloupe, and watermelon. Drizzle with the lime juice and zest and toss to coat. Garnish with spearmint leaves.

4 SERVINGS

NUTRITION INFORMATION PER SERVING:
Calories: 52
Fat: 0.5 g (4% calories from fat)
Protein: 1 g (7% calories from protein)
Carbohydrates (net): 13 g (89% calories from carbohydrates)

WHIPPED STRAWBERRY-BANANA MOUSSE

This mousse is creamy and light, but it falls quickly and should be served immediately for best results. Try experimenting with other fruits. If you do not use a high-water-content fruit (such as strawberries), then you will need to add some liquid (such as water, fruit juice, or skim milk).

1 banana, sliced

8 ounces fresh strawberries, hulled

4 ice cubes, crushed

4 tablespoons nonfat milk powder

2 teaspoons cocoa (optional)

In a food processor, combine the banana, strawberries, ice, milk powder, and cocoa (if using) and process for a few minutes until smooth.

Serve immediately.

4 SERVINGS

NUTRITION INFORMATION PER SERVING:
Calories: 72
Fat: 0 g (5% calories from fat)
Protein: 4 g (19% calories from protein)
Carbohydrates (net): 13 g (76% calories from carbohydrates)

DRINKS

Tips for all smoothies:

- Add wheat grass extract or other herbal extracts.

- Try using frozen fruit for a cooler, chunkier consistency.

- Add ice to any of the recipes for a frozen smoothie.

- Use your favorite fruits or toasted nuts to create your own recipe.

POME-GRAPEFRUIT SHAKE

Pomegranates are extremely high in antioxidants and boost the nutritional value of this shake. Note that grapefruit should not be consumed by individuals on statin drugs.

1 cup water
½ grapefruit, peeled, trimmed to remove white membrane, seeded, and cut into sections
2 tablespoons pomegranate juice
1 packet (2 scoops) Ray & Terry's™ Supreme Berry Nutritional Shake Mix

In a blender, combine the water, grapefruit, and pomegranate juice and blend. During mixing, add the shake mix and blend to desired consistency.

1 SERVING

NUTRITION INFORMATION PER SERVING:
Calories: 189
Fat: 3 g (15% calories from fat)
Protein: 23 g (48% calories from protein)
Carbohydrates (net): 16 g (37% calories from carbohydrates)

TRIPLE BERRY COOLER

Berries are a good fruit choice due to their nutrient density, low calorie levels, and high antioxidant levels.

> 1 cup unsweetened soy milk
> 1 packet (2 scoops) Ray & Terry's™ Supreme Berry Nutritional Shake Mix
> 1 cup ice
> $\frac{1}{2}$ cup fresh blueberries
> $\frac{1}{2}$ cup fresh strawberries
> $\frac{1}{2}$ cup fresh raspberries

In a blender, combine the soy milk, shake mix, and ice. Pulse to crush ice. Add the berries and blend to desired consistency. Pour and enjoy.

1 SERVING

NUTRITION INFORMATION PER SERVING:
Calories: 319
Fat: 9 g (25% calories from fat)
Protein: 33 g (41% calories from protein)
Carbohydrates (net): 21 g (34% calories from carbohydrates)

CHOCOLATE AVOCADO SHAKE

Using unsweetened soy milk instead of water lends an extra richness to this densely creamy shake.

1 cup water
½ avocado
⅛ cup walnuts
1 packet (2 scoops) Ray & Terry's™ Supreme Chocolate Nutritional Shake Mix

In a blender, combine the water, avocado, and walnuts and start blending. During mixing, add the shake mix. As all the ingredients blend together, you may need to add more water—this shake can get thick.

1 SERVING

NUTRITION INFORMATION PER SERVING:
Calories: 330
Fat: 24 g (63% calories from fat)
Protein: 22 g (25% calories from protein)
Carbohydrates (net): 5 g (12% calories from carbohydrates)

ANTIOXIDANT BLAST

Blueberries contain phytonutrients called anthocyanidins, which have strong antioxidant properties and support healthy veins and capillaries. They may also help prevent certain cancers, protect vision, and maintain a healthy brain.

> 1 cup unsweetened soy milk
> 1 packet (2 scoops) Ray & Terry's™ Supreme Chocolate Nutritional Shake Mix
> 1 cup ice
> 1 cup fresh blueberries

In a blender, combine the soy milk, shake mix, and ice. Pulse to crush ice. Add the berries and blend completely to desired consistency.

1 SERVING

NUTRITION INFORMATION PER SERVING:
Calories: 291
Fat: 9 g (29% calories from fat)
Protein: 28 g (38% calories from protein)
Carbohydrates (net): 21 g (33% calories from carbohydrates)

PATRIOTIC SMOOTHIE

Only unsweetened soy milk should be used because regular soy milk can contain a lot of sugar. To sweeten, simply add a little stevia.

8 ounces unsweetened soy milk
1 packet (2 scoops) Ray & Terry's™ Supreme Vanilla Nutritional Shake Mix
½ cup fresh raspberries
½ cup fresh blueberries

In a blender, combine the soy milk and shake mix. Blend slightly. Add the fruit and blend completely until the shake is frothy and the fruit is desired consistency. Pour into a tall glass. Garnish with fruit and enjoy!

1 SERVING

NUTRITION INFORMATION PER SERVING:
Calories: 293
Fat: 9 g (26% calories from fat)
Protein: 32 g (44% calories from protein)
Carbohydrates (net): 16 g (30% calories from carbohydrates)

CHOCOLATE MINT FROSTY

An excellent alternative to the tub of mint chocolate chip, this refreshing beverage may be enjoyed at breakfast or dessert. Any of our shakes can be blended with ice for a frosty treat.

> 1 packet (2 scoops) Ray & Terry's™ Supreme Chocolate Nutritional Shake Mix
> Handful of fresh mint leaves, torn into small pieces
> 8 ounces unsweetened soy milk
> ½ cup ice

Combine all the ingredients in a blender. Mix completely until the shake is frothy and the ice is crushed.

1 SERVING

NUTRITION INFORMATION PER SERVING:
Calories: 217
Fat: 9 g (37% calories from fat)
Protein: 28 g (51% calories from protein)
Carbohydrates (net): 4 g (12% calories from carbohydrates)

SMOOTHIE AND FRUITY

Chock full of potassium, bananas help maintain healthy blood pressure levels and lend a creaminess to this smoothie.

  8 ounces unsweetened soy milk
  1 packet (2 scoops) Ray & Terry's™ Supreme Berry Nutritional Shake Mix
  1 cup ice
  ½ medium banana
  ½ cup fresh blueberries

In a blender, combine the soy milk, shake mix, and ice. Blend slightly to crush the ice. Add the fruit and blend completely until the shake is frothy and the fruit is desired consistency.

1 SERVING

NUTRITION INFORMATION PER SERVING:
Calories: 316
Fat: 8 g (24% calories from fat)
Protein: 32 g (41% calories from protein)
Carbohydrates (net): 25 g (35% calories from carbohydrates)

PEACH DELIGHT SMOOTHIE

Commercially farmed stone fruits, such as peaches and plums, can be high in toxins, so make sure to use organic fruit for this simple summery shake.

    8 ounces unsweetened soy milk
    1 packet (2 scoops) Ray & Terry's™ Supreme Vanilla Nutritional Shake Mix
    1 organic peach, pitted and chopped (including skin)

In a blender, combine the soy milk and shake mix. Blend slightly. Add the peach chunks and blend completely to desired consistency.

1 SERVING

NUTRITION INFORMATION PER SERVING:
Calories: 278
Fat: 8 g (27% calories from fat)
Protein: 32 g (46% calories from protein)
Carbohydrates (net): 17 g (27% calories from carbohydrates)

CHOCOLATE CHERRY SMOOTHIE

Treat yourself to a protein boost as delightful as a bowlful of chocolate-covered cherries.

    1 packet (2 scoops) Ray & Terry's™ Supreme Chocolate Nutritional Shake Mix
    1 cup fresh pitted cherries
    8 ounces unsweetened soy milk

Combine all of the ingredients in a blender. Mix to desired consistency.

1 SERVING

NUTRITION INFORMATION PER SERVING:

Calories: 304
Fat: 9 g (28% calories from fat)
Protein: 29 g (37% calories from protein)
Carbohydrates (net): 26 g (35% calories from carbohydrates)

PIÑA COLADA SHAKE

This healthy twist on a healthy drink delights the taste buds and is easy on the waistline.

    1 cup water
    ½ cup fresh pineapple chunks
    1 inch of banana (or try 1 ounce coconut juice in place of banana)
    1 packet (2 scoops) Ray & Terry's™ Supreme Vanilla Nutritional Shake Mix

In a blender, combine the water, pineapple, and banana (or coconut juice) and blend. During mixing, add the shake mix.

All ingredients must be fresh. If using coconut juice, draining it directly from the coconut will yield best results.

1 SERVING

NUTRITION INFORMATION PER SERVING:
Calories: 184
Fat: 3 g (15% calories from fat)
Protein: 23 g (49% calories from protein)
Carbohydrates (net): 15 g (36% calories from carbohydrates)

COOL AS A CUCUMBER SHAKE

Cucumber pairs well with mint and soothes the palate in this frosty drink.

³⁄₄ cup cold water
¹⁄₂ cucumber, peeled
¹⁄₂ bunch fresh mint
1 packet (2 scoops) Ray & Terry's™ Supreme Vanilla Nutritional Shake Mix

In a blender, combine the water, cucumber, and mint and blend. During mixing, add the shake mix. For best results, use very cold water and blend extra well.

1 SERVING

NUTRITION INFORMATION PER SERVING:

Calories: 161
Fat: 3 g (20% calories from fat)
Protein: 24 g (58% calories from protein)
Carbohydrates (net): 5 g (22% calories from carbohydrates)

12

SUPPLEMENTS

"This has no vitamins or minerals but it's guaranteed not to cause cancer."

We feel that nutritional supplementation can serve as one of our key *TRANSCEND* strategies to "reprogram our biochemistry." It can help update and repair some of the glitches in our human genome version 1.0 that was "written" tens of thousands of years ago while we await biotechnology version 2.0 and nanotechnology update 3.0 slated for release about 15 and 25 years from now.

Like many people, you may feel a bit bewildered at times by the glut of conflicting information about nutritional supplements. Depending on the source—health providers, government agencies, supplement manufacturers—recommendations can vary widely regarding which supplements you should take and at what dosages, or whether you need to take any supplements at all. Even the nutritional guidelines developed by the U.S. Institute of Medicine can leave the average person scratching his head. Called the dietary reference intakes (DRI), these guidelines are broken into four subgroups: estimated average requirement (EAR), recommended dietary allowance (RDA), adequate intake (AI), and tolerable upper intake level (UL). In this chapter, we'll help you sort through the surplus of information and determine the right course of supplementation for you.

VITAMINS CAN DO MORE THAN PREVENT SCURVY AND RICKETS

While it has long been known that lack of certain nutrients leads to disease (for example, insufficient vitamin C causes scurvy, insufficient vitamin D causes rickets), until only recently, recommended levels of nutrients in the diet were set at the minimum required to prevent such deficiency diseases. And common wisdom had long held that a "balanced diet" was all that was needed to obtain all the nutrients required to be healthy. We now know that to be far from the truth.

In part, this stems from the fact that our food is not as nutrient rich as it used to be. Once-fertile soil has been stripped of essential nutrients through decades of intensive farming. Fruits, vegetables, and grains are transported thousands of miles and stored for months, and sometimes years. And the way vast quantities of our food is processed and prepared removes additional nutrients. As a result, even in developed countries with abundant food supplies, much of the population is not getting all the nutrients they need.

As research related to nutrition and disease continues to expand, so too does new evidence supporting the need for nutritional supplementation. For example, a study published in the journal *Nature Reviews Cancer* in 2002 found that deficiencies of vitamins C, B_6, and B_{12}; folic acid; iron; and zinc can lead to DNA damage and cause cancer. A 2003 study commissioned by Wyeth Consumer Healthcare found that a simple daily multivitamin, if taken by all Americans over age 65, would save Medicare an estimated $1.6 billion over 5 years by improving immune function and reducing risk of coronary artery disease. Nutritional supplementation has been found to improve memory, lower cholesterol levels, prevent prostate problems, relieve symptoms of menopause, reduce inflammation, and lower risk of cataracts.

Several recent studies have found beneficial effects from specific nutritional supplements on various disease processes.

• A study from the Netherlands of 4,400 people older than 55 years of age found that regular supplementation with beta-carotene slashed heart attack risk 45 percent over a 4-year period.

• A study of 11,000 elderly people between 67 and 105 years of age who participated in the Established Populations for Epidemiologic Studies of the Elderly found that supplementation with vitamin E decreased total mortality 34 percent and deaths from heart disease 47 percent.

• Supplementation with calcium and vitamin D can help prevent the bone loss of osteoporosis, and it has been estimated that more than 130,000 hip fractures could be prevented each year if everyone over the age of 50 consumed at least 1,200 milligrams of supplemental calcium per day.

• A study published in 2004 in the *Journal of the National Cancer Institute* of 1,000 men showed that those with higher blood levels of the mineral selenium had half the risk of developing advanced prostate cancer over a 13-year period of observation. As a result, the National Cancer Institute has enrolled 35,000 men over age 55 to participate in the follow-up Selenium and Vitamin E Cancer Prevention Trial (SELECT), which is ongoing.

• Many anecdotal reports have appeared in the medical literature suggesting that high-dose vitamin C given intravenously can help treat many different types of cancer. A recent animal experiment showed that cancer cells were killed by high-dose vitamin C treatments, and the first clinical trial of vitamin C in the treatment of human cancer patients is underway under the auspices of the Cancer Treatment Centers of America. Terry is involved in another research study involving the use of high-dose vitamin C as a possible treatment for hepatitis C thanks to a grant made possible by the Adolph Coors Foundation.

Yet, it seems that positive studies receive less media attention than studies suggesting hazards of vitamins and nutritional supplements. Even worse, in the cases we've seen, the questionable findings were often based on poorly designed or biased studies. For example, a highly publicized 2005 study in the *Annals of Internal Medicine* attacking use of vitamin E reported a slight increase in the risk of death from supplementing with vitamin E. But there were many serious methodological problems with the study, the most important being that the study was not about vitamin E at all; rather, it was about alpha-tocopherol, which is only one component of vitamin E. Vitamin E consists of eight components: four tocopherols (alpha, beta, gamma, and delta) and four tocotrienols (alpha, beta, gamma, and delta), yet this study addressed only one component, alpha-tocopherol. Many commercial brands of supplements that are labeled "vitamin E" are in fact just alpha-tocopherol, which is not what we recommend you take. The most important fraction in vitamin E, which contributes to vitamin E's well-known antioxidant benefits, is gamma-tocopherol, which constitutes about 70 percent of the vitamin E found naturally in food. Supplementing with just alpha-tocopherol actually depletes the body of gamma-tocopherol. Studies of vitamin E products that contain all of the tocopherols together show substantial health benefits. We recommend, therefore, that you take a blended form of vitamin E that contains all of the vitamin E components, especially the four tocopherols.

Another well-publicized negative study appeared in the February 28, 2007, issue of the *Journal of the American Medical Association* (*JAMA*) ("Mortality in Randomized Trials of Antioxidant Supplements for Primary and Secondary Prevention"), which denounced antioxidants in general. There were also numerous serious problems with this study. The study again used alpha-tocopherol for vitamin E rather than mixed tocopherols. For vitamin A they selected a strange study incorporating a single dose, which is not the recommended way to take vitamin A. Of 815 studies on these supplements that the authors of this study could have used, only 68 were included. Reviews of the selection of studies that were included showed substantial bias—well-designed major studies showing substantial benefits were excluded. For example, a study of 29,000 male smokers followed for 19 years that was excluded showed a 28 percent reduction in mortality for individuals with the highest levels of vitamin E compared with those who had the lowest levels. Another excluded study of over 3,000 people followed for

6 years showed significant reductions in mortality from heart disease and colorectal cancer in those with higher levels of vitamin A. Dozens of other excluded studies demonstrated substantial benefits from these supplements. Furthermore, this *JAMA* study looked only at vitamin A and C, alpha-tocopherol, and selenium, and ignored numerous other antioxidants that have antioxidant benefits and that we recommend in this book.

Below we outline a basic supplement program that we feel will be helpful to most people. As in all of medicine and technology, there is still much to learn, but these recommendations represent the cutting edge of what we know today.

FREE RADICALS VS. THE ANTIOXIDANTS

Not quite an NFL clash of titans, but this is a critical contest nonetheless. In fact, your life depends on it. When gasoline combines with oxygen in the cylinders of your automobile engine to provide power to the wheels, by-products are expelled through the exhaust pipe. In similar fashion, free radicals are by-products created when nutrients are oxidized in your cells to produce the energy you need to survive. You can also produce free radicals from experimental exposures, particularly if you are exposed to high levels of toxins. Regardless of the source, free radicals play a significant role in aging your body's systems and organs.

Compared with stable molecules, which have a set number of paired electrons, free radicals are missing an electron in their outer shell. Stability is the preferred state for molecules, so, to become stable again, a free radical steals an electron from another molecule to replace the one it is missing. But that leaves the donor molecule missing an electron, meaning it's now turned into a free radical itself, and it then needs to steal an electron from another molecule. All this petty theft can damage critical cell structures, hindering their ability to function properly or worse. For example, when a DNA molecule containing your genetic code is damaged in this way, genetic mutations can get passed on as the cell replicates itself; this can lead to cancer.

If production of free radicals were to continue unchallenged, they would pile up in your body and eventually kill you. That's where antioxidants come in. Also known as free-radical scavengers, antioxidants neutralize free radicals by giving up electrons to them so they don't steal from the crucial molecules in your cells. Antioxidants come from two sources, those you eat and those your body creates.

Antioxidants contained in food and nutritional supplements include vitamins A, C, and E, and the mineral selenium. Also on this list are vitamins B_2, B_3, and B_6, as well as alpha-lipoic acid, grape seed extract, coenzyme Q_{10}, and others. Diet alone cannot supply enough of these antioxidants to keep free radicals under control, especially as you age, so supplementation is required.

Your cells also produce free-radical scavengers known as *antioxidant enzymes*. Generally speaking, enzymes are proteins created to bring about chemical reactions that result in the formation of other substances your cells need to function. Your body produces thousands of different enzymes, each structured to perform a specific task, for example, the enzymes that aid in the digestion of fats, sugars, or proteins. To carry out their functions, however, many of these enzymes require vitamin or mineral cofactors; that is, these enzymes must bind to a specific vitamin or mineral or they can't do their job. Taking vitamin and mineral supplements ensures that enough of these cofactors are present in your body to meet the needs of your enzymes, including the antioxidant enzymes needed to battle those free radicals.

You may ask why you shouldn't just take antioxidant enzymes in supplement form. The problem with that approach is that antioxidant enzymes are poorly absorbed by the digestive system. The best way to be sure your antioxidant enzymes are functioning at peak capacity is to be sure sufficient quantities of vitamin and mineral cofactors are on hand by taking them as nutritional supplements.

GENETIC VARIATIONS

Unfortunately, it's not quite that simple (it's rarely simple when it comes to the human body). Researchers have discovered that genetic abnormalities can lead to production of enzyme molecules that are unable to bind correctly with their requisite cofactors because the structure of the connection sites just doesn't match up properly. In fact, a study published in the *American Journal of Clinical Nutrition* in 2002 estimates that as many as one-third of currently identified genetic variations result in defective enzymes. The authors also found that more than 50 diseases that involve these types of genetic abnormalities can be overcome when sufficiently large quantities of vitamin and mineral cofactors are available to stimulate the defective enzymes. But this requires supplementation, sometimes at hundreds of times the RDA amounts.

Genetic mutations aren't rare at all, and each of us has about a million of them. While most are not of major concern (in fact, gene variation is how we evolve), those that lead to defective enzymes can have serious consequences. Fortunately, genomic tests are available to identify a number of malformed enzymes, thus allowing you to determine which nutrients you may need to take in large doses to more effectively activate those enzymes.

SUPPLEMENT RECOMMENDATIONS

Results of scientific research often take time to move from laboratory to practice, and when it comes to nutritional supplementation, new discoveries can take quite some time indeed. It wasn't until 2001 that the *New England Journal of Medicine* suggested that most adults could benefit from a daily multivitamin supplement at the RDA level, with higher levels of folic acid, vitamins B_6, B_{12}, and D for those at risk of cardiovascular disease and vitamin D and calcium for bone loss. For the majority of physicians who had been taught that supplements were a waste of their patients' money, and that all supplements did was to create "expensive urine," this was a sea change. Meanwhile, dramatic new research continues to improve understanding of how vitamins and antioxidants work in the body, with numerous articles on the subject being published in scientific journals each year supporting nutritional supplementation to prevent specific diseases and promote optimal overall health.

But what do *you* need to take, and how much is enough? It depends on who you are, your genetic makeup, and your specific environment. It is clear, however, that you often need much more than RDA amounts as defined by the Institute of Medicine. For one thing, the RDAs don't take into account any variations you may have in your genetic code. And, as the government guidelines state, the RDA is based on "the average daily nutrient intake level sufficient to meet the nutrient requirement of nearly all (97 to 98 percent) *healthy* individuals in a particular life stage and gender group." That assumes, however, that the average American is "healthy," and clearly many of us are not. Do the two out of three Americans who are overweight or the one in four who are obese count as healthy? How about the tens of millions who are receiving medication for high blood pressure, diabetes, heartburn, head-aches, allergies, arthritis, depression, and erectile dysfunction? If these individuals enjoyed better nutrition thanks to eating better and taking nutritional supplements, perhaps they wouldn't need to take so much medicine and *then* could be counted as healthy. But this is not the case.

Therefore, rather than focusing on RDA amounts, we prefer to look at what we refer to as optimal nutritional allowances (ONAs), meaning dosages designed to optimize your health based on who you are, not merely to prevent deficiency diseases. Individual requirements necessarily vary because of differences in age, sex, genetic anomalies, and lifestyle, so our ONA recommendations cover a broad dosage range. As you try to decide what you should take, we suggest you start at the lower end of these recommendations and increase your dosages only if needed unless you know you have a specific health condition that calls for larger dosages.

THE THREE BASIC SUPPLEMENTS— RECOMMENDED FOR EVERYONE

There are three *universal* supplements recommended for almost everyone over the age of 30. Remember that's as long as we've been granted optimal health by our Stone Age genes. Up until the age of 30, it's pretty much a free ride, but after that, our caveman and cavewoman bodies are programmed to begin their downward slide and need additional help. Supplementation can play a vital role in helping overcome our outdated genetic programming. The three supplements that we feel should be taken by almost everyone over 30 are a **multiple vitamin/mineral formulation, fish oil, and extra vitamin D**.

Daily Multiple Vitamin/Mineral

A multiple vitamin/mineral formulation ("daily multiple") is necessary to ensure that the raw materials are always available to help your youth-sustaining enzymes function optimally. Numerous factors have conspired to create a situation in which most adults need supplementation. "Modern" farming methods have led to significant decreases in the vitamin and mineral content of our food, and almost no one eats enough fresh fruit and vegetables to get adequate amounts of vitamins and minerals without supplementation. In addition, digestive function decreases with age so that you don't absorb nutrients as well, which is another reason why you need more than what's available from food alone. Studies have shown that 90 percent of Americans are deficient in one or more vitamins or minerals, so by taking a daily multiple, your body can use what it needs and simply discard the rest.

There are many brands of multiple vitamin/minerals from which to choose. You want a daily multiple that contains enough vitamins and minerals to meet your needs for optimal nutrition, that is to say, ONA amounts. ONA amounts vary from person to person, but you want a daily multiple that contains enough of each ingredient to meet optimal health needs—and not just prevent deficiency diseases such as scurvy and rickets. One thing to realize is that these amounts cannot fit in one capsule or tablet—it would be too big. Therefore, a one-a-day vitamin won't fit the bill. Rather, you'll need to take a daily multiple that typically has two to six pills per day. That way, each tablet or capsule is a comfortable size for you to swallow. But, don't just take two to six one-a-day vitamins. Doing so would lead to your getting too much of one vitamin or mineral and not enough of another.

The ingredients in your daily multiple may vary from formulation to formulation, but most good products contain the following:

• **Vitamin A and beta-carotene** are crucial to proper eye function, support immune system function and resistance to infection, and are essential to bone growth and maintenance. ONA: 2,500 to 5,000 IU a day of vitamin A and a similar dose of beta-carotene is sufficient for most people. (Caution: Current smokers should avoid beta-carotene supplementation since studies have linked beta-carotene supplementation to higher lung cancer risk in smokers.)

• **B vitamins** are essential cofactors for enzymes that convert food to energy; promote healthy nerves, hair, skin, eyes, and gastrointestinal system; help combat stress; support healing from illness, injury, or surgery; and boost immune function. ONA amounts vary widely from individual to individual, but most good daily multiples have at least 25 milligrams of B_1, B_2, B_3, and B_6, and 25 micrograms of B_{12}, along with varying amounts of biotin, choline, inositol, PABA (para-aminobenzoic acid), and pantothenic acid (B_5).

• **Vitamin C** is our premier water-soluble antioxidant. It protects against heart disease and appears to reduce risk of cancers of the breast, lung, and gastrointestinal tract. Typical ONA is 500 to 2,000 milligrams per day. Some recent studies from the National Institutes of Health suggest that peak vitamin C levels in the bloodstream are achieved for most people with 500 milligrams daily. Note that this is several times the RDA amounts of 75 milligrams for women and 90 milligrams for men.

• **Vitamin D** is critical to the formation of healthy bones, and it appears to lower the risk of several types of cancer. It helps maintain a healthy immune system and regulates cell growth and differentiation. Currently, studies appear in the medical literature almost daily, touting the incredible benefits of vitamin D supplementation. Most daily multiples contain around 400 IU of vitamin D. We will discuss the value of taking additional vitamin D beyond what is contained in your daily multiple a little later.

• **Vitamin E** is your premier fat-soluble antioxidant that protects cells from toxins and carcinogens. Vitamin E also aids in the treatment of angina (reduced flow of blood to the heart muscle), arteriosclerosis (hardening of the arteries), and thrombophlebitis (blood clots in the legs). It helps prevent blood clots that can cause strokes; improves blood-flow to the extremities and relieves circulatory problems; increases high-density lipoprotein cholesterol while decreasing overall cholesterol levels; helps protect against cancers of the breast, cervix, lung, esophagus, and colon; and helps maintain levels of vitamin C. ONA is 400 to 800 international units of mixed tocopherols, which include several types of vitamin E, such as alpha-, beta-, delta- and gamma-tocopherol.

TABLE 12-1: OPTIMAL NUTRITIONAL ALLOWANCES FOR VITAMINS

VITAMIN	RDA	ONA
Vitamin A (international units)	2,333 (women)–3,000 (men)	2,500–5,000
Vitamin D (international units)	200	400–4,000
Vitamin E (international units)	15	400–800
Vitamin K (micrograms)	120	120
Vitamin C (milligrams)	90	500–2,000
Vitamin B_1 (milligrams)	1.2	10–100
Vitamin B_2 (milligrams)	1.3	10–100
Vitamin B_3 (milligrams)	16	20–35
Vitamin B_6 (milligrams)	1.3	10–100
Vitamin B_{12} (micrograms)	12	12–100
Folic acid (milligrams)	400	400–800

• **Minerals** function as cofactors in hundreds of different enzymes and are, therefore, essential to proper function of all organs and systems. Of the 92 naturally occurring elements, 14 are minerals that are essential to human health. We need more of some minerals than others. Minerals such as calcium and magnesium, which we need to take in milligram (1/1,000 of a gram) or even gram amounts daily are considered *macronutrients*, while some, such as selenium and chromium, are needed only in microgram (1/1,000,000 of a gram) amounts and are called *micronutrients* or trace minerals. We can get much of our ONA amounts of minerals by eating a healthy diet, but because of soil depletion, prolonged food storage, cooking, and the age-related decline in digestive function, many of us need supplementation. The 14 essential minerals are calcium, chromium, copper, fluorine, iodine, iron, magnesium, manganese, molybdenum, phosphorus, potassium, selenium, sodium, and zinc. A few minerals, such as boron, silicon, and vanadium, aren't considered essential to life, but have value and are often added to daily multiples.

Caution must be used when supplementing with minerals because of the greater risk of toxicity than with some other nutrients. For example, while 15 milligrams of zinc is within the ONA range, doses over 100 milligrams a day can be toxic. Iron and sodium are special cases of minerals that are essential but typically are found in adequate or excess amounts in most diets and are not included in our desired formulations. Excess sodium is a primary cause of elevated blood pressure and fluid retention, while excess iron has been linked to cancer, diabetes, heart disease, increased risk of infection, and worsening of rheumatoid arthritis. We don't recommend iron supplementation except under certain conditions, such as pregnancy, heavy menstrual bleeding, or chronic blood loss. Some recent studies suggest that calcium supplementation is associated with increased risk of heart attack in older women. As such, pending further research, we suggest women supplement with calcium until age 70, then stop. Much of your ONA for minerals can be gotten from dietary sources, and the ONAs listed below are the amounts we feel should come from supplements.

TABLE 12-2: OPTIMAL NUTRITIONAL ALLOWANCES FOR MINERALS

MINERAL	RDA	ONA (FROM SUPPLEMENTS)
Calcium	800–1,200 milligrams	400–800 milligrams (balance from diet)
Chromium	50–200 micrograms	50–200 micrograms
Copper	0.9–3 milligrams	0.9–3 milligrams
Fluorine	3–4 milligrams	3–4 milligrams
Iodine	80–150 micrograms	80–250 micrograms
Iron	10–18 milligrams	0 milligrams
Magnesium	300–400 milligrams	100–350 milligrams (balance from diet)
Manganese	2.5–5 milligrams	2.5–10 milligrams
Molybdenum	75 micrograms	75–800 micrograms
Phosphorus	800–1,200 milligrams	0 milligrams (balance from diet)
Potassium	2,000–3,000 milligrams	0 milligrams (balance from diet)
Selenium	40–70 micrograms	50–400 micrograms
Sodium	<2,400 milligrams	0 milligrams
Zinc	12–15 milligrams	12–30 milligrams

FISH OIL (EPA/DHA)

In addition to eating fish a few times each week, most adults can be helped by taking supplemental fish oil, which is a rich source of the omega-3 fatty acids eicosapentaenoic acid (EPA) and docosahexaenoic acid (DHA). EPA and DHA are precursors to chemicals in the body that help reduce inflammation. Recall that inflammation is the common pathway associated with many common and serious diseases, ranging from arthritis and asthma to cancer and heart disease.

Even conservative medical authorities have come out in support of fish oil supplementation in certain cases. The American Heart Association now recommends 1 gram of fish oil daily for patients with coronary artery disease. The National Institutes of Health considers fish oil of value in the treatment of heart patients as well, but also feels it has value for treating elevated triglycerides and high blood pressure. These three indications carry an A rating, meaning the National Institutes of Health feels that strong scientific evidence supports its use. Use of fish oil supplementation for primary prevention of heart disease and treatment of rheumatoid arthritis carries a B rating (good evidence supporting its use), while the use of fish oil for 27 other conditions ranging from cancer prevention to depression to schizophrenia carries a C rating (some evidence of value, but more study needed).

Fish oil is high in anti-inflammatory omega-3 fats. Most people today consume far more of their fat calories from omega-6 sources, which tend to increase inflammation. Years ago, before the days of processed food, the ratio of intake of omega-3 fats to intake of omega-6 fats was about equal. Today, it is not uncommon for people to get 25 times as much omega-6 fat than omega-3, increasing inflammation in the body and increasing health risks associated with inflammation. As we discussed in Chapter 2, inflammation underlies every step of the process leading to vulnerable plaque in the arteries and to heart attacks. It also underlies many other diseases, such as Alzheimer's disease, cancer, and arthritis. Reducing consumption of omega-6 fats (chiefly, vegetable oils) and increasing dietary and supplemental fish oil can help correct this imbalance.

There is no RDA for omega-3 fats, but the National Institutes of Health recommends 4 grams a day for healthy adults. Our ONA for EPA is 750 to 3,000 milligrams per day and for DHA, 500 to 2,000 milligrams per day. Vegetarians can get 2.5 grams of omega-3 fats from each teaspoon of flaxseed oil.

Some people worry about mercury contamination of their fish oil supplements. Thanks to California's Proposition 65, the mercury content of fish oil capsules must be less than 3 parts per million. Since most manufacturers want to be able to sell their products in California, this regulation has effectively had a national impact and most fish oil capsules now meet this standard.

Vitamin D

Studies showing benefits associated with having higher levels of vitamin D seem to be appearing almost daily. The evidence has been so overwhelming that even conventional physicians have taken notice and are now measuring vitamin D levels in their patients and recommending supplementation. We have found that, in addition to the vitamin D in your daily multiple, most people will benefit from taking an additional, separate vitamin D supplement as well.

Vitamin D is the one vitamin that you want to measure with a blood test that you can request from your doctor to determine your ONA. The test you want to ask for is called the 25-hydroxyvitamin D or 25(OH)D. The normal range for 25(OH)D at the reference lab we use is listed as 32 to 100 ng/dL, but you want your level to be at least 50. Most daily multiples contain about 400 international units (IU) of vitamin D, but we have found that most people need to take more vitamin D than is in their daily multiple to raise blood level into the desired range.

If your 25(OH)D level is 20 or less, we suggest you start with 5,000 IU of vitamin D per day. If your level is 21 to 30, then begin with 2,000 IU per day, and if it is 31 to 40, start at 1,000 IU daily. After 3 months have your level rechecked and adjust your dose accordingly. Don't be surprised if it takes 6 months or more for your level to reach the optimal range. After that, a dose of 1,000 to 2,000 IU per day is typically needed to maintain optimal levels, with continued blood monitoring to prevent excess build-up in your body. Vitamin D_3 (cholecalciferol) has been thought to be more effective than vitamin D_2 (ergocalciferol), although some recent studies suggest they are equivalent.

Vitamin D supplementation has long raised concern about toxicity because vitamin D is fat soluble and can be stored in fat tissues, with excess levels leading to elevated levels of calcium in the blood. Newer research, however, suggests that this is rarely a problem and that the current RDA of 400 IU is far too low.

A natural way to increase vitamin D levels is to get direct sunlight on the skin. Exposure to sunlight enables you to make vitamin D from cholesterol in the skin, but use of sunscreen prevents this conversion. In addition to taking vitamin D supplements, we recommend that you try to get direct sun exposure to your skin *without sunscreen* before 10:00 a.m. or after 4:00 p.m. for up to 30 minutes a day. At these hours, the most dangerous and damaging ultraviolet rays are reduced, so you can get the benefit of sunlight while minimizing the risk of skin damage. Vitamin D is not naturally found in significant amounts in foods, but is added as a supplement to milk and some other "fortified" foods.

> **Terry2023:** We now have implantable devices the size of a grain of rice that can monitor the levels of nutrients such as vitamins in our bloodstream and release just the right amounts that we need on a continuous basis.

> **Reader:** I thought I read there were some technologies like that today.

> **Ray2023:** Actually, back in your day, there was a fully artificial pancreas being tested that monitored glucose levels and gradually released insulin using the same method as a real pancreas. That's now an approved therapy, and these new implantable devices for nutrient levels are based on the same idea.

> **Ray2034:** Now in 2034, the nanobots take care of this. These blood cell–size robots are continuously monitoring our blood for nutrients, hormone levels, glucose, and everything else. If levels are not where they should be, they send an e-mail to other nanobots to correct the levels. People take a spoonful of nanobots each day, which are programmed to respond to the latest internal measurements of what your body needs.

> **Reader:** So I won't have to watch what I eat?

> **RayandTerry2034:** Basically, that's right.

> **Reader:** Isn't there some value in taking responsibility for the biological consequences of our actions?

> **Terry2034:** Well, with birth control, you've already separated at least some biological consequences from your actions back in your day. So overcoming the limitations of biology is not a new idea.

ADDITIONAL SUPPLEMENTS

In addition to our universal supplements that we feel have value to almost everyone over 30, there are several other supplements that we feel have significant benefits and should be considered as part of an optimal supplementation program.

Coenzyme Q_{10} or Ubiquinol

Coenzyme Q_{10} is needed in the production of ATP (adenosine triphosphate), the primary fuel for cellular function. It is also a powerful antioxidant that aids in the regeneration and recycling of vitamins C and E; helps protect the body from heart disease and various types of cancer; is helpful in preventing toxic effects of some types of chemotherapy; and aids in the treatment of cardiovascular conditions such as angina, high blood pressure, and congestive heart failure.

Free radicals can combine with oxygen to form reactive oxygen species, which are quite harmful, especially to your DNA, and damaged DNA can lead to cancer. A study published in 2000 in the journal *Clinical Biochemistry* found that coenzyme Q_{10} helps prevent cancer by neutralizing reactive oxygen species before they can damage DNA molecules.

While coenzyme Q_{10} is produced naturally by the body, most people can benefit from supplementation. For some people, however, coenzyme Q_{10} supplements aren't optional—they're necessary. For example, we feel that patients who are receiving statin drugs must take supplemental coenzyme Q_{10} because these drugs interfere with coenzyme Q_{10} production in the body.

ONA amounts of coenzyme Q_{10} are 30 to 150 milligrams twice a day with meals. Coenzyme Q_{10} is also available in its chemically reduced form as *ubiquinol*, which appears to be a more potent form. ONA for ubiquinol is 50 milligrams once or twice daily.

Grape Seed Extract

Grape seed extract is rich in proanthocyanidins, extraordinarily powerful free radical scavengers. Grape seed extract promotes healthy cell growth, reduces inflammation, increases blood vessel strength and elasticity, and protects against heart disease, strokes, and cancer. It removes amyloid, a factor in Alzheimer's disease and age spots on the skin; aids in collagen repair and reverses the appearance of aging; and helps protect liver and kidney cells in cases of acetaminophen (Tylenol) overdose.

Grape seed extract is a more powerful antioxidant than vitamin E, vitamin C, or beta-carotene. In fact, recent studies have shown grape seed extract to be 20 times more effective than vitamin C and 50 times more effective than vitamin E at scavenging free radicals.

ONA is 50 to 100 milligrams of grape seed extract twice a day.

ALPHA-LIPOIC ACID

Alpha-lipoic acid (ALA) is a potent free radical scavenger that can neutralize the most dangerous reactive oxygen species of all, the OH- (hydroxyl) free radical, among others. ALA aids in the recycling of other antioxidants, such as vitamins C and E, glutathione, and coenzyme Q_{10}, and enhances glucose utilization and uptake of insulin into the cells. It inhibits formation of age-related AGEs (advanced glycation end products), which can cause a number of conditions, such as high blood pressure and age spots on the skin.

Although ALA is formed naturally in the body in small amounts, supplementation is recommended to reach effective levels. No serious side effects have been identified in more than 40 years of research, and ALA has been found to be safe over a wide range of dosages. Our ONA of 50 to 100 milligrams once or twice a day is adequate for most healthy adults. We recommend 100 to 300 milligrams a day for anyone trying to correct glucose intolerance or the metabolic syndrome, and 300 to 600 milligrams a day for patients with diabetes.

RESVERATROL

It is difficult to attend a meeting or open a magazine on breakthroughs in longevity research without hearing about resveratrol, the so-called "red wine extract." Resveratrol is produced naturally in plants in response to attacks by pathogens such as viruses or fungi and is found in small amounts in red wine. Resveratrol is one of a group of compounds that affect *sirtuins*, proteins in the body that help control aging. Studies by David Sinclair at Harvard have shown that resveratrol supplementation can increase the lifespan of yeast, fruit flies, and fish; and a 2008 study showed that resveratrol was able to counter the harmful effects of a high-fat diet in mice. Resveratrol appears to mimic some of the effects of caloric restriction, the only proven method for life extension in animal experiments. While it is still too early to know if supplemental resveratrol will translate to significant life extension in humans, both Ray and Terry take 50 milligrams of trans-resveratrol twice daily.

ACETYL-L-CARNITINE

Carnitine is a naturally occurring molecule that helps transport fatty acids across the membranes of the mitochondria, the energy powerhouses of the cells, so they can be burned as fuel to generate ATP. Acetyl-L-carnitine (ALC) is a more bioavailable form of carnitine that has particular benefit for brain activity and mental function. ALC appears to work synergistically with ALA to rejuvenate aging mitochondria. The combination of fat-burning and enhanced brain activity makes ALC a very desirable nutrient. Typical doses range from 300 milligrams once a day to 1,000 milligrams three times a day.

ACETYL GLUTATHIONE

Acetyl glutathione is one of the most important and versatile new supplements we have seen in several years. Glutathione is one of the body's main endogenous or built-in antioxidants. The body makes it naturally from three amino acids: cysteine, glycine, and glutamic acid. Glutathione works in harmony with vitamin C and vitamin E to recycle these nutrients, but glutathione levels are easily depleted by increased stresses to the body, such as infection or toxin exposure. Glutathione itself is not well absorbed when taken orally, and intravenous injections of glutathione have a very short half-life in the bloodstream. Acetyl glutathione is made in the laboratory by attaching an acetyl group (chemical formula $COCH_3$) to glutathione molecules. This changes the molecule so that it is easily absorbed and has a much longer half-life in the bloodstream. As an easily absorbable and durable form of glutathione, acetyl glutathione can circulate throughout the body and reach all the tissues and easily enter the cells. Increased glutathione can help with numerous important processes, such as removal of toxic heavy metals. It can improve lung and liver function, improve immunity, and help repair injured tissues. Studies have also shown that glutathione decreases as a direct consequence of the aging process. We recommend that healthy individuals over 50 years of age consider taking one 100-milligram acetyl glutathione tablet daily and that people over 70 consider taking one twice daily.

SAMPLE SUPPLEMENT REGIMENS

Our ONA recommendations provide broad guidelines for nutritional supplementation, but they are not "one size fits all." Your specific needs depend on numerous factors, including your sex, age, weight, occupation, stress level, health condition, and genetic predisposition. To help you develop a supplementation regimen that will be most effective in meeting your needs, the following sample regimens are suggested as basic programs of supplementation. Work in collaboration with your healthcare provider to help with appropriate testing at regular intervals to ensure that you are taking the supplements you need at the correct dosages.

Ages 30–40
(In general, supplementation is optional for individuals younger than 30 years of age, but these amounts may be safely taken by people 20–30.)

Two-a-day daily multiple	1 twice per day
EPA 480 milligrams/	
DHA 240 milligrams (fish oil)	1 twice per day
Vitamin D	as needed to reach 25(OH)D > 50
Calcium (women only)	400–800 milligrams per day, so total including food = 1,200 milligrams

---

Ages 40–50

Daily multiple (full dose = 6 per day)	2 twice per day
EPA 480 milligrams / DHA 240 milligrams (fish oil)	1 twice per day
Vitamin D	as needed to reach 25(OH)D > 50
Calcium (women only)	400–800 milligrams per day, so total including food = 1,200 milligrams
Ubiquinol 50 milligrams	1 per day
Grape seed extract 100 milligrams	1 twice per day

Ages 50 +

Daily multiple (full dose = 6 per day)	3 twice per day
EPA 480 milligrams/ DHA 240 milligrams (fish oil)	1–2 twice per day
Vitamin D	as needed to reach 25(OH)D > 50
Calcium (women only)	400–800 milligrams per day, so total including food = 1,200 milligrams
Ubiquinol 50 milligrams	1 to 2 per day
Grape seed extract 100 milligrams	1 twice per day
Alpha-lipoic acid 100 milligrams	2 per day
Acetyl-L-carnitine 250 milligrams	2 per day
Acetyl glutathione 100 milligrams	1 per day (after 70 years, 2 per day)

These guidelines are intended as a basic program of supplementation to help protect against the type of changes associated with the aging process. But they are not the whole answer. Other factors may suggest the need for additional protection. For instance, if your lifestyle exposes you to higher amounts of free radicals, you should increase your total consumption of antioxidants. If you live in the mountains, where the air is clean and water is clear, and grow your own organic vegetables, your needs would be less than if you live in a large metropolitan area and work at a highly stressful job in an office located next to a heavily trafficked thoroughfare.

As you perform the testing recommended elsewhere in this book, additional supplements may come to mind. If your level of inflammation is higher than desired, as measured by an elevated high-sensitivity C-reactive protein, increasing the amount of fish oil and adding curcumin would be of benefit. Elevations of cholesterol might suggest the need for pharmacologic doses (as much as 1,500 to 3,000 milligrams) of vitamin B_3 (niacin) (see Chapter 2).

You should also round out your program based on any genetic information you have, either as a result of formal genomics testing or on the basis of your family history. In summary, you shouldn't rely on RDA-based supplement recommendations; rather, use your supplement program to reprogram your personal biochemistry to avoid disease and slow down aging.

EARLY DETECTION
Vitamins

Other than vitamin D, we generally don't recommend regular blood monitoring of most vitamin levels. Terry will sometimes perform a blood analysis of fat-soluble vitamins A, D, and E as well as coenzyme Q_{10} on his patients as part of a comprehensive evaluation. We don't feel there is as much value in checking levels of water-soluble vitamins such as vitamin C and the B vitamins because their levels change so frequently.

Minerals

Semi-quantitative information can be obtained about mineral status through the use of hair analysis. In addition to measuring the levels of toxic heavy metals, such as lead, mercury, and cadmium, hair mineral analysis also provides data about levels of essential minerals, such as calcium, magnesium, and selenium. Two common patterns that we see are patients who have low levels of most minerals or patients with a specific pattern of high calcium, magnesium, and strontium. Patients in this first group, with low levels of many of the common minerals, often have *hydrochlorhydria*, or inadequate stomach acid. One of the most common reasons for these low minerals is prolonged usage of acid-suppressive medications such as H2 blockers (Zantac or Pepcid, for example) or PPIs (proton pump inhibitors) such as Prilosec, Nexium, or Prevacid. To absorb minerals properly, you need adequate amounts of stomach acid, so when we see this pattern of many low minerals, we'll usually recommend supplementation with betaine hydrochloride, a pill form of hydrochloric acid. This will enhance absorption of minerals from your supplements and diet. In addition, we have also found that acid reflux in patients with GERD (gastroesophageal reflux disease) will often improve with the use of supplemental hydrochloric acid, so we consider this therapy rather than suppressing acid production with the acid-blocking drugs. Hydrochloric acid supplementation should not be taken if you have a history of peptic ulcer disease.

Another common pattern that we see is the combination of high levels of calcium, magnesium, and strontium in the hair. This often occurs in menopausal or perimenopausal women with varying degrees of bone loss. Calcium, magnesium, and strontium are found in large amounts in the bones, and when increased bone turnover exists, such as in cases of osteopenia or osteoporosis, these minerals can get deposited in the hair. When we see this pattern, we'll recommend that women undergo further osteoporosis screening and appropriate treatment.

13

CALORIE REDUCTION AND WEIGHT LOSS

"More die in the United States of too much food than of too little."
—JOHN KENNETH GALBRAITH

"To lengthen your life, shorten your meals."
—PROVERB

The National Health and Nutrition Examination Survey 2001 to 2004 (NHANES) found that two-thirds of American adults are overweight and that obesity rates had more than doubled from about 15 percent in 1974 to almost 34 percent in 2006. In a troubling indicator of future trends, a 2006 report from the Department of Health and Human Services found that the number of overweight children ages 10 to 17 had tripled between 1980 and 2006 to about 15 percent. The U.S. Surgeon General described this as an "obesity epidemic" for good reason. A 2002 report in the *New England Journal of Medicine* found that being just 20 percent overweight triples your risk of high blood pressure and type 2 diabetes and boosts your risk of heart disease by 60 percent. Being overweight or obese also increases your risk of stroke, osteoarthritis, sleep apnea, and a number of cancers. The financial cost of excess weight is also enormous—the National Institute of Diabetes and Digestive and Kidney Diseases estimates annual weight-related medical costs in the United States to be as high as $92.6 billion, with additional lost productivity costs as high as $3.9 billion (2002 dollars).

Most people know that excess weight is unhealthy—not to mention unflattering—but losing weight and keeping it off is often a major struggle. According to a 2007 survey by the International Food Information Council, during any given year, 70 percent of Americans change their diets in an effort to lose weight. And in 2005, *Forbes* reported that $46 billion is spent each year in the United States on diets and diet aids. Yet overweight and obesity rates continue to rise.

As a society we clearly need to go in the other direction—towards reducing calories. The reason we have been gaining so much weight as a society is obvious—we're eating much more than we used to. Total calories consumed by Americans in 2004 averaged 2,750 per day, 500 calories more per day than consumed daily in 1970. This has translated into drastic increases in our average weight. According to figures from the Centers for Disease Control and Prevention, the average man in 1960 was 5 feet 8 inches tall and weighed 166 pounds. By 2002 he had grown an inch and a half taller but gained 25 pounds to an average weight of 191. In 1960 women averaged 5 feet 4 inches in height and 140 pounds. By 2002 they were 5 feet 4 inches and 164 pounds.

Perhaps the biggest problem in losing weight is that most people approach "going on a diet" as a temporary change in eating habits, a period of deprivation to be followed by a return to their normal routine once the extra pounds have been dropped. This type of dieting is almost completely ineffective; two-thirds of dieters regain whatever weight they lost within a year, and 97 percent gain it all back within 5 years. This continuing cycle of weight loss and gain has come to be known as yo-yo dieting, and it can be even more detrimental to your health than maintaining a consistent number of excess pounds. Obviously, yo-yoing will not help you lose excess weight and maintain your optimal (most healthy) body weight for the *long term*. The solution lies in changing the way you eat for the rest of your life. We will describe below how you can make one single change to your diet that will allow you to achieve your optimal weight—and keep it there—for good.

Walk the aisles of the typical American supermarket, and the vast majority of what you see on the shelves is laden with sugars, refined carbohydrates, and saturated fats. The same is true of most restaurant food, especially fast-food restaurants. These unhealthy ingredients lie at the heart of the obesity epidemic as well as many of the other life-shortening conditions we discuss elsewhere in this book. At the center of our weight loss plan is a diet that drastically reduces or eliminates these killer ingredients (like sugar, the "white Satan"!) and instead emphasizes a specific balance of low-starch vegetables, lean proteins, and healthy fats. In Chapter 11, we discussed these nutritional recommendations and the reasons why developing these new eating habits is so crucial, not just to trimming your excess body fat but also to your overall well-being and longevity.

Calorie Reduction, the C in our *TRANSCEND* plan, helps you to lose weight gradually over time without feeling hungry or deprived and to keep that weight off for the long term. (The plan also works if you are underweight and want to add some pounds.) The key word is *gradually*. Any excess body fat you may be carrying took a while to accumulate, and shedding it for good will take some time as well. If you are patient and stick to these new eating habits, you will be rewarded with a leaner and healthier body for the rest of your life.

STEP ONE: GETTING TO YOUR GOAL WEIGHT

By making one simple change to your diet, you can gradually approach your goal weight and remain there permanently. **All you need to do is consume as many calories each day as you would if you were *already* at your goal weight.** That's all there is to it! This is your *maintenance calorie level* (at your target weight). Regardless of your current weight, you will gradually reach your optimal weight by consistently eating as if you were already at your desired maintenance calorie level. There is no need to have any sensation of hunger doing this, because as you follow the diet we suggest, you will be replacing calorically dense sugary and high-fat foods with healthier choices that are lower in calories. This will actually enable you to eat more than you ever have before and never experience any hunger. The key is proper food choices. The following four steps will help you determine your maintenance calorie level based on your body frame, height, and activity level:

1. **Determine your body frame size.** Measure your wrist circumference (use a tape measure or a piece of string that you then measure with a ruler).

TABLE 13-1: DETERMINING FRAME SIZE FROM WRIST CIRCUMFERENCE

	SMALL FRAME	MEDIUM FRAME	LARGE FRAME
	WRIST MEASUREMENT (INCHES)		
Adult males	Under 6¼	6¼–7	Over 7
Adult females	Under 5¼	5¼–6	Over 6

2. **Determine your optimal weight.**

TABLE 13-2: IDEAL WEIGHT RANGE IN POUNDS FOR WOMEN (INDOOR CLOTHING)

HEIGHT	SMALL FRAME	MEDIUM FRAME	LARGE FRAME
4' 10"	102–111	109–121	118–131
4' 11"	103–113	111–123	120–134
5' 0"	104–115	113–126	122–137
5' 1"	106–118	115–129	125–140
5' 2"	108–121	118–132	128–143
5' 3"	111–124	121–135	131–147
5' 4"	114–127	124–138	134–151
5' 5"	117–130	127–141	137–155
5' 6"	120–133	130–144	140–159
5' 7"	123–136	133–147	143–163
5' 8"	126–139	136–150	146–167
5' 9"	129–142	139–153	149–170
5' 10"	132–145	142–156	152–173
5' 11"	135–148	145–159	155–176
6' 0"	138–151	148–162	158–179

Women between 18 and 25 should subtract 1 pound for each year under 25.

Weight in pounds according to frame (wearing indoor clothing weighing 3 pounds and shoes with 1-inch heels). Courtesy of Metropolitan Life Insurance Company.

TABLE 13-3: IDEAL WEIGHT RANGE IN POUNDS FOR MEN (INDOOR CLOTHING)

HEIGHT	SMALL FRAME	MEDIUM FRAME	LARGE FRAME
5' 2"	128–134	131–141	138–150
5' 3"	130–136	133–143	140–153
5' 4"	132–138	135–145	142–156
5' 5"	134–140	137–148	144–160
5' 6"	136–142	139–151	146–164
5' 7"	138–145	142–154	149–168
5' 8"	140–148	145–157	152–172
5' 9"	142–151	148–160	155–176
5' 10"	144–154	151–163	158–180
5' 11"	146–157	154–166	161–184
6' 0"	149–160	157–170	164–188
6' 1"	152–164	160–174	168–192
6' 2"	155–168	164–178	172–197
6' 3"	158–172	167–182	176–202
6' 4"	162–176	171–187	181–207

Weight in pounds according to frame (wearing indoor clothing weighing 3 pounds and shoes with 1-inch heels). Courtesy of Metropolitan Life Insurance Company.

3. **Determine your activity level.**

   ○ *Sedentary:* You have no regular exercise routine, walk only occasionally, and most of your activity is typical of daily life.

   ○ *Moderately active:* Combined activity from your normal routine (work, recreation, etc.) or regular exercise program is equivalent to walking or running 1.5 to 3 miles per day in addition to physical activities typical of routine daily life.

   ○ *Very active:* Combined activity from your normal routine (work, recreation, etc.) or regular exercise program is equivalent to walking or running more than 3 miles per day in addition to physical activities typical of routine daily life.

4. **Determine your maintenance calorie level.**
   o Find your optimal weight in the first column of Table 13-4.
   o On that line, find your maintenance calorie level in the column that best describes your activity level. (If your weight falls between two lines in the table, you can calculate the difference between the two corresponding maintenance calorie levels.)
   o Remember that if you consume the number of calories each day that you would if you were already at your goal weight, you will gradually achieve your goal weight.
   o There are a number of excellent online resources to help you keep track of your calories. Ray has been keeping a spreadsheet diary of his daily caloric intake for over 15 years. Terry hasn't been doing this for quite as long, but uses an application downloaded to his iPhone to keep track of his calories.

TABLE 13-4: DETERMINING YOUR MAINTENANCE CALORIE LEVEL

TARGET WEIGHT	SEDENTARY	MODERATELY ACTIVE	VERY ACTIVE
90	1,170	1,350	1,620
100	1,300	1,500	1,800
110	1,430	1,650	1,980
120	1,560	1,800	2,160
130	1,690	1,950	2,340
140	1,820	2,100	2,520
150	1,950	2,250	2,700
160	2,080	2,400	2,880
170	2,210	2,550	3,060
180	2,340	2,700	3,240
190	2,470	2,850	3,420
200	2,600	3,000	3,600
210	2,730	3,150	3,780
220	2,860	3,300	3,960
230	2,990	3,450	4,140
240	3,120	3,600	4,320

PUTTING THE PLAN TO WORK

Your maintenance calorie level is the key to gradually reaching and sustaining your optimal body weight for the long term. Not just any calories will do, however. You need to eat a variety of healthy foods in the right proportions. If weight loss has been difficult for you in the past or to lose weight more quickly, you may want to follow our recommendations in Chapter 11 for the Low-Carbohydrate Corrective Diet, which limits carbohydrates to 20 percent of calories. Crucial information is provided in that chapter regarding the specific kinds of proteins, carbohydrates, and fats you should eat to meet all your nutritional needs and lose excess body fat. And be careful not to go too far in limiting your daily calories. At a minimum, you should eat 13 calories per day for each pound of your optimal weight; in general, that means no fewer than 1,000 calories each day for women or 1,200 calories for men.

As you put our weight loss plan into action, here are a few things to keep in mind:

- *Make good health your ultimate goal, not weight loss.* Maintaining optimal body weight is just one aspect of living healthy, living well, and living long. If you focus on making sustainable lifestyle changes that lead to better overall health for the *long term*, weight loss will follow automatically. A narrow focus on losing pounds can obscure the larger objective, and it's a primary reason many people get discouraged by diet plans and give up when their weight levels off. Paying too close attention to the bathroom scale can be misleading. Your weight will naturally fluctuate because of water retention, bowel movements, menstruation, and changes in medications. Also, building muscle through strength training (see Chapter 14) can actually add pounds because muscle tissue is denser than fat tissue, but this is a desirable change.

- *Gradual weight loss is healthier and more sustainable.* Preliminary research suggests that yo-yo dieting, the continuing cycle of weight loss and gain that plagues so many dieters, may be explained by the ghrelin hormone that is secreted by the stomach. Ghrelin levels increase when the stomach is empty, stimulating the appetite and slowing metabolism, while ghrelin levels drop when the stomach is full. A 2002 study in the *New England Journal of Medicine* found that ghrelin levels increase significantly after rapid weight loss. On the other hand, gradual weight loss does not appear to trigger a similar spike in ghrelin, so hunger cravings are much less intense. Losing weight gradually also puts far less stress on the organs and systems of your body than does rapid weight loss.

• *Choose foods you can live with for the long term.* A weight loss plan is doomed to fail if it leaves you impatiently counting the days (or pounds) until you reach some arbitrary goal where you can get off the diet and return to your old eating habits. Successful long-term weight loss comes from embracing a new way of eating that you can continue for the rest of your life. The magic of our program is that once you achieve your target weight, there is no further change to make. You are already eating at the maintenance level for your new desirable weight.

The transition to new eating patterns is not something you can expect to do overnight. You will need to experiment, adding new healthy foods to your diet as you eliminate those that are unhealthy. The sample menus and recipes in Chapter 11 can help you discover delicious healthy alternatives to foods you currently eat.

• *Eliminate sugar and refined carbohydrates.* If you currently consume a lot of sugar and refined carbohydrates, cutting those foods from your diet can be particularly challenging. These empty carbohydrates don't just contribute to excess body fat, they also have addictive properties similar to tobacco and narcotics, and eliminating them from your diet can actually cause withdrawal symptoms. However, those cravings will dissipate in a week or two if you go "cold turkey" and drop sugar and refined carbohydrates completely from your life. Eliminating high-glycemic-index and sugary carbohydrates is the single most important thing you can do to lose weight.

• *Exercise is your friend.* Being sedentary contributes to weight gain, while being active contributes to weight loss. You should strive to be at least moderately active. Engaging in a regular routine of both aerobic and strength-building exercise not only burns calories while you are exercising but also raises your metabolic rate while you are inactive, so you burn more calories all the time. It will also make you feel good by releasing endorphins, which are natural morphinelike chemicals. Strengthening your muscles (resistance training) is particularly important to losing weight. Your cells contain mitochondria, structures that burn nutrients to produce energy, and there are more mitochondria in muscle cells than in fat cells. As a result, muscle tissue is much more effective at burning calories. See Chapter 14 for our exercise recommendations.

STEP TWO: HAVING LESS TO GET MORE

Step One helps you achieve your goal weight when you consume as many calories as if you were already at your goal weight. Once you get to your goal weight, you can take it to the next level and consider refining your program by engaging in a deliberate program of *calorie reduction*—eating even less than you need for your goal weight.

First reported in 1982 in *Journals of Gerontology*, calorie restriction grabbed the public's attention with its tantalizing promise of dramatic life extension. In that study, laboratory rats fed a normal diet lived a normal life span of about 1,000 days. Rats fed similar amounts of vitamins, minerals, protein, and essential fatty acids, but with one-third fewer calories, lived for about 1,500 days, meaning that eating less extended their lives by 50 percent. That extended life was also good life (to the extent that laboratory rats can have a good life!). The calorie restriction rats were more energetic, less feeble, and more adept at running mazes than the control rats. Their immune systems were stronger, and they had reduced rates of cancer, diabetes, and cataracts.

To date, more than 2,000 studies where calories have been reduced have confirmed these results with numerous other animal species. And while not enough years have passed to determine whether calorie restriction can extend the human life span by 50 percent, studies of people practicing calorie restriction have found the same signs of decelerated aging that are evident in calorie restriction lab animals. Calorie restriction suggests a theory that each species has a fixed number of calories it can burn during the course of a normal life span, so that the fewer calories burned each day, the more days it takes to use up the total calories that can be burned before disease and aging mechanisms kick in.

HARA HACHI BU

Yet, it appears you don't need to restrict calories 30 to 35 percent to experience many of the benefits. Studies of long-lived human populations, such as the Japanese, have confirmed that reducing calories by a lesser amount also translates into increases in health and longevity.

Before beginning a meal, it is common for the Japanese to say, "Hara hachi bu," which means "stomach eight parts (80 percent) full." The Japanese try to remember not to overeat each time they sit down to eat, and by reducing calories about 20 percent, they have achieved radical increases in both youthfulness and longevity. Stories abound in Japan of 85-year-old physicians still in practice and 100-year-old tai chi instructors. Also, mean life expectancy in Japan, currently at 82.1 years, is among the longest in the world, 4 years longer than the 78.1-year average in the United States, so rather than recommending full-blown caloric restriction wherein calories are reduced 30 to 35 percent, we recommend "hara hachi bu" calorie reduction of 10 to 20 percent less than the maintenance recommendations.

The health and longevity characteristics of reducing calories have been attributed to several factors. Calorie restriction animals have significantly lower levels of blood glucose and body fat, which may protect against conditions associated with higher levels of glucose and fat, such as type 2 diabetes, metabolic syndrome, and cardiovascular disease. With less food consumed, fewer free radicals are produced as the food is metabolized. This decreases the overall level of free radical damage, including DNA mutations that lead to cancer. Indeed, calorie restriction animals also have more robust DNA-repairing enzymes.

There is no doubt that calorie restriction offers some exciting prospects, but a bit of caution is advised before you adopt a reduced-calorie diet. There are limits to how far calories can be restricted before the effect becomes life-limiting. Cutting too deep can cause deficiencies in crucial vitamins, minerals, proteins, essential fats, and other nutrients, which can lead to illness and even death. With lab animals, that limit appears to be two-thirds the number of calories they would normally consume if allowed to eat without restraint. You won't be anywhere near that by following our program of calorie reduction of up to 20 percent below your maintenance requirements. While the weight loss associated with calorie restriction is beneficial up to a point, humans who adhere to the two-thirds calorie level used in animal experiments tend to lose so much weight that they look gaunt and too thin. In light of the potential hazards of nutrient deficiencies, and with calorie restriction research on humans still quite limited, we recommend the following guidelines for our more moderate 10 to 20 percent form of *calorie reduction.*

• Most people can tolerate a 10 to 20 percent reduction in calories with relative safety as long as optimal amounts of nutrients are obtained. You can use online software to keep track of nutrients as well as calories. We recommend a calorie/nutrient counter developed by the late Dr. Roy Walford, a pioneer in caloric restriction. The DWIDP (Dr. Roy Walford Interactive Dietary Planner) is available to paid members of the Calorie Restriction Society (www.calorierestriction.org). By joining the Calorie Restriction Society (and paying the nominal fee of $35 per year), you have full access to the DWIDP. Using the DWIDP can help ensure that you are getting adequate amounts of all nutrients when following a reduced-calorie diet.

• Eat foods that are high in fiber and nutrients but low in calories, such as low-starch vegetables (broccoli, spinach, zucchini, etc.), and avoid high-starch foods (potatoes, rice, pasta, etc.).

• Be sure to get all the nutrients you need by maintaining a healthy balance of carbohydrates, lean proteins, and good fats (see Chapter 11) and by taking appropriate nutritional supplements (see Chapter 12).

By following a modest 10 to 20 percent reduction in daily calories, you will inevitably lose some weight. For example, Table 13-4 above tells us that a 130-pound moderately active woman would consume about 1,950 calories per day to maintain her weight. If she reduces her calories by 10 percent or 195 calories, her weight will gradually settle at around 118 pounds, a weight loss of 12 pounds. A moderately active 200-pound man who decides to consume 20 percent fewer calories would end up weighing about 160 pounds. The table below shows the number of calories associated with a reduction of 20 percent.

TABLE 13-5: DETERMINING YOUR WEIGHT LOSS CALORIE LEVEL

WEIGHT	SEDENTARY		MODERATELY ACTIVE		VERY ACTIVE	
	NORMAL	20% CALORIE REDUCTION	NORMAL	20% CALORIE REDUCTION	NORMAL	20% CALORIE REDUCTION
90	1,170	936	1,350	1,080	1,620	1,296
100	1,300	1,040	1,500	1,200	1,800	1,440
110	1,430	1,144	1,650	1,320	1,980	1,584
120	1,560	1,248	1,800	1,440	2,160	1,728
130	1,690	1,352	1,950	1,560	2,340	1,872
140	1,820	1,456	2,100	1,680	2,520	2,016
150	1,950	1,660	2,250	1,800	2,700	2,160
160	2,080	1,664	2,400	1,920	2,880	2,304
170	2,210	1,768	2,550	2,040	3,060	2,448
180	2,340	1,872	2,700	2,160	3,240	2,592
190	2,470	1,976	2,850	2,280	3,420	2,720
200	2,600	2,080	3,000	2,400	3,600	2,880
210	2,730	2,184	3,150	2,520	3,780	3,024
220	2,860	2,288	3,300	2,640	3,960	3,168
230	2,990	2,392	3,450	2,760	4,140	3,312
240	3,120	2,496	3,600	2,880	4,320	3,456

Reader: You mentioned earlier a gene that causes every unused calorie to be stored in my fat cells.

Ray: Yes, the fat insulin receptor (*FIR*) gene.

Reader: And a way to turn particular genes off.

Ray: Yes, RNA interference. It basically destroys the RNA that expresses a gene.

Reader: Okay, so how about if I turn off that gene? Then I can eat this hot fudge sundae?

Ray: Yes, well, turning off the fat insulin receptor gene is a great idea. When that gene was suppressed in animals (using another that also turns genes off), those animals got most of the benefit of calorie restriction, including weight loss, while eating a lot.

Terry: Of course, like all of these other interventions, it needs to be fully tested and go through the approval process. There are biotech companies working on that.

Ray: And there is at least one further consideration, which is that the *FIR* gene needs to be turned off only in the fat cells. If it were turned off in, say, the muscle cells, that would lead to muscle wasting. There are ways of directing RNA interference to specific types of body tissues, but it's a complication.

Terry2023: Well, I'm pleased to say that here in 2023, this concept has now been perfected.

Reader: Terrific, I'm not sure this particular hot fudge sundae will last until then, but I'm glad I'll be able to enjoy them then.

Terry2023: Well, you still need to be concerned about the quality of the calories you consume. The *FIR* inhibitor drugs will protect you from the weight gain of excess calories, but they won't protect you from all of the sugar and saturated fat.

Ray2034: For that you need to wait a bit longer. Here in the 2030s, the nanobots monitor the nutrients in your blood and remove unwanted substances, such as toxins or excess sugar, add nutrients you may not be eating, maintain optimal hormone levels, and so on.

Reader: So, uh, that hot fudge sundae?

Terry2034: Well, if you can wait a quarter century, you'll be able to eat as many as you want.

Ray2023: Yes, now all you need to worry about is taste fatigue!

14

EXERCISE

"Your exercise program—weights, yoga, walking—has helped us keep our bodies strong and flexible with a minimum of time. In about an hour a day, we've strengthened our bodies significantly. When we've had an 'accident' or a fall, our bodies recover quickly. We'll continue using it for our lifetimes. We like to know that our 'numbers' are good, but we especially like how good we FEEL on this program!"

—BEVERLY (65) AND KEN (66), MASSACHUSETTS

Exercise will not only make you stronger—it is one of the strongest pillars of your program for prevention of disease. Regular exercise is a critical component of the *TRANSCEND* program to help you live long enough to live forever. Yet, few people seem able to make the time or generate the energy needed to remain on a consistent exercise program. Even though only a fraction of us exercise regularly, no one seems to doubt the health benefits of exercise. In fact, the news from the medical literature just keeps getting better and better. For instance:

• A 2007 study of 16,000 male veterans followed for 7 and a half years showed that exercise was able to decrease their risk of death by 50 to 70 percent depending on the amount of exercise done. The authors of the study concluded that exercise appears to be an even more important risk factor for early mortality than cholesterol, high-density lipoprotein (HDL), and low-density lipoprotein (LDL).

• A 2007 study by Italian researchers showed that the more exercise you do, the less the risk of Alzheimer's disease.

• A study published in 2008 of 12,000 Danish adults showed that combining regular physical activity with moderate alcohol consumption slashed the risk of heart attack by 50 percent and decreased all-cause mortality by one-third.

• The January 28, 2008, issue of the American Medical Association journal *Archives of Internal Medicine* reported a study conducted by researchers in England and the United States that found leisure time physical activity in men and women is associated with increased leukocyte (white blood cell) telomere length compared with that of sedentary individuals. Longer telomere length is associated with increased longevity.

In an effort to combat the sedentary lifestyles of so many adults, the American College of Sports Medicine has joined with the American Medical Association to promote the "Exercise Is Medicine" program. The goal is to encourage physicians to regard physical activity as another *vital sign* that they measure in their offices along with blood pressure and pulse.

WHY WE DON'T WANT TO EXERCISE

If we know that exercise is so good for us, and we look and feel so much better when we do it, why don't we? Even though almost everyone realizes the proven importance of regular exercise for maintaining health, only a small fraction of us remain successful on a consistent basis. According to the U.S. Department of Health and Human Services, 74 percent of Americans over 18 don't exercise regularly and nearly 40 percent don't get any leisure time exercise *at all*. The reason is one of those facts that is so obvious that it seems to have been almost completely ignored. The elephant in the room is the fact that *most people don't like to exercise.*

Recall that we are still stuck with caveman genes. Scientists speculate that preferring rest to exercise may have led to a survival advantage for our ancestors in the distant past. Long ago, getting enough food simply to stay alive from day to day was a major struggle. Cavemen and women hadn't learned yet how to grow crops or how to store food for more than a few days. Their existence was hand-to-mouth at all times, and there were no refrigerators to store food. Taking it easy and resting any time they could—that is, any time that lack of food or the need to escape danger didn't require that they get up and do something about it—enabled our ancestors to conserve calories, which helped the oftentimes marginal food supply last longer and enhanced their prospects for survival.

But, there is also no question that the human body was designed to be exercised vigorously and regularly. This was never a problem in prehistoric times, as life in the Stone Age necessarily included plenty of vigorous physical activity. Walking and running long distances, climbing, and throwing were an inherent part of the way our hunter-gatherer ancestors acquired their food. They also got quite a bit of exercise avoiding danger. Much of the time, they were either running *after* their next meal or away *from* becoming a predator's next meal. Our Cro-Magnon ancestors didn't need willpower or self-control to make them *want* to get regular exercise; their empty bellies and impending danger saw to that.

Understanding that we are still cavemen and women who are genetically programmed to enjoy those rare moments when we could rest, while also understanding that our bodies need regular exercise, can help to generate the motivation to exercise regularly. Moreover, being aware of the profound benefits of exercise on every aspect of our biology is the best motivation. In our work with thousands of friends, colleagues, and patients over the years, we have found that simply coming to the full realization of these facts can be liberating and empowering.

Terry2023: This discussion about our genetic bias against exercise is very interesting from a historical perspective, but we're starting to have other alternatives today. Thanks to biotech drug breakthroughs, we can magnify the benefits of exercise. Even couch potatoes are in pretty good shape nowadays.

Reader: You don't need to do strength training or cardio exercise to maintain fitness?

Terry2023: There's still some controversy, but the "exercise mimetic" drugs are looking promising, and are becoming increasingly popular.

Reader: You're able to run marathons without training?

Terry2023: No, the drugs aren't that powerful, but they do allow people who don't want to exercise to maintain about the same level of endurance as folks who jog for half an hour three times a week. Their muscle strength is about what they would have by lifting weights twice a week. High-level fitness, such as being able to run more than 26 miles in a marathon, still requires regular training, although taking a fitness pill along with regular exercise can make your training easier and more effective.

Reader: How does it work?

Terry2023: In 2008 two drugs were found to reproduce many of the beneficial effects of a program of regular exercise. These drugs came to be known as exercise mimetics, meaning they mimicked the effects of exercise on the body. The first class of these drugs was known as peroxisome proliferator activated receptor delta (PPARd) agonists. PPARd drugs were initially found to increase HDL or good cholesterol, which of course was a good thing in and of itself. But, doctors knew that regular exercise was one of the few ways of increasing HDL cholesterol levels, so they wondered if these drugs might also mimic other beneficial effects of exercise as well.

Ray2023: First, they tried the PPARd drug GW1516. They found that just taking the drug didn't increase fitness, but when it was combined with regular exercise, laboratory animals were able to run 70 percent longer and farther.

Terry2023: And this led to the second class of exercise mimetic drugs, the AMPK (5'AMP-activated protein kinase) activators. The first of the AMPK drugs, commercially available as AICAR, was able to increase running endurance 45 percent without any exercise at all. Safety concerns emerged with the AMPK drugs, so successor drugs were designed that have recently been FDA approved. Today, most people take successors to one or the other of these types of drugs.

Ray2023: Couch-potato types take a successor to the AMPK-type drug and achieve a reasonable level of fitness without doing any exercise, while people who enjoy exercise often take a PPARd-class drug and experience far greater levels of fitness than they could achieve with exercise alone.

Reader: So exercise has become pretty much optional, and the only people who do it in the future do so simply because they enjoy it?

Terry2023: Well, the ideal approach is exercise plus the enhancement drugs, but coach potatoes today are in much better shape than they used to be.

Ray2023: We also have the fat insulin receptor gene inhibitor drugs.

Reader: Yes, we discussed that in Chapter 13.

Ray2023: A substantial portion of the population today takes an RNA-interference based drug to turn off the fat insulin receptor gene in the fat cells. Obesity has become a rare condition.

Reader: What does the International Olympic Commission have to say about all of this?

Ray2023: Athletes who want to compete are still barred if they take these drugs, although it's creating quite a bit of controversy. Athletes who take the drugs—and are thus prohibited from competing in the Olympics—can outrun, outlift, outswim, outjump the Olympians. So, some people have suggested having two sets of Olympics—one that is drug-free and one that is not.

Ray2034: The strongest argument was that unlike the steroid drugs early in this century, these drugs are actually good for your health, although not everyone accepts that. Drug-enhanced Olympics actually came into existence beginning with the 2028 summer games. The drug-free Olympics were held in Toronto, Canada, and the drug-enhanced games were held in Warsaw, Poland. By the 2030 Winter Olympics and the 2032 summer games, twice as many people were watching the drug-enhanced games as the drugless ones. The enhanced Olympics were far more interesting and exciting to most people as world records were broken in almost every event.

Reader: You said these drugs were first used in 2008. Where can I get them today?

Terry2023: I'm afraid you're going to have to keep exercising for a while longer as it took over a decade for them to be perfected and approved.

RayandTerry: Which brings us—appropriately enough—to the next part of this chapter: what you need to do to get (and stay) fit.

Terry2023: Until a pill a day can do it for you . . .

Nowadays, with a few simple modifications, it is possible to turn your exercise sessions into events that you look forward to—rather than drudgery that you need to force yourself to do. People are often amazed at how quickly they begin to look and feel better and how easy it is to stay on the program we describe below. One key is using your exercise to raise your levels of *endorphins*, chemicals in your body associated with intense pleasure. Another key is to combine your exercise program with activities that you naturally enjoy, such as exploring a new scenic park or walking with a friend. By making exercise an activity that generates pleasure, it's possible to overcome your Stone Age genetic preference for rest over exercise.

LIVE 10 YEARS LONGER WITH EXERCISE

One of the theories for why we age relates to the length of the telomeres on the ends of our chromosomes. Telomeres are highly repetitive sequences of DNA at the ends of chromosomes that keep the double-stranded DNA from unraveling. They can be likened to the plastic tips on the ends of your shoelaces that prevent them from becoming frayed. Each time a cell replicates, one of the telomere "beads" (repeated DNA sequences) drops off. When all of the telomeres have dropped off, the cell can no longer replicate and dies. A recent study from King's College in London compared telomere length in a group of 2,400 twins. The mean difference in telomere length between the most and least physically active people was 200 nucleotides. In other words, the most active participants had the telomere length of people 10 years younger. The least active people got only 16 minutes of exercise a week, while the most active about 200 minutes. Three hours of exercise a week translated into the reversal of 10 years of aging—as far as telomere length is concerned. The preservation of your telomeres is just one in a long list of benefits from exercise.

We recommend three types of exercise: aerobic exercise, strength training, and flexibility training. In the remainder of this chapter, we'll describe a comprehensive fitness program that incorporates each of these in turn.

AEROBIC EXERCISE

The term *aerobics* was coined by Dr. Ken Cooper in 1968 and refers to exercise done in the presence of adequate oxygen. The basic idea behind aerobic exercise is to perform physical activity that gets the heart and lungs working at a level of intensity that increases the flow of oxygen to the body. This type of sustained activity results in improved cardiovascular fitness and is essential to maintaining cardiovascular health. Examples of aerobic exercise include fast walking, jogging, vigorous bicycling, in-line skating, rowing, and cross-country skiing.

Although aerobic exercise is done mainly to improve and maintain cardiovascular health, almost every tissue in your body benefits from it. Aerobic exercise is especially beneficial to get rid of abdominal fat and is also one of the best ways to increase your natural supply of endorphins. Endorphins, chemically related to morphine, are responsible for the "runner's high," and they help explain why some long-distance runners seem to become addicted to running. Endorphins are the reason some runners keep going even when their doctors tell them to stop or take a break for a few weeks or months to let tissues heal—just like an addict hooked on drugs. When you perform aerobic exercise regularly, you increase your endorphin level, which can play a key role in helping you overcome the natural tendency to not want to exercise. Indeed it is an effective treatment for mild depression.

Types of Aerobic Exercise

Exercise experts remind us that the best aerobic exercise for you is the one you are most likely to keep doing. Therefore, pick something that you enjoy and is convenient for you. Fast walking, jogging, swimming, and cycling are good aerobic exercises. They elevate your heart rate for a sustained period, and they don't require any specialized equipment. Golf and bowling aren't aerobic since they don't increase your heart rate enough. Downhill skiing doesn't count since it doesn't elevate your heart rate continuously. Cross-country skiing, on the other hand, is aerobic. Competitive sports such as racquetball and basketball can be aerobic exercise if you play continuously.

Some types of activity, such as running, jogging, or jumping rope, involve repetitive pounding or jarring to the joints. We do need to note that people who run or jog on a regular basis often develop problems with their joints and tendons. With this in mind, you may want to consider activities such as fast walking, cross-country skiing, bicycling, swimming, or rowing that are less stressful to your joints.

Warning—Before You Begin!

Speak to your health professional. If you are over 40 years of age and haven't been exercising on a regular basis previously, you should speak with your health professional before starting on an exercise program. This is particularly important if you have a history of heart disease or other serious illness or have been expressing any unusual symptoms, such as chest pain or pressure, lightheadedness, dizziness, or back or joint pain.

Stress test. The American Heart Association recommends you get an exercise stress test before beginning exercise if:

- You are a man over 40 or a woman over 50 (heart disease typically strikes men 10 years earlier than it does women).

- You have two or more coronary risk factors, such as family history of heart disease, cigarette smoking, hypertension, elevated cholesterol, diabetes, or sedentary lifestyle (see Chapter 2).

- You have had abnormal results at a physical.

GUIDELINES FOR AEROBIC EXERCISE

Frequency

The Centers for Disease Control and Prevention (CDC) and the American College of Sports Medicine (ACSM) suggest that all adults perform aerobic exercise for at least 30 minutes on as many days of the week as possible. You can exercise every day, but three times a week is regarded as a minimum. You should also try not to let there be more than 3 days between aerobic sessions.

Intensity and Duration

To achieve maximal training benefits, you want to keep your heart rate in a range between 65 and 85 percent of your maximum predicted heart rate (MPHR). MPHR is calculated as 220 minus your age, so MPHR would be 190 for a 30-year-old and 170 for a 50-year-old. The chart on the next page lists the ranges for various age groups.

Your Aerobic Session

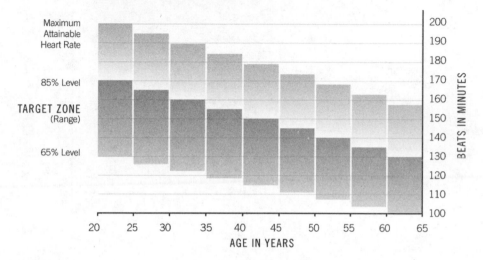

MAXIMUM HEART RATE
AND TRAINING HEART RATE RANGE

The maximum heart rate shows the peak heart rate for average individuals. Any individual may vary from these averages, which apply to about two-thirds of the population. Shown below the maximum heart rate are the corresponding training heart rate ranges for each age.

There are four phases to an aerobic session.

Warmup. Spend 3 to 5 minutes walking and warming your muscles. This increases bloodflow to the muscles and prepares them for the next aerobics stage. As part of your warmup, you also want to do in a slow, gentle fashion the neck, back, and abdominal stretches described below.

Aerobics exercise. This is where you get your heart rate into your desired training range of 65 to 85 percent of your MPHR. Training benefits to the heart begin after 20 minutes of continuous exercise, so sessions should last at least this long, but may extend to 30 to 40 minutes or longer to really take advantage of the aerobics phase of your exercise. If walking is your preferred form of aerobic exercise, you'll need to walk at a rate of at least 4 miles an hour to get close to your training range.

Cooldown. After completing your aerobics phase, walk slowly for about 3 to 5 minutes to allow your heart rate to return to normal and prevent pooling of blood in your lower body. This is the stage of exercise during which arrhythmias (abnormal heartbeats) are most likely, and cooling down slowly reduces their likelihood.

TARGET EXERCISE PATTERN

The diagram shows a desirable pattern of heart rate
for an aerobic exercise session

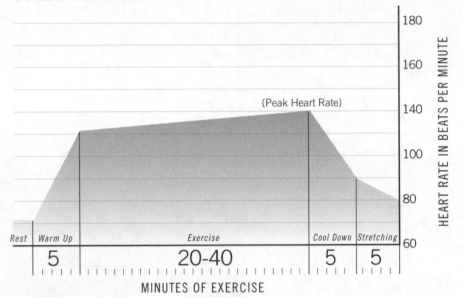

Stretching. Tendons, ligaments, and muscles tend to tighten after a vigorous exercise session, so it is important to engage in a few minutes of gentle stretching of the muscles which were just used.

How to Exercise Safely

Begin slowly. Once you have the go-ahead from your health professional, the next step is to ease into exercise. The objective is to exercise on a regular basis and build this activity into a predictable routine. Endurance and speed will come naturally as your fitness improves.

Don't overdo it. Forget about "No pain, no gain" and replace it with "No pain, no pain." Aerobic exercise should be associated with a sense of exertion, but not to the point that you feel pain. Should you develop discomfort in your ankles, shins, knees, hips, or back, slow down or change what you're doing. *If you ever feel pain in your chest, stop immediately and consult your doctor!* If the pain persists for more than a few minutes, it could be a sign of an impending heart attack, and you should chew a full-strength 5-grain (325-milligram) aspirin tablet and seek medical care on an emergency basis.

Build up gradually. One of the main reasons people discontinue exercise programs is that they overdo it early on. Going too fast too soon leads to soreness and injuries and makes further exercise difficult or impossible. Begin slowly and build up gradually. Remember: *"No pain, no pain."* The important thing is to do it; don't worry about how quickly you are progressing.

Make it a habit. The most common reason that people give for why they don't exercise regularly is lack of time. Very few people these days have any extra time, but consider that all of the U.S. presidents for the past several decades have continued to exercise regularly throughout the course of their presidencies. They *made* time. "Failing to plan is planning to fail" applies more so to exercise than most other activities. Making exercise a priority and establishing a regular time for it in your schedule is critical to success. Most people find that exercising first thing in the morning before breakfast works best. You may prefer to exercise at other times of the day, but you want to wait 30 to 45 minutes after eating to exercise, and working out and exercising before bed can interfere with sleep.

What You Will Need

Shoes and clothes. It is important to choose clothing and shoes that are appropriate for both climate and activity. High-performance clothing including synthetic blends such as Lycra have the advantage of wicking sweat away from your body. This keeps you warmer and is useful for exercise in the cold. Layering of clothing is a good plan in cold weather, as you tend to warm up with progressive exercise and can add or remove layers as needed.

Heart rate monitor. It is important to monitor your heart rate when first beginning aerobic exercise to ensure that you are keeping your heart rate within your target training zone. Use of a stopwatch, a wristwatch with a second hand, or some type of heart rate monitor that has electrodes attached to the chest or elsewhere can be helpful. After you have been exercising for a period of time, you'll develop a sense of what your target heart rate feels like and you won't need to monitor your pulse rate as often.

Water. If you're exercising within your target heart rate for 30 minutes or more, you may lose a lot of your body water through sweating. Drinking water before, during, and after workouts can help keep you hydrated. Keep an eye on the color of your urine. If it becomes dark, drink two more glasses of water. One 8-ounce glass of water is about right for every 15 minutes of sustained aerobic exercise. Exercise in hot or humid conditions increases your need for water even further.

Take It to a Higher Level—Add Interval Training

Interval training refers to a type of aerobic exercise in which you vary the intensity of your workout periodically. In a typical interval training workout, you exercise in your normal training range at 65 to 85 percent of your maximum heart rate most of the time, but periodically pick up the pace so that you are at 95 percent of your maximum (or more) for 15 seconds to 2 minutes. If jogging is your preferred aerobic activity, you would jog at your normal rate for, say, 3 minutes, then run as fast as you can for 1 minute, then slow back down for 4 minutes, then sprint again for, say, 45 seconds, and keep varying both the length of time and intervals between the bursts of maximal exertion.

These bursts of near maximum exertion add several additional benefits to your aerobics program. Fat loss is greater, cardiovascular conditioning occurs more quickly, and injuries occur less often. You will obtain many benefits from a standard aerobics program. Adding interval training will take you to an even higher fitness level.

STRENGTH TRAINING

Strength training is a specific type of exercise in which muscular force is exerted against an opposing object to build strength. Few enough people perform aerobic exercise regularly, but even fewer engage in regular strength training. Among individuals over 65, which is the group that needs to maintain their muscle mass the most, only 12 percent perform regular strength training. People who don't do this type of exercise will lose as much as 40 percent of their muscle mass between 20 and 80 years of age. While the primary benefit of aerobic exercise is keeping your cardiovascular system healthy, strength training can keep some of your hormone levels higher. Our original blueprint written in the Stone Age only planned for our hormones to remain at optimal levels into our forties. Strength training is designed to counteract the natural tendency of your muscles to shrink with age and will keep some of your hormones higher.

Most important of all, strength training can help you feel better, look better, and age better. You will reap a host of health benefits from your strength training workouts. You will increase your muscle mass, enabling you to do your activities of daily living more easily. Your ligaments and tendons will be stronger, which will protect your joints from arthritis. Strength training increases bonc mass, reducing the risk of osteoporosis. You'll be able to eat more, while weighing less, since muscle mass burns so many calories. But, most important of all, strength training will increase your levels of anabolic hormones—the hormones that build you up, including testosterone, growth hormone, and DHEA—naturally and safely. A study published in 1999 in the *Journal of Applied Physiology* showed that men averaging 62 years of age who performed strength training exercise twice a week for 12 weeks were able to raise their levels of testosterone to the average levels of 30-year-old men. Since our levels of these hormones begin a steady decline after age 30, strength training is truly "the biggest no-brainer since the Stone Age." Men and women of any age can literally erase some of the harmful effects of constantly declining hormone levels by simply using their muscles to stretch some resistance tubes or lift some dumbbells for a total of 60 to 90 minutes each week. Hopefully these benefits will motivate you to get started, so let's talk about how to do a strength training workout.

STRENGTH TRAINING WORKOUT

Objects used in strength training include free weights such as barbells and dumbbells, specialized weight machines, and elastic resistance bands or tubes. Resistance bands are flat hollow rubber bands, and resistance tubes are hollow tubes. They come in various thicknesses, which determine the

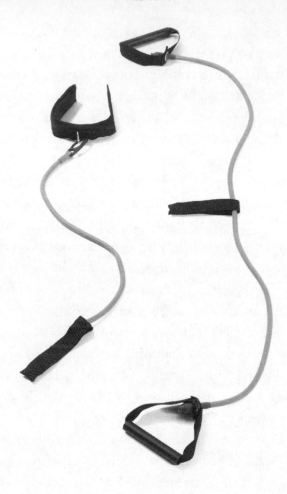

amount of effort needed to stretch them. We will describe a strength training program below that uses resistance tubes.

Convenience is the key to strength training, and the reason we like using tubes is because they're so convenient. They are very inexpensive, can be used at home, and are lightweight and portable enough to take with you while traveling. You can get started for a few dollars, and using these devices doesn't require you to drive to the gym or pay monthly dues. Because they are so safe, easy to use, inexpensive, portable, and effective for all body types and all muscle groups, we have chosen resistance tubes as our strength training equipment of choice. They can easily be found on the Internet or at most sports and department stores. The series of exercises shown below can also be done by using handheld dumbbells if you prefer. By making a few modifications, you can also easily do them on weight machines at a gym.

The simplest way to complete these exercises is by using two types of resistance tubes along with door attachment clips. These will be an arm band with handles at each end for upper-body exercises and an ankle band with an ankle strap for lower-body training. Special door clips attach the tubes to a door for some of these exercises. Kits of these resistance tubes and clips are available very inexpensively at sporting goods stores or online.

Resistance tubes are color coded—yellow, green, red, blue, black—depending on the level of difficulty and amount of tension desired. There is no consistency among manufacturers; for one, green is an easier band, where for another it is the most difficult, and the same applies to the other colors. Tubes are so inexpensive that you might want to start with two or three adjacent colors. As you get stronger, you will have the next color on hand to use. Additionally, you might be stronger on some exercises than others and want to use several different tubes during your workout for maximum effect. Women who haven't been doing strength training may want to begin with the lightest tubes, and men, with intermediate ones. Men and women who have been doing strength training previously may need the tighter ones, which offer considerable resistance.

Guidelines for Resistance Tube Strength Training

• In each exercise you will be using resistance and breathing to control all movements.

• Resistance is achieved by using very slow and controlled movements. Maintain continuous tension in the tube throughout the exercise. Do not let the tube go slack or snap back.

• Maintain the same rhythm while resisting and releasing.

• In upper-body exercises, keep your wrist straight so it is in line with the forearm at all times.

• Consciously squeeze the working muscle with every repetition.

• Exhale during the exertion part of the exercise and inhale on the release. Inhaling or holding your breath during the exertion phase of strength training can increase muscle fatigue and raise your blood pressure excessively and should be avoided. *Always* exhale while exerting.

• Form is key to this type of exercise. Optimal results are achieved by keeping abdominal muscles engaged (tightened) at all times and by keeping shoulders back and down, relative to your head and neck. Good posture encourages good form.

• Muscles grow stronger during rest days. Do not work the same muscle groups 2 days in a row.

• Be sure to warm up for 5 minutes before your workout by walking in place, going up and down stairs, dancing, or any continuous movement that involves the big muscles of the legs.

• Always place the door attachment along the hinged side of the door. Never hook it along the side with the door handle as it can pull loose from this side, possibly causing injury.

If using tubes for your workouts, as you improve in strength, you can increase the intensity of your workout by adjusting your body (e.g., standing farther from the door), doubling your tubes, or using thicker tubes. If you are using dumbbells or weight machines, you will want to gradually lift heavier weights, increase the number of repetitions, or change the speed of each repetition. Slowing the speed of your repetitions can add intensity to your workout.

Start with a resistance level that enables you to safely and effectively perform the exercise. If muscle strength is your goal, find a resistance band that enables you to perform 8 to 12 repetitions, with the last several repetitions being somewhat difficult. If endurance is your goal, find a band that enables you to perform 12 to 15 repetitions, with the last several reps being challenging. No matter your goals, you will be stronger and have more muscle tone with this workout.

For optimal fitness, it is important to work the muscles of both the upper and lower body and core. The workout on the following pages, including upper- and lower-body and core exercises followed by stretches, should take about 40 minutes. If you prefer to concentrate on one area at a time, you could simply double the number of sets of upper-body exercises one day (but no lower body) and then do just the opposite the next time. Core-body exercises should be done at each session.

UPPER-BODY EXERCISES

There are six upper-body exercises designed to train the major muscle groups, including the chest (pecs), back (traps, lats), shoulders (delts), back of the upper arm (triceps), and front of the upper arm (biceps).

You will need an arm band with handles at each end and an attached door clip.

CHEST PRESS
Muscles worked: pecs, triceps

Attach resistance band to door at chest height.

Facing away from the door, place one handle in each hand, hands at shoulders, palms facing one another.

Step away from the door so that your arms feel some tension as you begin.

Engage your abdominal muscles and press arms out in front of you straight away from chest. Keep wrists straight and palms facing one another and keep arms at shoulder height.

Hold for 2 seconds and return slowly to start position.

Do one set of 8 to 12 reps. Rest 30 to 60 seconds, then repeat a second set.

TRICEPS KICKBACK
Muscles worked: triceps

Starting with right arm at side, step on band with right foot at a point to create tension.

Step forward with left foot, lean body forward slightly, keeping your back straight, and place your left hand on thigh for balance. Pull abs in. Hold tube in hand with elbow bent at a 90-degree angle. Keep your elbow in the same spot throughout the exercise.

Extend forearm behind you, keeping your elbow still as you pass your hip in a fluid motion.

Squeeze back of upper arm as you hold at the top for 2 seconds, and then return slowly to the starting position.

Do one set of 8 to 12 reps. Rest 30 to 60 seconds, then repeat a second set.

Do the same with your other arm.

OVERHEAD PRESS

Muscles worked: lats, delts, and triceps

Holding band with both hands, step on band with one foot (feet are parallel).

Bring hands to shoulder height with palms facing forward, making a goal post position.

Completely extend hands above head, hold for 2 seconds, then return slowly to goal post position.

For even more intensity, stand on band with feet shoulder-width apart.

Do one set of 8 to 12 reps. Rest 30 to 60 seconds, then repeat a second set.

Caution: Individuals over 50 years of age should perform this exercise with caution and with less than maximum resistance to avoid shoulder injuries. If you feel any discomfort while performing this exercise, exclude it from your routine or try doing one arm at a time.

UPRIGHT ROW
Muscles worked: delts

Holding band with both hands, step on band with one foot (feet are parallel).

Starting with arms fully extended, palms facing towards body, bring hands to just below the chin, extending elbows out, but not higher than shoulders. Hold for 2 seconds and slowly return to start position.

For even more intensity, stand on band with both feet shoulder-width apart.

Do one set of 8 to 12 reps. Rest 30 to 60 seconds, then repeat a second set.

BENT-OVER ONE-ARM ROW

Muscles worked: traps and shoulders

Starting with right arm at side, step on tube with right foot at a point to create tension. Take a slight step forward with your left foot.

Lean body forward slightly, keeping your back straight and right arm extended. Place your left hand on thigh for balance. Pull abs in.

Pull your elbow back until your hand is at your rib cage. Hold for 2 seconds and return slowly to start position. Do one set of 8 to 12 reps. Rest 30 to 60 seconds, then repeat a second set.

Do the same with your other arm.

BICEPS CURL
Muscles worked: biceps

Holding tube with both hands, step on tube with one foot (feet are parallel). To increase tension, stand on tube with both feet, either together or up to shoulder-width apart.

Starting with arms fully extended at your sides, palms facing back, pull up on tube while rotating your hands up towards the ceiling, elbows glued to your side. Without bending your wrists, bring hands up to your shoulder. Hold for 2 seconds, release and slowly return to starting position.

Do one set of 8 to 12 reps. Rest 30 to 60 seconds, then repeat a second set.

Upon completion of your series of upper-body resistance exercises, you will do the six upper-body stretches beginning on page 392.

Lower-Body Exercises

You will perform four exercises designed to train the major muscles of the lower body: front of thigh (quads), back of thigh (hamstrings), rear end (glutes), outer thigh (hip abductors), and inner thigh (hip adductors). You will need an ankle tube with ankle strap and door clip, and a sturdy chair.

You will attach the ankle strap to your right ankle, complete all four exercises on the right side, then switch the ankle strap to your left side and repeat.

SEATED LEG EXTENSION
Muscles worked: quads

Set up clip tube in door at ankle height, place chair facing away from door.

Sit on chair, attach ankle cuff to left ankle, with clip at back of ankle.

Sitting up straight, engage abdominals, plant both feet firmly on the ground, extend leg until it is parallel to floor, hold for 2 seconds, and slowly release.

Do one set of 8 to 12 reps. Rest 30 to 60 seconds, then repeat a second set.

Move chair away from the door to increase tension as your strength improves.

STANDING LEG CURL
Muscles worked: hamstrings

Stand up and face the door. Ankle clip will be towards the door. Pull abs in.

Move chair to your side so you can use it for balance.

Keeping your upper body still and upright, bend your knee to 90 degrees, keeping your thigh stationary. Keep knee pointing down. Hold for 2 seconds.

Allow your leg to return slowly to starting position.

Do one set of 8 to 12 reps. Rest 30 to 60 seconds, then repeat a second set.

OUTER-THIGH LIFT

Muscles worked: glutes, hip abductors

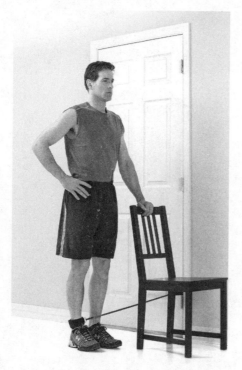 

Stand perpendicular to the door with attached ankle farthest from the door and clip pointing towards the door.

Move chair so that it is in front of you.

Slowly, move your ankle in an arc away from the door, hold for 2 seconds, then let it return to starting position.

Do one set of 8 to 12 reps. Rest 30 to 60 seconds, then repeat a second set.

INNER-THIGH LIFT

Muscles worked: hip adductors

Face sideways with attached ankle closest to the door and clip towards door.

Move chair so that your leg can move freely in front of your body.

Hold on to the back of the chair for balance.

Move leg across the front of your body, bringing the heel of the leg you are working to your opposite toe (or beyond). Hold for 2 seconds and slowly release.

Do one set of 8 to 12 reps. Rest 30 to 60 seconds, then repeat a second set.

After completing the series of exercises with one leg, attach the ankle tube to your opposite leg and repeat the entire series. Upon completion of the lower-body resistance exercises, you will do the lower-body stretches beginning on page 386.

CORE-BODY EXERCISES

A strong core is central to linking upper and lower body. Most of your movement, such as walking, bending, and lifting, is powered by the core musculature—the muscles of the abdomen, back, and pelvis. As you get stronger in your arms and legs, it is critical to have a powerful core to assist in your exercise and other activities of daily life. Although ignored for many years, core strengthening has come to occupy—appropriately enough—a central role in most fitness programs. Benefits of a strong core include:

- Reduced risk of injury (particularly low back)
- Improved ability to perform daily activities
- Better performance in all sporting activity
- Reduction in back pain
- Improved posture

Guidelines for Core Exercises

The two exercises you will do to strengthen your core are abdominal crunches and the side plank. In addition, Pilates exercises (not described here) are very effective for core strengthening and can be added to your program if desired. It is important while doing any abdominal exercise to pull your belly inward (like you're trying to suck your belly in to fit into a tight pair of jeans) during the exhalation phase of each exercise.

ABDOMINAL CRUNCHES

Lie with your back flat on the floor, knees bent at a right angle with your feet on a chair or on a wall. Place your hands gently behind your head.

Exhale while pulling belly in and do a situp to 45 degrees, then return to the floor and inhale.

Try to keep your abs tight throughout, but do NOT hold your breath.

Shoulders should touch the floor at the end of each rep, but not necessarily your head.

Work up to 30 to 60 reps or more.

SIDE PLANK

Lie on your right side with your upper body supported by your right arm, as shown in the photograph.

With your left hand on your hip or placed in front of you for balance, exhale and lift your body, keeping your hips in line with your shoulders.

Tighten your abs, continue to breathe, and hold 10 to 60 seconds. Do not let your hips sag.

Repeat 8 to 12 times on the right side, then roll over and repeat on the left side.

FLEXIBILITY TRAINING

There are few enough people who like to do aerobics, and fewer still who do regular strength training. Stretching has even fewer devotees, although yoga is a relaxing way to maintain flexibility, and many are discovering it every day. People who are stiff don't like to stretch because it is uncomfortable. People who are limber don't feel that they need it. Another problem with stretching is that, unlike aerobic exercise and strength training, which lead to rapid improvements in fitness or physique, the benefits seen from stretching are not as apparent and may seem to take forever to achieve.

But after exercise and as a result of the aging process, there is a tendency for individual muscle fibers to get shorter. With time, this muscle shortening puts increased stress on ligaments and joints, increasing the risk of arthritis. Stretching or performing flexibility training after each exercise session helps prevent this damage.

Unless you stretch these muscles regularly, they will continue to shorten, which can lead to poor posture, decreased range of motion, muscle stiffness, and other problems. With stretching, "old age posture" can be prevented and movement can become more smooth and comfortable. Many people find another profound benefit of stretching—a feeling of overall mental and physical relaxation as you breathe deeply and as the tension leaves the muscles during the stretch.

We describe a series of basic stretches below. In addition, you may want to consider joining a yoga class for additional flexibility training.

Guidelines for Stretching

• Stretching can be a little uncomfortable, but shouldn't hurt. If it does, you're pushing too hard or doing something wrong.

• Stretch each muscle to the point of tension, but not to the point of pain. This sense of tension or lengthening of the muscle should be felt the most in the middle of the muscle.

• Muscles tend to shorten after they are exercised, and they must be lengthened after either an aerobic or strength training workout in order to maintain flexibility.

• Never try to stretch cold muscles.

• Hold each stretch for up to 30 seconds, but don't bounce.

• Approach your flexibility training as a time to slow down and relax. Breathe slowly, and inhale and exhale deeply.

• If you only do flexibility training, do it at least three times a week. Try it after a busy workday or a stressful event. You can try it while watching TV.

THE STRETCHES

There are three neck, back, and abdominal stretches, six upper-body stretches, and six lower-body stretches. After aerobic exercise, perform at least the six lower-body stretches and, if possible, the full series of stretches. The full series of stretches should be done as a course of your strength training workout with the resistance tube. Before beginning to use the tubes, do the three neck, back, and abdominal stretches. After completing your lower-body strength training exercises, do the series of lower-body stretching exercises. Then, after completing your upper-body strength training exercises, do the corresponding six upper-body stretches. Keep in mind you should not strain or force any of these stretches.

NECK, BACK, AND ABDOMINAL STRETCHES

NECK STRETCH

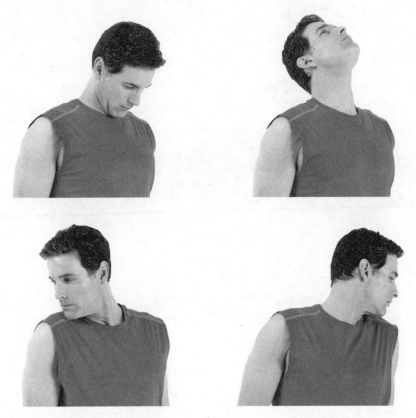

Sit or stand and relax your shoulders.

Let your head drop down slowly towards your chest until you feel a gentle pull in the muscles at the back of your neck. Hold for 8 to 10 seconds.

Very gently push your head backwards until you feel a slight pull at the front of your neck. Don't go back as far as you can go. Hold for 8 to 10 seconds.

Do the same exercise towards each side, letting your head gently drop towards your shoulders. Hold each side for 8 to 10 seconds.

LOWER-BACK STRETCH

Sit on the floor with your legs straight in front of you.

Bend your right knee to 90 degrees, then move your right foot to the outside of your left knee, so that your right leg crosses over your left.

Place your left elbow to the outside of your right knee and gently twist to the right.

Hold for 10 to 20 seconds.

Repeat on the other side.

ABDOMINAL STRETCH

Kneel on the floor with your back straight.

Place your hands on your lower back and gently bend your upper body back. You should feel the stretch in your abdomen. Keep your abs engaged so as not to put stress on your lower back.

Hold for 8 to 10 seconds. Repeat.

LOWER-BODY STRETCHES

CALF STRETCHES

Stand about 1 or 2 feet away from a wall with your toes pointing straight ahead.

Hold your hands against the wall and lean your body toward the wall at a 45-degree angle as you press your heels down toward the floor as shown.

You should feel the stretch in your calf muscles.

Hold for 15 to 30 seconds.

ANKLE STRETCHES

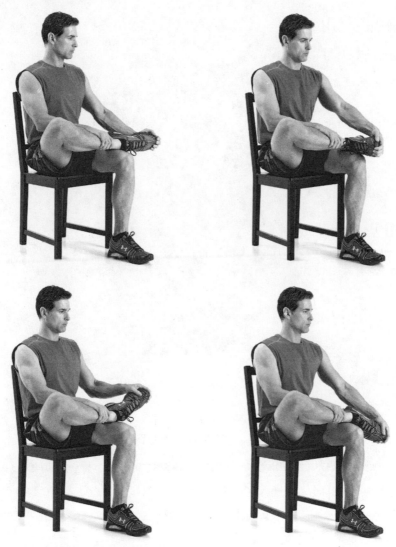

Sit on a sturdy chair and lift your right leg as shown.

Use your right hand to hold your ankle and your left hand to hold your heel.

Use your left hand to pull your foot towards you. Hold for 5 seconds.

Then, push your foot away and hold for 5 seconds.

Finally, rotate your foot first clockwise a few times, then counterclockwise a few times.

Switch sides and repeat on your other ankle.

HAMSTRING STRETCH

Stand with your feet shoulder-width apart.

Bend at the hips (not the waist), letting your upper body hang.

Reach your hands towards the floor until you feel a slight stretch in your hamstrings.

If needed, bend your knees slightly.

If this stretch causes discomfort in your low back, keep your back straight and place your hands on your thighs.

For a deeper stretch, place your palms flat on the ground.

Hold for 15 to 30 seconds.

ANTERIOR-THIGH (QUADS) STRETCH

Balancing yourself with your left hand on the wall, take hold of your right foot or ankle and bring it behind you.

Keep your right knee pointing down and your rear end tucked and not sticking out.

Bring your heel as close to your buttock as possible without pain.

Hold for 10 to 15 seconds, then repeat on the other side.

BUTTOCKS STRETCH

Lie on your back and gently pull your right knee towards your chest.

Keep your left leg straight.

Hold for 15 to 30 seconds, then repeat on the other side.

GROIN STRETCH

Sitting comfortably on the floor, touch the soles of your feet together.

Take hold of your feet or ankles with your hands.

Gently pull your upper body forward, bending at the waist.

Try to keep your back straight.

Hold for 15 to 30 seconds.

UPPER-BODY STRETCHES

CHEST STRETCH

Stand in a doorway facing perpendicular to the wall.

Bend your arm to 90 degrees and place your forearm against the door jam.

Rotate your body away from the door until you feel a stretch in your chest.

Hold for 10 seconds, then repeat on the other side.

UPPER-BACK STRETCH

Extend your arms in front of you with your fingers interlaced and your palms facing forward.

Push your hands forward while exhaling, allowing your back to arch slightly.

Hold for 10 to 15 seconds.

SHOULDER STRETCH

Place your left arm across the front of your body.

Hook your right elbow in front of your left elbow.

Pull slightly to the right while turning your head slightly to the left.

Hold for 15 to 30 seconds, then repeat on the other side.

DELTOIDS STRETCH

With your hands behind your back, take hold of your right wrist with your left hand.

Pull gently towards the left until you feel the stretch in the front of your shoulder.

Hold for 10 to 15 seconds, then repeat on the other side.

TRICEPS STRETCH

Lift your left arm straight overhead and then allow elbow to bend.

Allow fingers to touch back between shoulder blades.

Place your right hand across the top of your head and take hold of left elbow. Stand up straight and feel the stretch in left triceps.

Hold for 10 to 15 seconds, then repeat on the other side.

BICEPS STRETCH

Face away from wall and bend over.

Place hands close together, palms on wall as high on wall as possible. Point fingers towards ceiling.

Squat down slowly.

Hold for 15 to 30 seconds.

PUTTING IT ALL TOGETHER

Exercise plays a critical role in enhancing all parts of your *TRANSCEND* program. An effective exercise program need not take very much time or cost a lot of money. A good pair of walking or running shoes, some rubber tubes, and your complete fitness center is ready to go—no monthly fees, no fancy equipment. And all it takes is about 3 hours a week of your time.

SUGGESTED BASIC EXERCISE PROGRAM

Aerobics—pick an exercise you enjoy, such as fast walking, jogging, inline skating, swimming, or bicycling, that will get your heart rate into your aerobic training range of 65 to 85 percent of your maximum predicted heart rate (220 minus your age). Don't start a program if you're over 40 until you've been cleared by your doctor. Perform this activity three times a week (or more) for at least 30 minutes. Do a series of stretches afterwards.

    Strength training—Follow the program described on the preceding pages using resistance tubes, dumbbells, or weight machines two or three times a week.

    Flexibility—Perform the lower-body stretches we've described after each aerobic exercise session that uses the lower body, such as walking or jogging. Do both upper- and lower-body stretches after each aerobic exercise session that uses both upper and lower body, such as cross-country skiing, tennis, or racquetball. Perform a series of upper, lower, and core stretches after each strength training session.

A TYPICAL WEEK OF EXERCISE

MONDAY	TUESDAY	WEDNESDAY	THURSDAY	FRIDAY	SATURDAY	SUNDAY
Aerobics 30 min + Targeted stretches	Off	Strength 45 min + All stretches	Aerobics 30 min + Targeted stretches	Off	Aerobics 60 min + Targeted stretches	Strength 45 min + All stretches

You will notice we've included a longer aerobics workout on the weekend. With exercise, the more the better, so feel free to increase the amount of time you spend doing any of the above. By following this program, you will satisfy (and exceed) the minimum requirements set forth by the U.S. Department of Health and Human Services that adults get a minimum of 150 minutes of moderate exercise or 75 minutes of vigorous exercise weekly along with muscle strengthening twice a week. According to their definitions, moderate exercise includes brisk walking, water aerobics, general gardening, and ballroom dancing, while racewalking, jogging, swimming, jumping rope, or hiking uphill or with a heavy backpack are examples of vigorous exercise.

The people to whom we've taught this program constantly give us feedback saying that when they exercise consistently, they have much more energy throughout the day. They report being happier and less stressed when faced with the challenges of everyday life.

If you already exercise, vary your routine from time to time so it doesn't get boring. If your main aerobic exercise is jogging, add occasional cross-country skiing, swimming, bicycling, or some other activities available in your area. If you're walking, try different routes—explore historic and scenic paths in your community. If you usually use resistance tubes, try an occasional workout with dumbbells or weight machines.

If you have never exercised before or if you have a medical challenge, you can safely exercise as long as you first consult with your doctor, start slowly, and use good common sense. We have advised people who are 80 years old who do this workout on a regular basis and didn't begin exercising until they were in their seventies.

Above all, keep safe and have fun—it will help you to get to the future bridges in good shape.

15

NEW TECHNOLOGIES

"I want to know how forever feels."
—KENNY CHESNEY

Up until just recently there was no conceivable way to cheat death, so our philosophies rationalized death as actually a good thing. "Death gives meaning to life" is an age-old adage, but in our view death does exactly the opposite. It is a great robber of relationships, knowledge, wisdom, and skill, all the things that make life worth living. The authors believe that it is life itself—and all the things we can do with our lives—that gives life meaning.

Although we are not yet at the point where we can adopt a fixed program to forestall death indefinitely, we do have the means right now to slow down disease and the aging process sufficiently so that even older baby boomers like the authors of this book can not only be alive but *well* when we do gain the knowledge and means to stop and even reverse the processes that lead to disease and death. We firmly believe that we'll get to a point in a few decades where we can literally create back-up copies of the information in our bodies and brains that makes us who we are. Then, if we're careful to make frequent back-ups, we'll become immune even to the proverbial bus in the road (and, of course, we'll have replaced buses with vehicles that fly with nano wings by that time).

Reader: So if I just follow the nine steps in your *TRANSCEND* program, it really is possible that I can live forever?

RayandTerry2023: We're still here and doing fine.

RayandTerry2034: And we're still here as well and doing even better. The prospects for human immortality have never been brighter.

Terry: Yes, but remember that one of the nine *TRANSCEND* steps is "New Technologies."

Reader: So how exactly do I take that step?

RayandTerry2023: You take it by keeping yourself in good health until the new technologies we've been talking about throughout this book become available in the near future.

Reader: I knew there was a hitch.

Terry: The only hitch is that you'll need to stay in good health while you wait.

Reader: And that's what the other eight steps are for.

Terry2034: Precisely, but we wanted you to know what's coming, so you'll be motivated to stay healthy the old-fashioned way a while longer.

Ray2034: And because of the explosive nature of exponential growth, "a while" is going to be a lot quicker than most people assume.

BEYOND SPAGHETTI ON THE WALL

As we pointed out at the beginning of this book, the key point with regard to new technologies is that health and medicine have now been transformed from a hit-or-miss affair (throwing spaghetti on the wall to see what will stick) into an information technology. A key feature of information technologies is that they inevitably proceed at an exponential pace. Several developments, all of which have occurred at the same time, are responsible for this grand transformation.

We finished collecting the human genome in 2003. As we showed in the Introduction, genetic sequencing is an outstanding example of smooth exponential growth in which the amount of genetic data collected doubled every year, and the cost came down by about half each year.

We have means of turning genes off by using RNA interference, a method that was quickly recognized with the Nobel Prize. Numerous genes promote disease, aging, weight gain, and other undesirable effects that we would like to turn off, and there are over 1,000 drugs in the development and testing pipeline that use RNA interference methods.

We now have effective ways to add new genes using new forms of gene therapy—for example, taking cells out of the body, adding a new gene, replicating that cell a millionfold, and injecting the now gene-enhanced cells back into the body. These techniques avoid the problems of earlier forms of gene therapy that triggered adverse reactions by the immune system or ended up inserting the new gene in the wrong place. So we'll not just create designer babies but designer baby boomers—essentially changing our adult genes to encourage health and to inhibit aging.

We have new means of reprogramming the information processes that underlie biology by turning proteins and enzymes on or off and changing biochemical processes. These interventions (such as new drug developments) can now be designed on computers rather than the time-consuming method of trying out existing or new chemicals one at a time.

Drugs and other interventions can now be tested on increasingly sophisticated biological simulators, which are scaling up in complexity at an exponential pace. New drug testing will also be done much more quickly when we're able to use reliable biological simulators that can operate much faster than real time. We can already transform one type of cell into another. Recently it was discovered that by adding four genes to ordinary skin cells, we can change them into pluripotent stem cells that can be transformed into many other types of cells. There are already experimental treatments to rejuvenate a variety of body tissues, such as the heart, using adult stem cells, which we'll ultimately be able to create from our skin cells. The promise of this line of research is to rejuvenate all of the cells in our bodies using cells with our own DNA. And these cells can be DNA corrected.

New cells used to rejuvenate your tissues can also be used to extend or lengthen their telomeres (the repeated sequences of DNA at the end of chromosomes that indicate aging). There have been recent breakthroughs in the understanding of the telomeres. For example, it has recently been discovered that both cancer cells and germ-line cells (such as sperm and egg cells) produce an enzyme known as *telomerase*, which rebuilds the telomeres after each cell division, allowing these cells to replicate indefinitely. Research into the workings of telomeres and telomerase holds great promise for life-extending therapies as a technique to stop aging. Since cancer cells also produce telomerase, work is being directed at how to block telomerase to help stop the growth of malignant cells.

LONGEVITY ESCAPE VELOCITY

Aubrey de Grey, by training first an engineer and now a prominent biomedical gerontologist, is a leading advocate of applying the engineering approach to stopping and reversing the aging process. Our bodies and brains are designed to age and die at a young age because that was in the interest of the species as a result of evolution during an era of scarcity. We can now reprogram these processes, an approach he calls "strategies for engineered negligible senescence" (SENS). According to de Grey, we don't need to repair the damage done by aging perfectly, just well enough to buy more time than is going by. Then, as progress continues, we can perfect these age-reversal methods. He calls this tipping point "longevity escape velocity," and he believes that with sufficient funding we are only a couple of decades from achieving this threshold in humans. His Methuselah Foundation is working to show the feasibility of longevity escape velocity in mice. He believes (as do we the authors) that such a demonstration in a mammal that shares 99 percent of our genes will cause a sea change in the public's appreciation for the feasibility of dramatically reengineering the life span of complex animals such as humans. De Grey predicts that the average age at death of people being born at the end of this century will probably be at least 5,000 years.

De Grey has identified seven principal aging processes that can be modified to achieve his goal of longevity escape velocity. He has articulated strategies based on what we already know about aging in humans and other advanced species. Some species, such as worms and fruit flies, have already had their life spans increased four- or five-fold through genetic manipulations. Some of de Grey's strategies include repairing the following:

- *Chromosomal mutations*—We can delete the genes for telomerase and other enzymes that cancer cells need in order to kill us.

• *Toxic cells*—We can eliminate cells that have become toxic by activating the immune system to recognize them as undesirable and to attack them or by targeted "suicide gene therapy" to make them die on their own.

• *Mitochondrial mutations*—We can use gene therapy to move the mitochondrial genes to the nucleus, where they would be more protected. Evolution has already done this with most of the mitochondrial genes, so we can essentially finish evolution's job by moving the rest of these genes to the nucleus.

• *Cell loss and atrophy*—We can use stem cells to replace worn-out or lost cells in crucial organs and systems in which the capacity for cell replacement diminishes as we age.

Now that health and medicine is an information technology, it will be subject to Ray's law of accelerating returns, which is a doubling of capability every year. As we noted earlier, although these technologies are now in an early stage, they will be a million times more powerful in 20 years. And at that point, it will be an entirely new era.

THE BRIDGES AHEAD

That new era will be the full flowering of the biotechnological revolution, which we have referred to as *Bridge Two*. However, even a reprogrammed and optimized biology will ultimately be limited compared to what we can create if we step outside of biology. *Bridge Three* will be based on nanotechnology, which is also an information technology. Nanotechnology involves applying massively parallel computerized processes to reorganize matter and energy at the molecular level to create new materials and new mechanisms even more intricate and powerful than biology.

The nanotechnology revolution is in an even earlier stage, but it will also progress at an exponential pace. There are already dozens of experiments with nanoengineered blood-cell-size devices in the bloodstreams of animals that are performing therapeutic functions. Today, proto-nanoscale devices known as *BioMEMS* (biological microelectromechanical systems) have been developed for a wide range of very precise diagnostic and therapeutic applications. Today's BioMEMS release blood-clotting factors in hemophiliac patients, monitor blood insulin levels in diabetic patients, release dopamine into the brain of patients with Parkinson's disease, monitor electrical activity of patients with neurologic diseases, and directly target malignant tumors with cancer drugs. A team at Sandia National Laboratories has built a micromachine that can grab an individual cell in its "jaw" and implant it with DNA, medications, proteins, or other therapeutic substances. There are already conceptual designs for nanobots to replace red blood cells, white blood cells, and platelets and to do their jobs thousands of times more effectively than our natural cells.

Ultimately we will have billions of nanobots in our bloodstreams augmenting and replacing our biological blood cells, keeping our bloodstreams filled with exactly the right level of nutrients, hormones, and other substances. At the same time, they will remove toxins; destroy pathogens such as bacteria, viruses, cancer cells, and prions; and generally keep every one of our cells vital and in good health.

Nanobots will eventually perform a range of surgical procedures from inside the body, requiring no incisions and leaving no scar. Injected into a patient by the millions and operating synergistically, each nanosurgeon will perform its work one cell at a time, removing cancerous tumors, mending broken bones, or clearing blocked arteries with more precision than any human surgeon ever could. Advances in the field of artificial intelligence will enable individual members of such "nanosurgical teams" to communicate and coordinate their actions. Nanobots will even be able to work inside cells to repair or replace damaged or faulty structures, such as mitochondrial or nuclear DNA.

Nanobots will also continually monitor the status of various bodily systems, allowing us to rapidly adjust treatment regimens to address changes in condition and alerting us to impending emergencies before they happen. These sensors will patrol inside our bodies and, when threats are detected, will initiate treatment, such as dispensing drugs and performing microscopic surgical procedures. A team at Harvard is already developing nanosensors that can test any bodily fluid (such as blood, saliva, or urine) to detect problems such as viral infections, cancers, genetic defects, and potential drug interactions. Researchers at Edinburgh University are developing spray-on nano-health monitors that combine a computer, wireless communications, and sensors for heat, pressure, light, magnetic fields, and electrical currents. And a research team at Pittsburgh's Allegheny Singer Research Institute is developing sensor robots that will work inside the body to detect infection, identify its source, and dispense the appropriate antibiotic to counter the specific pathogen.

Instead of repairing our genetic code, we will eventually be able to completely replace our DNA with microscopic computers whose code could be wirelessly reprogrammed to quickly address threats, such as a viral infection or cancer. Researchers at several universities have already modified cells to perform a number of basic computer functions and have been able to wirelessly communicate with computers embedded inside of cells. Such developments will ultimately lead to the ability to control gene expression by remote control.

Reader: If we're going to have all this technology swimming around in our bodies, it sounds like we won't be human anymore.

Ray2034: It's fair to say that almost everyone is a cyborg here in 2034.

Terry: But, there are cyborgs among us today in 2009—a patient with Parkinson's disease who has neural implants or a patient with diabetes who has an implanted artificial pancreas, for example.

Ray: But we still consider both of these types of patients to be fully human.

Terry2034: And we still consider ourselves to be fully human in the 2030s with our billions of nanobiotic implants. Our technology has always been part of who we are.

16

DETOXIFICATION

"There's so much pollution in the air now that if it weren't for our lungs there'd be no place to put it all."

—ROBERT ORBEN

Your world and your own body are awash in chemicals, many beneficial, some benign, but others are quite hazardous to your health. Billions of pounds of toxic waste are dumped into the environment every year, leaching into our drinking water, accumulating in our food, and carried on the air we breathe. Many common household and personal products such as toilet cleaners, hairsprays, and air "fresheners" contain chemicals that, if not outright deadly, can over time affect systems in our bodies, leading to a host of ailments from hormone imbalance and respiratory disorders to birth defects and cancer. Some jobs may require that we work directly with hazardous materials or radiation. If not, we're still exposed in the workplace and at home to toxins in the carpeting, wall coverings, furniture, and electronic equipment. Even our children's toys may expose them to lead or other dangerous chemicals. And then there's the toxic waste our own bodies produce in the natural course of living.

The fact is we can't avoid exposure to toxins. We can, however, greatly reduce that exposure while strengthening our body's defenses and removing toxins that have accumulated. For this reason, we include *Detoxification* as a crucial component of our *TRANSCEND* program for living long enough to live well forever.

HOW MUCH CLEAN AIR DO WE NEED?

Many of us wonder what Lee Iacocca, former CEO of Chrysler, had in mind when he said, "We've got to pause and ask ourselves: How much clean air do we need?" There's no question the air we breathe is hazardous to our health. From factories and power plants to planes, trains, and automobiles, we pump an enormous amount of toxins into the sky every day. With the increasing emphasis on reducing the energy used to heat and cool our homes and work-places, we are insulating and caulking every last crevice, sealing in even more indoor air pollutants, including outgassing from plastics and other materials, vapors from solvents and cleaners, and tobacco smoke.

Your best defense is to reduce your exposure. Some simple things you can do to limit airborne toxins:

- Conventional "all-purpose," oven, and toilet cleaners, glass cleaners, and detergents contain toxic ingredients. Lemon juice, baking soda, borax, and vinegar can be used instead and are nontoxic. (See "Home and Work Haz-ards" below for specific suggestions.) Commercially available products from Earth Friendly, Ecover, and Seventh Generation are also safe.

- Keep insects at bay with botanical-based pest control products.

- Use stand-alone air filters. Indoor air contains bacteria, viruses, dust mites, pollen, smoke, dead skin, and more. You want a device that removes both large particles that you can see as well as microscopic par-ticles such as bacteria, viruses, and fungi.

- Keep lots of houseplants—plants are ideal for reducing airborne toxins such as benzene and formaldehyde that air filters can't remove.

- Keep printers, copiers, and fax machines as far as possible from your work-space and ventilate the area where these machines are used. The toner car-tridges in laser printers, for example, release ultrafine particles that are inhaled into the lungs at a rate comparable to that from a burning cigarette.

- Wear water-washable clothing, and if you must dry clean, let clothes "outgas" outdoors or in your garage before bringing them into your home. The main solvent used in dry cleaning, perchloroethylene, is on the Environmental Protection Agency (EPA)'s list of suspected carcinogens. People who live in the neighborhoods of dry cleaners appear to have increased rates of cancer.

• Install a gas fireplace to avoid toxins put out by wood-burning fireplaces and wood- or coal-burning stoves.

• Install a radon mitigation system if tests show infiltration through your foundation. Radon gas is formed from the radioactive decay of uranium found in rock and soil. The EPA has found that one in 15 homes has unsafe levels of radon, with higher levels in the Midwest and Rocky Mountain regions. Inhalation of radon gas is the second leading cause of lung cancer (after cigarette smoking) and has been linked to over 21,000 deaths per year in the United States.

• Most important, don't smoke or allow others to smoke in your home, and avoid venues where smoking is allowed.

IS IT SAFE TO DRINK THE WATER?

The water you drink can also hide dangerous pollutants. A bout of stomach "flu" can be brought on by microscopic bacteria, viruses, or parasites carried in your drinking water. A miserable but temporary inconvenience for most of us, such infections can be life-threatening for those with a weak immune system. More serious diseases such as hepatitis and cholera can also be spread through drinking water.

Chemicals from industrial production and farm runoff enter our water supply. Thousands of different toxins, from pesticides to arsenic to radioactive elements, find their way into our streams, rivers, and groundwater, ultimately to reach our taps. A 2008 Associated Press survey showed residues of pharmaceutical drugs such as birth control pills, antibiotics, heart and seizure medications, and more in tap water coming out of municipal water treatment plants. No one knows the long-term effects of decades of exposure to trace amounts of these chemicals. Water-processing facilities add additional chemicals such as chlorine to kill bacteria, viruses, and parasites.

Long-term exposure to these chemicals has been linked to cancer and liver, kidney, and reproductive tract problems.

Often, bottled water isn't a better alternative since 40 percent of bottled water is bottled tap water. In addition, it takes 1.5 million barrels of oil to make plastic water bottles, 86 percent of which are not recycled. Of course, you can't stop drinking water, so how do you avoid the contaminants? One word: Filter! You can start immediately by filtering all the water you drink and use for cooking. A number of relatively inexpensive countertop, tap-attachment, or under-sink filter systems are available. A bit of online research can help you determine which system would be best for your use. For a few hundred dollars you can install a household system that filters the water coming from all the taps in your house—not just your drinking and cooking water but also your bathing water. This prevents chlorine from being absorbed through your skin in the bath or shower.

HOME AND WORK HAZARDS

Some jobs are just plain dangerous, and not just in the obvious industries like logging and construction. Agricultural workers who are exposed to pesticides, fungicides, herbicides, or fertilizers are at higher risk of non-Hodgkin's lymphoma, neurologic illness, and brain and prostate cancer. Frequent exposure to petroleum-based products is associated with higher rates of leukemia and cancers of the lip, stomach, liver, pancreas, connective tissues, prostate, eye, and brain. If your work brings you in contact with hazardous materials, make sure you understand and follow all of the most up-to-date safety procedures.

Around the home, you can also reduce your exposure. Follow label directions carefully and wear appropriate safety gear (e.g., coveralls, rubber gloves, face mask) when handling solvents, pesticides, etc. Better yet, safely dispose of your hazardous cleaning supplies through your municipal recycling program and begin using nontoxic alternatives instead, such as:

- All-purpose household cleaner—½ cup borax dissolved in 1 gallon of hot water. Add a teaspoon of lemon juice.

- Household disinfectant—mix equal amounts of isopropyl (rubbing) alcohol and water.

- Laundry bleach alternative—use baking soda or borax rather than bleach.

- Glass cleaner—mix equal parts white vinegar and water.

WATCH WHAT YOU EAT

Consumers want unblemished, perfectly shaped produce all year long. Producers want to increase their yields, ship their products over great distances, and store them for long periods of time. As a result, modern commercial food production has come to rely on tons of toxic chemicals.

Pesticides kill bugs and rodents that can devour a crop. Fungicides prevent fungi (organisms such as mold and mushrooms) from damaging food plants. Herbicides clear fields of "weeds" that compete with crops for soil, water, and sunlight, and get in the way when it comes time to harvest. Produce is often coated in wax and stored in warehouses filled with toxic gases to extend "shelf-life" and maintain an appealing "fresh from the field" appearance. As we eat such conventionally produced foods, those chemicals accumulate in our tissues. Whether you eat a bright red apple straight off the tree or dig into a bowl of your favorite breakfast cereal, toxins are present in all types of plant-based foods.

Meat, poultry, and fish contain higher levels of toxins than grains, fruits, and vegetables. Being higher up the food chain, these animals concentrate chemicals from the plants or other animals they eat, primarily in their fatty tissues. Some livestock are also fed antibiotics and hormones to prevent disease and promote growth.

As with air and water, you have to eat. Here are some things you can do to reduce your risks:

- Eat organic whenever possible (this applies to meat and poultry, too).

- When organic is not available, steer clear of produce that tends to be the most contaminated (bell peppers, spinach, celery, potatoes, peaches, nectarines, strawberries, apples, pears, cherries, imported grapes, raspberries) and reach for items that tend to be less contaminated (sweet corn, avocado, cauliflower, asparagus, onions, sweet peas, broccoli, pineapples, mangoes, kiwifruit, papaya, bananas), a practice that is estimated to reduce pesticide exposure by 90 percent.

• Use a nontoxic, plant-based fruit and veggie cleanser to remove wax and other contaminants from the outside of your produce. You can soak produce in a solution of ¼ cup of 3 percent food-grade hydrogen peroxide in a sink full of water for 20 minutes as well.

• Eat lower on the food chain, with more plant-based foods and fewer animal products on your plate, which is consistent with our nutritional recommendations (see Chapter 11).

• Trim fat from red meat and remove the skin from poultry.

• Boil meat and poultry to allow fat to rise to the water's surface so it can be skimmed before eating. Environmental pollutants such as pesticides found in animal feed and hormones given to livestock to hasten growth are fat-soluble and have higher concentrations in fatty tissue.

RADIATION, ALL DAY, EVERY DAY

Concentrated exposure to high doses of electromagnetic radiation can be lethal, but research remains inconclusive as to the dangers associated with lower doses. Evidence suggests, however, that ongoing environmental exposure to low-dose electromagnetic radiation can produce changes in biological tissue function and may lead to a wide variety of adverse health effects, including DNA damage. While waiting for conclusive proof one way or the other, why take a chance?

You are constantly barraged by electromagnetic radiation across a wide frequency range. Your body absorbs radiation 24/7 from satellite transmitters, cell-phone towers, and radio, television, and radar antennae. And closer to home, you are radiated on a daily basis by your cell phone (a miniature radio transmitter) and other electronic devices such as TVs, computer monitors, hair dryers, electric razors, waterbed heaters, and electric blankets.

Short of living in a lead box, you can't completely avoid electromagnetic exposure, but you can limit it. Here are some tips:

- Use a Bluetooth earpiece when using your cell phone. The radiation from the Bluetooth device is far lower than the amount coming from the body of the cell phone itself.

- Change any tube-based computer monitors for flat screen versions, which emit almost no radiation.

- Limit time spent using hair dryers, electric shavers, and other high-power electrical devices.

- Sit at least 10 feet away from wide-screen TVs (especially important for the kids).

- Don't use an electric blanket or sleep on a heated water bed.

- Don't live next to high-voltage electrical lines or transmission (radio/TV/cell phone/radar) towers.

DEATH BY HEAVY METAL

We're not talking head-banger music here. Toxic heavy metals are far more dangerous. They can be deadly in high concentrations and extremely damaging to your health even at low exposure. A toddler teething on a lead-painted windowsill can suffer severe developmental disabilities. Arsenic, beryllium, cadmium, chromium, cobalt, and nickel can cause cancer. Toxic heavy metals have been shown to increase free-radical activity and can contribute to premature aging and age-related diseases. Abnormal immune function, mood disorders, neurodegenerative diseases, fatigue, poor concentration, and hair loss can all result from exposure to toxic heavy metals.

Certain occupations and hobbies can bring you into contact with toxic heavy metals. Naturally occurring sources can leach into groundwater supplies, of particular concern to those who drink well water. Of broader concern is mercury, which accumulates in fish species throughout the world. Mercury suppresses immune system function and is a known neurotoxin (Lewis Carroll's character the Mad Hatter references the high incidence of mental illness among hat makers of the time who handled mercury in the course of their work).

You can reduce your exposure to toxic heavy metals. For example:

• Test your home for lead paint, especially if you have young children in the house.

• Educate yourself about toxic heavy metals used on the job or in pursuit of a hobby and always follow the most up-to-date safety procedures.

• Instead of amalgam dental fillings, which are more than 50 percent mercury, ask your dentist to use gold or less toxic composite materials (polymers of glass or ceramic) to fill cavities and for other dental reconstruction.

• Don't use aluminum cookware or aluminum foil, and if you use antiperspirant, look for one that does not contain aluminum.

• Filter your drinking water.

• Limit your consumption of fish and other seafood to species with lower concentrations of mercury. According to the U.S. Department of Health and Human Services and the U.S. Environmental Protection Agency (EPA), fish and other seafood can be divided into three groups depending on their level of mercury contamination:

Lowest-Mercury Fish and Seafood (Can Be Eaten Anytime)

Anchovies

Butterfish

Catfish

Clams

Cod

Crab (blue, king, snow)

Crawfish/crayfish

Croaker (Atlantic)

Haddock (Atlantic)

Hake

Herring

Jacksmelt

Lobster, spiny

Mackerel (North
   Atlantic, Pacific Chub)

Mullet

Oysters

Perch (ocean)

Pollock

Salmon (canned, fresh,
   frozen)

Sardines

Scallops

Shad (American)

Shrimp

Squid

Tilapia

Trout (freshwater)

Whitefish

Whiting

Intermediate-Mercury Fish and Seafood (Can Be Eaten Periodically)

Bass (saltwater, striped, black, Chilean)
Bluefish
Carp
Cod (Alaskan)
Croaker (white Pacific)
Grouper
Halibut

Lobster (North America)
Mackerel, Spanish
Marlin
Monkfish
Orange roughy
Perch (freshwater)
Sablefish

Skate
Snapper
Tuna (canned albacore)
Tuna (fresh, frozen)
Tuna (yellowfin)
Weakfish (sea trout)

Highest-Mercury Fish and Seafood (Should Not Be Eaten)

King mackerel
Shark

Swordfish

Tilefish (Gulf of Mexico)

NUTRITION AND SUPPLEMENT RECOMMENDATIONS

Reducing your exposure to the myriad pollutants in your world will go a long way toward protecting your health and extending your life. However, our *TRANSCEND* recommendations for *Nutrition*, *Supplementation*, and *Exercise* can also strengthen your body's ability to remove accumulated toxins. The following suggestions will help:

• Make sure your diet includes lots of cruciferous vegetables (broccoli, cauliflower, kale, cabbage, Brussels sprouts, bok choy) for the detoxification properties of their many antioxidants.

• Eat lots of garlic, onions, lemon, rosemary, and green tea to help your liver eliminate heavy metals.

• The flavorful herb cilantro is a natural chelator of heavy metals.

• Strengthen liver function with milk thistle (silymarin) and alpha-lipoic acid.

• Vitamin C, magnesium, selenium, and many of the B vitamins help optimize detoxification enzyme function.

• N-acetylcysteine supplements help boost levels of glutathione, one of the liver's most important phase II (see below) detoxifiers.

• There are new products available (see the Resources section) that allow glutathione to be taken orally and to survive the digestive process. Supplementary glutathione provides a powerful boost to your liver's ability to detoxify the body.

• Maintain a regular aerobic exercise routine and treat yourself to the occasional sauna; heavy metals and fat-soluble toxins may be partially excreted in sweat.

EARLY DETECTION— TOXIN ACCUMULATION AND DETOXIFICATION STATUS

According to the EPA, we are now releasing over 7 billion pounds of 650 different industrial compounds into the environment each year, and, unavoidably, some ends up in our bodies—with adverse health consequences. Here are a few simple tests you can use to determine the levels of some of the more common toxins in your body and evaluate how well your liver is performing its job of toxin removal.

TOXIC HEAVY METALS ANALYSIS

Hundreds of enzymes in the body rely on essential minerals to work properly. It is possible for toxic heavy metals to attach to these enzymes instead of the proper minerals, making the enzymes useless. For example, if an enzyme is supposed to bind with magnesium but instead becomes bound up to arsenic, that enzyme is unable to function properly.

Because of the increasing amounts of pollution present as part of modern-day life, many of us have accumulated significant amounts of toxic heavy metals, which interfere with proper enzyme function and promote processes associated with aging. *Hair mineral analysis* is a simple, noninvasive, and

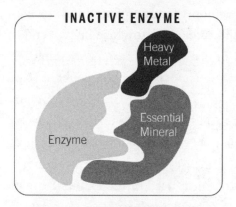
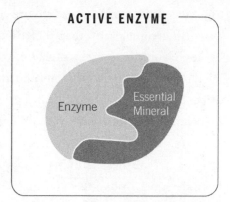

Heavy metals will bind to the enzyme, causing the shape of the enzyme to change, and preventing the enzyme and essential mineral from fitting together exactly.

inexpensive way to measure the levels of toxic metals in the body. These toxins include aluminum, arsenic, lead, mercury, and others. Hair testing provides a semi-quantitative method of assessing your exposure to these toxins, particularly over the previous few months. One gram of hair is taken from the nape of the neck and sent to the laboratory to be analyzed by a mass spectrophotometer. The specimen can be collected by a nutritionally oriented healthcare practitioner, or you can do it yourself at home. If the results suggest significant accumulations of these toxins, detoxification strategies such as oral *chelation therapy* may be done to safely remove these toxins from your body. In cases of more severe toxic heavy metal accumulations, intravenous chelation therapy may be needed.

Environmental Pollutants Panels

Environmental pollutants panels are screening urine tests that measure the levels of several common environmental pollutants in the body. We recommend a panel that tests for benzene, parabens, phthalates, styrene, toluene, and xylene, available from complementary physicians and clinics such as Terry's clinic, Grossman Wellness Center. *Benzene* is an industrial solvent and is used in the manufacture of some plastics. It exerts toxic effects on the brain and bone marrow, and long-term exposure has been associated with anemia and leukemia. *Parabens* are commonly found in foods and cosmetics, skin creams, sunscreen lotions, and shampoos, and are known endocrine disruptors, which interfere with hormone function. *Phthalates* are used to soften plastic products and are also endocrine interrupters that can affect breast tissue in women and prostate tissue in men. *Styrene* is found in fast-food ware such as Styrofoam cups and bowls. Long-term exposure to even small amounts of styrene can cause nervous system problems such as fatigue, nervousness, and insomnia. It depresses the bone marrow and has been declared a possible carcinogen. *Toluene* is used in the manufacture of rubber products, oils, resins, adhesives, inks, detergents, dyes, and explosives. It is toxic to the nervous system, and long-term low-level exposure can result in headaches, fatigue, loss of appetite, disturbances in menstruation, and reductions in IQ and psychomotor skills. *Xylene* is a solvent used in the manufacture of gasoline and polyester fibers. It is used to make dyes, paints, and lacquers. Unlike other pollutants, xylene does not accumulate in the body, but long-term exposure has been associated with disturbances to the nervous system. Symptoms such as headaches, irritability, depression, insomnia, agitation, extreme tiredness, tremors, and impaired concentration and memory have been linked to long-term xylene exposure. Specific detoxification strategies can be undertaken if accumulations of the above toxins and pollutants are found. Many of these toxins will come out in sweat, so vigorous exercise that leads to sweating and the use of saunas can help with toxin removal.

HEPATIC DETOXIFICATION PROFILE

In addition to measuring the actual levels of toxins, it is useful to see how good a job your body is doing at eliminating these toxins. The liver has the greatest responsibility for toxin decontamination, and it possesses hundreds of specialized detoxification enzymes to assist with this task. This occurs in two steps, referred to as phase I and phase II detoxification reactions. In phase I, the liver converts mostly fat-soluble toxins into water-soluble compounds that can be more easily excreted in the urine and stool. In phase II, a water-soluble molecule is attached to the partially processed toxin, which makes it much less toxic and easier for the body to excrete.

The products of the phase I reactions can still be quite toxic, so it is important that phase I and phase II reactions work in harmony or else phase I toxins can accumulate. The *hepatic detoxification profile* is a urine test that shows how well these phases are working and also how well they are synchronized. A common problem is for phase I to work too quickly and phase II to be too slow. An easy remedy is to drink a small amount of grapefruit and pomegranate juice each day. Grapefruit juice contains a compound known as *naringinen*, which slows down phase I, while pomegranate juice is rich in ellagic acid, which enhances phase II activity. If you do a hepatic detoxification profile and find your phase I is too fast for phase II to keep up, a cocktail of 3 ounces each of grapefruit and pomegranate juice each day can help normalize your liver's detoxification function.

Reader: Hey future guys, I suppose the nanobots are going to keep me detoxified in the future?

Ray2034: Well, that's exactly right. The nanobots monitor all of the substances in your blood and make sure you have optimal amounts of needed nutrients and an absence of toxins.

Terry2034: A new generation of nanobots is now being tested that examine each cell and make sure the right nutrients actually reach the cells, which is really the goal of your biology, and that the cells remain toxin free.

Reader: Hmmm, sounds like I won't need a lot of my organs in the future.

Terry2034: Here in 2034, there are already a lot of artificial organs in use such as an artificial pancreas, kidneys, and various gland organs such as the thyroid.

Ray2034: In some cases, swarms of nanobots are replacing the organs.

Reader: It sounds like you could replace just about every organ.

Ray2034: Well, those organs that basically put things into the bloodstream or remove them are being replaced or are likely to be replaced by nanobots very soon.

Terry2034: There are artificial hearts already being used, and new artificial blood cells that provide their own mobility are likely to replace the heart's function in the near future.

Reader: Sounds like I'll want to keep my brain, though.

Ray2034: Yes, good idea. But here in the 2030s it is already routine to have nanobots travel into the brain through the capillaries. The nanobots interact with your biological neurons and thereby provide an extension to your brain providing full-immersion virtual reality from within the nervous system and generally expanding your memory and thinking ability. So our brains are already partly nonbiological.

Reader: Well, biology was good while it lasted.

Terry2034: Well put.

EPILOGUE

Ray: Looks like this is the last chapter of the book. And you know what that means?

Reader: It makes me a little sad when books I'm enjoying get close to ending.

Terry: It makes us sad, too. We've enjoyed getting to know you.

RayandTerry2023: It also means it's nearly time for us to disappear.

RayandTerry2034: Us, too. Back to the future, to coin a phrase.

Reader: None of you can stay around?

RayandTerry2023, RayandTerry2034: Unfortunately, no. According to the rules of our poetic license, as soon as this book is finished, we have to return to our age-appropriate time period. We're history . . . future history, that is.

Reader: All that you've talked about has given me a lot of hope for what lies ahead.

Terry: Yes, that is the main point we've tried to make. The conventional wisdom has been if you adopt various lifestyle changes, you can add some years to your remaining life expectancy. But if those additional years get you to a point in time where we have dramatically more advanced methods for life extension from biotech and nanotech, then you'll get even more years.

Ray: And that will get you more advances for yet more years, and so on.

Reader: Sounds good to me, but good-byes still make me sad.

Ray2034: We have a couple people we want you to meet first. We think they'll cheer you up.

Reader: Hmmm, these two look familiar.

Reader2023: I should hope so! I'm the future you.

Reader2034: Me too.

Reader: Nice to meet . . . uhhh . . . me. You guys look younger than I do, especially you, uh, me, in 2034.

Terry2023: Rejuvenation medicine using the stem cells you stored back in 2009 helped you to become younger.

Ray2034: And nanotherapies finished the job during the late 2020s.

Reader: This really makes me want to do what I can to get to these future bridges in good shape.

Terry: Well, you can start by **T**alking with your doctor. Discuss your commitment to true preventive medicine and early detection of disease with your medical care provider. Arrange to have regular exams including occasional comprehensive evaluations so nothing sneaks up on you unaware.

Ray: Be sure to get plenty of **R**elaxation. All work and no play is not just no fun, it's bad for your health. Take out time for hobbies and vacations and allow ample time for friends and family. Your anti-aging program should be enjoyable and relaxing—red wine, dark chocolate, blueberries, green tea, massages, and regular sex should be regular aspects of your program.

Terry2023: Perform thorough health **A**ssessments. Do self-tests that will help you assess your body for both fitness and early detection of disease on a regular basis.

Ray2023: Follow good **N**utrition. Providing your body with healthful meals such as those listed in the Recipes section shows that eating properly needn't be boring. Use your imagination to create additional healthy, tasty food, and use a bit of self-control to eliminate all the junk food and sugary foods from your diet.

Terry2034: Don't forget to take your **S**upplements. The health of virtually everyone can be improved by taking some basic nutritional supplements such as a multiple vitamin-mineral, fish oil, and extra vitamin D. Your need for additional supplements can be determined by the results of your ongoing testing. You need to reprogram your biochemistry, especially as you get older.

Ray2034: Reduce your **C**alories. You don't need to follow severe caloric restriction to obtain health and longevity benefits. Follow the example of the Japanese, who have the longest life expectancies on Earth, and consider reducing your calories 10 to 20 percent.

Reader2023: Keep **E**xercising regularly. Just 180 minutes a week of combined aerobic exercise, strength training, and stretching will improve your health significantly. Work up to 1 hour per day for even more benefits.

Reader2034: Adopt **N**ew technologies as they become available. Several genomics tests and a few stem cell therapies are already available to you in 2009. Much more comprehensive and less costly diagnostic testing such as "the $500 genome" and telomere length testing to accurately assess true biological age will be available to you soon. New therapeutic interventions such as organ replacement, genetic modification through RNA interference, gene therapy, and personalized medications—to name just a few—are only several years away. The pace of these new developments will continue to accelerate.

Terry2023: And, finally, don't forget about **D**etoxifying toxic substances that cause you to age more quickly when they accumulate in your body. Eat organic food, use "green" household cleansers, and minimize your exposure to other environmental toxins at your place of employment. Measure the level of toxins in your body, and perform regular detoxifying strategies to clear them.

Terry: These are the tools you'll need to help you *TRANSCEND* the limits of your outdated Stone Age genes that want your body to begin deteriorating at an ever-increasing rate starting at 30 years of age.

Ray: So that you can live long enough and remain healthy long enough to take full advantage of the technological breakthroughs of the Information Age in the decades ahead.

Reader: Makes me want to go back to the beginning and start rereading the book.

RayandTerry2034: Good idea, then we won't have to leave right away.

Reader: Yes, I think I'll keep you guys around until I am able to transcend my biology.

RayandTerry2023: Well put. To *TRANSCEND* is what we humans do well.

RESOURCES AND CONTACTS

FINDING A DOCTOR

After reading this book, many readers will want to find a doctor they can "talk with." Three physician organizations maintain lists of physicians and other healthcare practitioners who are open to new ideas and willing to work with their patients on optimal wellness programs and early detection of disease.

- **The American College for Advancement in Medicine (ACAM)** lists more than 1,000 physicians trained in complementary, alternative, and integrative medicine. www.acam.org, 800-532-3688

- **The American Academy of Anti-Aging Medicine (A4M)** is an organization of more than 17,000 physicians and describes itself as "the only non-profit, non-commercial, medical society in the world devoted to eradicating the degenerative diseases of aging." www.worldhealth.net, 800-558-1267

- **The American Holistic Medical Association (AHMA)** is "the leading advocate for the use of holistic and integrative medicine by all licensed healthcare providers." www.holisticmedicine.org, 216-292-6644

PERSONALIZED ONLINE HEALTH RECOMMENDATIONS

We have developed an online personalized program that closely tracks the *TRANSCEND* Program. We invite you to take a few minutes to fill out this interactive questionnaire and immediately receive a set of personalized recommendations regarding diet and detox, stress management and supplementation, weight loss, and more.

www.rayandterry.com/personalprogram

GROSSMAN WELLNESS CENTER

Patients from all regions of the United States and every continent have come to Terry's clinic, the Grossman Wellness Center, in Denver to experience comprehensive wellness and executive health evaluations based on the TRANSCEND program. www.grossmanwellness.com, 303-233-4247 or 877-548-4387

KURZWEILAI.NET

We have tried to make this book as up-to-date as possible, yet changes in medicine and technology occur on a daily basis. Ray's free daily newsletter, available at KurzweilAI.net, is one of the best ways to keep abreast of the latest breakthroughs in medicine, biotechnology, nanotechnology, computer science, information technology, artificial intelligence, and more.

BOOK WEB SITE

Visit us at our book Web site, www.rayandterry.com/transcend. We've gathered a comprehensive set of additional information, including product recommendations, reader Q&A, glossary, test information, health resources, health research and news, excerpts, press, and much more.

CONTACTING THE AUTHORS

The authors remain committed to spreading the *TRANSCEND* message of wellness and health. Ray can be reached at ray@rayandterry.com and Terry at terry@rayandterry.com. Contact information for Ray2023, Terry2023, Ray2034, andTerry2034 will be available in the future.

SELECTED REFERENCES

Introduction

1. Key TJ, Schatzkin A, Willett WC, Allen NE, Spencer EA, Travis RC (2003) Diet, nutrition and the prevention of cancer. *Annu Rev Physiol* 65:313–332.

Chapter 1: Brain and Sleep

1. Bielak AA, Hughes TF, Small BJ, Dixon RA (2007) It's never too late to engage in lifestyle activities: significant concurrent but not change relationships between lifestyle activities and cognitive speed. *J Gerontol B Psychol Sci Soc Sci* Nov; 62(6):P331–339.
2. Lydic R, McCarley RW, Hobson JA (1984) Forced activity alters sleep cycle periodicity and dorsal raphe discharge rhythm. *Am J Physiol* 247(1 Pt 2): R135–145.
3. Stanford Report, "Remediation training improves reading ability of dyslexic children," February 25, 2003 (http://news-service.stanford.edu/news/2003/february26/dyslexia-226.html).
4. McDaniel, MA (2003) Brain-specific nutrients: a memory cure. *Nutrition* 11-12: 957–975.
5. Bower, B (1998) To dream, perchance to scan. *Science News* 153:44.

Chapter 2: How to Keep Your Heart Beating . . .

1. Liem AH, Jukema JW, van Veldhuisen DJ (2003) Secondary prevention in coronary heart disease patients with low HDL: What options do we have? *Int J Cardiol* 90(1): 15–21.
2. National Center for Health Statistics, "Deaths—leading causes," http://www.cdc.gov/nchs/FASTATS/lcod.htm.
3. Lagergvist B, James S, Stenestrand U, Lindbäck M, Nilsson T, Wallentin L (2007) Long-term outcomes with drug-eluting stents versus bare-metal stents in Sweden. *NEJM* 356(10):1009–1019.
4. Enos WF, Holmes RH, Beyer J (1953) Coronary disease among United States soldiers killed in action in Korea. *JAMA* 152: 1090–1093.
5. The Society for Women's Health Research, "What really claims women's lives?" http://www.womenshealthresearch.org/site/PageServer?pagename=hs_whatclaims.
6. Becker D, Gordon R, Morris P, Yorko J, Gordon J, Li M, Iqbal N (2008) Simvastatin vs therapeutic lifestyle changes and supplements: Randomized primary prevention trail. *Mayo Clin Proc* 83:758–764.
7. Lu Z, Kou W, Du B, Wu Y, Zhao S, Brusco O, Morgan J, Capuzzi D (2008) Effect of Xuezhikang, an extract from red yeast Chinese rice, on coronary events in a Chinese population with previous myocardial infarction. *Amer J of Card* 12(101):1689–1693.
8. Gouni-Berthold J, Berthold H (2002) Policosanol: Clinical pharmacology and therapeutic significance of a new lipid-lowering agent. *Amer Heart J* 143(2):356–365.

9. Mörike EM (1996) Vitamin E treatment of patients with coronary disease. *Deutsch Med Wochenschr* May, 121(21):A9.

10. Brook JG, Linn S, Aviram M (1986) Phosphatidlylinositol raises HDL cholesterol levels in humans. *Biochem Med Metabol Biol* (35):31–39.

11. Katsuyuki M, Daviglus ML, Dyer AR, Kiang L, Garside DB, Stamler J, Greenland P (2001) Relationship of blood pressure to 25-year mortality due to coronary heart disease, cardiovascular diseases, and all causes in young adult men: The Chicago Heart Association detection project in industry. *Arch Intern Med* 161:1501–1509.

12. Bolotin HH (2007) DXA in vivo BMD methodology: An erroneous and misleading research and clinical gauge of bone mineral status, bone fragility, and bone remodelling. *Bone* 41(1):138–154.

Chapter 3: Digestion

1. Drossman DA, Li Z, Andruzzi E, Temple RD, Talley NJ, Thompson WG, Whitehead WE, Janssens J, Funch-Jensen P, Corazziari E et al. (1993) U.S. householder survey of functional gastrointestinal disorders. Prevalence, sociodermography, and health impact. *Digestive Disease Science* 38(9): 1569–1580.

2. Krasinski SD, Russell RM, Samloff IM, Jacob RA, Dallal GE, McGandy RB, Hartz SC (1986) Fundic atrophic gastritis in an elderly population. *J Am Geriatr Soc* 34(11): 800–806.

3. Scrimshaw NS, Murray EB (1988) The acceptability of milk and milk products in populations with a high prevalence of lactose intolerance. *Am J Clin Nutr* 48(Suppl 4): 1079–1159.

4. Swagerty DL Jr, Walling AD, Klein RM (2002) Lactose intolerance. *Amer Fam Physician* 65(9):1845–1850.

Chapter 4: Hormone Optimization

1. Canaris GJ, Manowitz NR, Mayor G, Ridgeway EG (2000) The Colorado thyroid prevalence study. *Arch of Int. Med* (160):526–534.

2. Svec F, Porter JR (1998) The actions of exogenous dehydroepiandrosterone in experimental animals and humans. *Proceedings of the Society for Exp Bio and Med* (218):174–191.

3. Lemon JA, Boreham DR, Rollow CD (2003) A dietary supplement abolishes age-related cognitive decline in transgenic mice expressing elevated free radical processes. *Exp Bio Med.* (228):800–810.

4. Butler RN, Fossel M, Mitchell H, Heward CB, Olshansky SJ, Perls TT, Rothman DJ, Rothman SM, Warner HR, West MD, and Wright WF (2002) Is there an anti-aging medicine? *J of Gerontology Series A: Bio Sci Med Sci* (57):B333-B3385.

5. Straub RH, Lehle K, Herfarth H, Weber M, Falk MW, Preuner J, Scholmerich J (2002) Dehydroepiandrosterone in relation to other adrenal hormones during an acute inflammatory stressful disease state compared with chronic inflammatory disease: Role of interleukin-6 and tumour necrosis factor. *Eur J Endo* 146:365–374.

6. National Sleep Foundation, "Epidemic of Daytime Sleepiness Linked to Increased Feelings of Anger, Stress, and Pessimism," http://www.sleepfoundation.org/site/c.huIXKjM0IxF/b.2417359/k.3028/Epidemic_of_Daytime_Sleepiness_Linked_to_Increased_Feelings_of_Anger_Stress_and_Pessimism.htm.

7. Leger D, Laudon M, Zisapel N (2004) Nocturnal 6-sulfatoxymelatonin excretion in insomnia and its relation to the response to melatonin replacement therapy. *Am J Med* 116:91–95.

8. The Writing Group for the WHI Investigators (2002) Risks and benefits of estrogen plus progestin in healthy post-menopausal women: Principal results of the Women's Health Initiative randomized controlled trial. *JAMA* 288(3):321–333.

9. The Women's Health Initiative Steering Committee (2004) Effects of conjugated equine estrogen in postmenopausal women with hysterectomy. The Women's Health Initiative randomized controlled trial. *JAMA* 291:1701–1712.

10. Rossouw JE, Prentice RL, Manson JE, Wu L, Barad D, Barnabei VM, Ko M, LaCroix AZ, et al. (2007) Postmenopausal hormone therapy and risk of cardiovascular disease by age and years since menopause. *JAMA* 297:1465–1477.

11. Fournier A, Berrino F, Riboli E, Avenel V, and Clavel-Chapelon F (2005) Breast cancer risk in relation to different types of hormone replacement therapy in the E3N-EPIC cohort. *Int J of Cancer* 3(114):448–454.

12. Rudman D, Drinka PJ, Wilson CR, Mattson DE, Scherman F, Cuisinier MC, Schultz S (1994) Relations of endogenous anabolic hormones and physical activity to bone mineral density and lean body mass in elderly men. *Clin Endo* 40:653–661.

13. Stephenson K, Price C, Kurdowska A, Neuenschwander P, Stephenson J, Pinson B, Stephenson D, Alfred D, Krupa A, Mahoney D, Zava D, and Bevan M (2004) Topical progesterone cream does not increase thrombotic and inflammatory factors in post-menopausal women. ASH Annual Meeting Abstracts 104:5318.

Chapter 5: Metabolic Processes

1. Das UN (2006) Exercise and inflammation. *The Euro Heart J.* 27(11):1385.

2. Fenech M, Aitken C, Rinaldi J (1998) Folate, vitamin B_{12}, homocysteine status and DNA damage in young Australian adults. *Carcinogenesis* 19(7):1163–1171.

Chapter 6: Cancer

1. American Cancer Society, "Cancer Facts and Figures," http://www.cancer.org/docroot/stt/stt_0.asp.

2. American Cancer Society, "Costs of Cancer," http://www.cancer.org/docroot/MIT/content/MIT_3_2X_Costs_of_Cancer.asp.

3. Key TJ, Schatzkin A, Willett WC, Allen NE, Spencer EA, Travis RC (2004) Diet, nutrition and the prevention of cancer. *Pub Health Nutr* Feb 7(1A):187–200.

4. Trichopoulou A, Costacou T, Bamia C, Trichopoulos D (2003) Adherence to a Mediterranean diet and survival in a Greek population. *N Engl J Med* 348(26):2599–2608.

5. Dominique MS, Simin L, Giovannucci E, Willett WC, Graham CA, Fuchs CS (2002) Dietary sugar, glycemic load and pancreatic cancer risk in a prospective study. *J of Natl Can Inst* 94(17):1293–1300.

6. Higginbotham S, Zhang ZF, Lee IM, Cook N, Giovannucci E, Buring JE, Liu S (2004) Dietary glycemic load and risk of colorectal cancer in the Women's Health Study. *J of Natl Can Inst* 96(3):229–233.

7. Calle EE, Rodrigue C, Walker-Thurmond K, Thun M (2003) Overweight, obesity, and mortality from cancer in a prospectively studied cohort of U.S. adults. *NEJM* 348(17):1625–1638.

8. Zheng QS, Zheng RL (2002) Effects of ascorbic acid and sodium selenite on growth and redifferentiation in human hepatoma cells and its mechanisms. *Pharmazie* 57(4):265–269.

9. Duffield-Lillico AJ, Reid ME, Turnbull BW, Combs GF, Slate EH, Fischbach LA, Marshall JR, Clark LC (2002) Baseline characteristics and the effect of selenium supplementation on cancer incidence in a randomized clinical trial. *Cancer Epidemiol Biomarkers Prev* 11: 630–639.

10. Lappe JM, Travers-Gustafson DT, Davies KM, Recker RR, Heaney RP (2007) Vitamin D and calcium supplementation reduces cancer risk: Results of a randomized trial. *Amer J of Clin Nutr* 85(6):1586–1591.

11. Eichholzer M, Luthy J, Moser U, Fowler B (2001) Folate and the risk of colorectal, breast and cervix cancer: The epidemiological evidence. *Swiss Med Wkly* 131:539–549.

Chapter 7: Genomics

1. Kathiresan S, Melander O, Anevski D, Guiducci C, Burtt NP, Roos C, Hirschhorn JN, Berglund G, Hedblad B, Groop L, Altshuler DM, Newton-Cheh C, Orho-Melander M (2008) Polymorphisms associated with cholesterol and risk of cardiovascular events. *N Engl J Med.* 358(12):1240–1249.

2. Love-Gregory L, Sherva R, Sun L, Wasson J, Schappe T, Doria A, Rao DC, Hunt SC, Klein S, Neuman RJ, Permutt MA, Abumrad NA (2008) Variants in CD36 gene associate with the metabolic syndrome and high-density lipoprotein cholesterol. *Human Molecular Genetics* 17(11):1695–1704.

Chapter 8: Talk with Your Doctor

1. National Center for Health Statistics, "Deaths—leading causes," http://www.cdc.gov/nchs/FASTATS/lcod.htm.

Chapter 9: Relaxation

1. Holmes T, Rahe R (1967) The social readjustment rating scale. *J Psychosom Res* 11(2):213–218.

2. Weber G (2004) Lost time: Vacation days go unused despite more liberal time-off policies. *Workforce Man.* Dec:66–67 http://www.workforce.com/section/02/feature/23/89/82/index.html.

3. Haskell CF, Kennedy DO, Milne AL, Wesnes KA, Scholey AB (2008) The effects of l-theanine, caffeine and their combination on cognition and mood. *Bio Psych* 77(2):113–122.

4. Zaret BL, Moser M, Cohen LS (1992) *Yale University School of Medicine Heart Book* www.med.yale.edu/library/heartbk/p.97.

5. Moyer CA, Rounds J, Hannum JW (2004) A meta-analysis of massage therapy research. *Psych Bull* 130(1):3–18.

6. Benson H, Beary JF, Carol MP (1974) The relaxation response. *Psychiatry* 37(1):37–46.

7. Klag MJ, Wang NY, Meoni LA, Brancati FL, Cooper LA, Liang KY, Young JH, Ford DE (2002) Coffee intake and risk of hypertension: The Johns Hopkins precursors study. *Arch Intern Med* 162(6):657–662.

8. El Yacoubi M, Ledent C, Parmentier M, Costentin J, Vaugeois JM (2000) The anxiogenic-like effect of caffeine in two experimental procedures measuring anxiety in the mouse is not shared by selective A(2A) adenosine receptor antagonists. *Psychopharmacology* (Berl) 148(2):153–163.

9. Haskell CF, Kennedy DO, Milne AL, Wesnes KA, Scholey AB (2007) The effects of L-theanine, caffeine and their combination on cognition and mood. *Biol Psychol* 77(2):113–122.

Chapter 10: Assessment

1. van Dam RM, Li T, Spiegelman D, Franco OH, Hu FB (2008) Combined impact of lifestyle factors on mortality: Prospective cohort study in US women. *BMJ* 337:a1440.

2. Kösters JP, Gøtzsche PC (2003) Regular self-examination or clinical examination for early detection of breast cancer. Cochrane Database of Systematic Reviews 2: CD003373. DOI: 10.1002/14651858.CD003373

Chapter 11: Nutrition

1. O'Keefe JH, Gheewala NM, O'Keefe JO (2008) Dietary strategies for improving postprandial glucose, lipids, inflammation, and cardiovascular health. *J Am Coll Cardiol* 51:249–255.

2. MedWireNews, September 28, 2007, "Alcohol amount, not type, affects breast cancer risk," http://www.medwire-news.md/265/69846/The_Week_in_Medicine/Alcohol_amount,_not_type,_affects_breast_cancer_risk.html.

3. Sesso HD, Paffenbarger Jr. RS, Lee IM (2001) Alcohol consumption and risk of prostate cancer: The Harvard Alumni Health Study. *Int J of Epi* 30:749–755.

4. Johnston CS, Kim CM, Buller AJ (2004) Vinegar improves insulin sensitivity to a high-carbohydrate meal in subjects with insulin resistance or type 2 diabetes. *Diabetes Care* 27:281–282.

Chapter 12: Supplements

1. Ames BN, Wakimoto P (2002) Are vitamin and mineral deficiencies a major cancer risk? *Nature Reviews Cancer* 2:694–704.

2. University Medical Center Rotterdam, "The Rotterdam Study," http://www.epib.nl/ergo.htm.

3. Losonczy KG, Harris TB, Havlik RJ (1996) Vitamin E and vitamin C supplement use and risk of all-cause and coronary heart disease mortality in older persons: The Established Populations for Epidemiologic Studies in the Elderly. *Am J Clin Nutr* 64(2):190–196.

4. Li H, Stampfer MJ, Giovannucci EL, Morris JS, Willett WC, Gaziano JM, Ma J (2004) A prospective study of plasma selenium levels and prostate cancer risk. *J Natl Cancer Inst* 96(9):696–703.

5. Miller ER, Pastor-Barriuso R, Dalal D, Riemersma RA, Appel LJ, Guallar E (2005) Meta-analysis: High-dosage vitamin E supplementation may increase all-cause mortality. *Ann Intern Med* 142:37–46.

6. Ames BN, Elson-Schwab I, Silver EA (2002) High-dose vitamin therapy stimulates variant enzymes with decreased coenzyme binding affinity (increased K_m): Relevance to genetic disease and polymorphisms. *Amer J of Clin Nut* 75(4):616–658.

7. Chen Q, Espey MG, Sun AY, Pooput C, Kirk KL, Krishna MC, Khosh DB, Drisko J, Levine M (2008) Pharmacologic doses of ascorbate act as a prooxidant and decrease growth of aggressive tumor xenografts in mice. *Proc Natl Acad Sci U S A.* 105(32):11105–11109.

8. Bjelakovic G, Nikolova D, Gluud LL, Simonetti RG, Gluud C (2007) Mortality in randomized trials of antioxidant supplements for primary and secondary prevention: Systematic review and meta-analysis. *JAMA* 297(8):842–857.

9. Willett WC, Stampfer MJ (2001) Clinical practice. What vitamins should I be taking, doctor? *N Engl J Med* 345(25):1819–1824.

10. Portakal O, Ozkaya O, Erden Inal M, Bozan B, Koşan M, Sayek I (2000) Coenzyme Q10 concentrations and antioxidant status in tissues of breast cancer patients. *Clin Biochem* 33(4):279–284.

11. Bagchi D, Bagchi M, Stohs S, Ray SD, Sen CK, Preuss HG (2002) Cellular protection with proanthocyanidins derived from grape seeds. *Ann N Y Acad Sci* 957:260–270.

12. Baur JA, Pearson KJ, Price NL, Jamieson HA, Lerin C, Kalra A, Prabhu VV, Allard JS, Lopez-Lluch G, Lewis K, Pistell PJ, Poosala S, Becker KG, Boss O, Gwinn D, Wang M, Ramaswamy S, Fishbein KW, Spencer RG, Lakatta EG, Le Couteur D, Shaw RJ, Navas P, Puigserver P, Ingram DK, de Cabo R, Sinclair DA (2006) Resveratrol improves health and survival of mice on a high-calorie diet. *Nature* 444 (7117): 337–342.

Chapter 13: Calorie Reduction and Weight Loss

1. National Institute of Diabetes and Digestive and Kidney Disease, Statistics Related to Overweight and Obesity, http://www.win.niddk.nih.gov/statistics/index.htm.

2. Cummings DE, Weigle DS, Frayo RS, Breen PA, Ma MK, Dellinger P, Purnell JQ (2002) Plasma ghrelin levels after diet-induced weight loss or gastric bypass surgery. *NEJM* 346(21):1623–1630.

3. Allison DB, Miller RA, Austad SN, Bouchard C, Leibel R, Klebanov S, Johnson T, Harrison DE (2001) Genetic variability in responses to caloric restriction in animals and in regulation of metabolism and obesity in humans. *J Gerontol A Biol Sci Med Sci* 56 Spec No 1:55–65.

4. Cummings DE, Weigle DS, Frayo RS, Breen PA, Ma MK, Dellinger EP, Purnell JQ (2002) Plasma ghrelin levels after diet-induced weight loss or gastric bypass surgery. *N Engl J Med* 346:1623–1630.

5. Masoro EJ, Yu BP, Bertrand HA (1982) Action of food restriction in delaying the aging process. *Proc Natl Acad Sci U S A* 79(13):4239–4241.

Chapter 14: Exercise

1. Kokkinos P, Myers J, Kokkinos JP, Pittaras A, Narayan P, Manolis A, Karasik P, Greenberg M, Papademetriou V, Singh S (2007) Exercise capacity and mortality in black and white men. *Circulation* 117:589–591.

2. Cherkas LF, Hunkin JL, Kato BS, Richards B, Gardner JP, Surdulescu GL, Kimura M, Lu X, Spector TD, Aviv A (2008) The association between physical activity in leisure time and leukocyte telomere length. *Arch Int Med* 168(2):154–158.

3. Reeves ND, Narici MV, Maganaris CN (2006) Myotendinous plasticity to ageing and resistance exercise in humans. *Exp Physiol* 91(3):483–498.

4. Kraemer WJ, Häkkinen K, Newton RU, Nindl BC, Volek JS, McCormick M, Gotshalk LA, Gordon SE, Fleck SJ, Campbell WW, Putukian M, Evans WJ (1999) Effects of heavy-resistance training on hormonal response patterns in younger vs. older men. *Appl Physiol* 87(3):982–92.5.

5. Cornelissen VA, Fagard RH (2005) Effect of resistance training on resting blood pressure: A meta-analysis of randomized controlled trials. *J Hypertens* 23(2):251–259.

6. Henriksson J (1995) Influence of exercise on insulin sensitivity. *J Cardiovasc Risk* 2(4):303–309.

7. Engelke K, Kemmler W, Lauber D, Beeskow C, Pintag R, Kalender WA (2006) Exercise maintains bone density at spine and hip EFOPS: A 3-year longitudinal study in early postmenopausal women. *Osteoporosis Int* 17(1):133–142.

Chapter 15: New Technologies

1. De Grey A, Rae M. (2007) *Ending Aging: The Rejuvenation Breakthroughs That Could Reverse Human Aging in Our Lifetime.* New York: St. Martin's Press.

2. Kurzweil, R. (2005) *The Singularity Is Near, When Humans Transcend Biology.* New York: Viking.

Chapter 16: Detoxification

1. Sundell J (2004) On the history of indoor air quality and health. *Indoor Air* 4:14 Suppl 7:51–58.

2. Yoder J, Roberts V, Craun GF, Hill V, Hicks LA, Alexander NT, Radke V, Calderon RL, Hlavsa MC, Beach MJ, Roy SL; Centers for Disease Control and Prevention (CDC) (2008) Surveillance for waterborne disease and outbreaks associated with drinking water and water not intended for drinking—United States, 2005–2006. MMWR Surveill Summ 57(9):39–62.

3. Hyland GJ (2000) Physics and biology of mobile telephony. *Lancet* 356(9244):1833–1836.

4. Beyersmann D, Hartwig A (2008) Carcinogenic metal compounds: Recent insight into molecular and cellular mechanisms, *Arch Toxicol* 82(8):493–512.

INDEX

Underscored page references indicate sidebars and tables. **Boldface** references indicate illustrations and photographs.

AA (arachidonic acid), 116, 227
Abdomen, examination of, 161
Abdominal crunches, 379, **379**
Abdominal fat, increasing cancer mortality, 123
Abdominal stretch, 385, **385**
Acesulfame-K, 223
Acetyl glutathione, 331–32
Acetyl-L-carnitine, 14, 15, 331
Activity level, determining, 340
Addictions, 11–13
  as response to stress, 187–89, 190–91
Additives, avoiding, 242
Adrenal glands, 76, 77
Adrenal stress index, 100, 101, **101**
Advanced glycation end products. See AGEs
Aerobic conditioning, assessing, 204, 205
Aerobic exercise
  caution about, 355, 356
  definition of, 354
  detoxification from, 416
  guidelines for, 356–60
  health benefits from, 355
  stretching after, 382
  suggested program for, 398, 398, 399
  types of, 354, 355
Age, as risk factor for heart disease, 34, 36
AGEs (advanced glycation end products)
  from excess blood sugar, 212, 221
  glycation and, 108–9
Age spots, from AGE production, 212
Aging
  genetic code and, xxiii–xxiv
  vs. growing older, 3–4
  hormone levels and, 73, 82–83
  hormones of, 75–78
  processes causing, 4
  reversing process of, xviii, xxi, 403–4
  slowing process of, xii–xiii, xviii, xxi
Agricultural chemicals, 124–25, 235. See also Pesticides

Airborne toxins, reducing, 408–9
Air filters, 408
ALA. See Alpha-linolenic acid; Alpha-lipoic acid
Alcohol
  addiction to, 13
  elevating triglycerides, 38
  increasing mortality, 194–95
  moderate use of
    decreasing heart attack risk, 350
    decreasing homocysteine levels, 119
    definition of, 236
    vs. excessive use, 187–88, 190
    health effects of, 236–37
    increasing longevity, 12
Alpha-linolenic acid (ALA), 114, **115**, 226
Alpha-lipoic acid (ALA), 330, 415
AlphaStim, for stress management, 185
Aluminum, avoiding use of, 414
Alzheimer's disease
  genomics testing and, 147, 152, 169
  from inflammation, 112, 114
  from methylation, 117, 119
  preventing, with
    curcumin, 241
    exercise, 349
  screening for, 164, 169
Amalgam dental fillings, alternatives to, 414
Amino acids
  essential, 233, 234, 235
  for increasing growth hormone production, 81
Anabolic steroids. See Hormones, anabolic
Anger, health risks from, 45, 186
Angina, 29, 30, 48–49
Angioplasty, studies on effectiveness of, 29, 30
Angiotensin II antagonists, for high blood pressure, 45
Ankle stretches, 387, **387**

433